Prion Protein Protocols

METHODS IN MOLECULAR BIOLOGY™

John M. Walker, SERIES EDITOR

Prion Protein Protocols

Andrew F. Hill
Editor

Department of Biochemistry and Molecular Biology,
Bio21 Molecular Science and Biotechnology Institute;
Department of Pathology and Mental Health Institute of Victoria,
University of Melbourne,
Melbourne, Australia

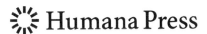 Humana Press

Editor
Andrew F. Hill, Ph.D.
Department of Biochemistry and Molecular Biology
Bio21 Molecular Science and Biotechnology Institute
University of Melbourne
30 Flemington Road
Parkville, VIC 3010
Australia
a.hill@unimelb.edu.au

Series Editor
John Walker, Ph.D.
28 Selwyn Avenue
Hatfield, Hertfordshire, UK

ISBN: 978-1-58829-897-3   e-ISBN: 978-1-59745-234-2
DOI: 10.1007/978-1-59745-234-2
ISSN 1064–3745

Library of Congress Control Number: 2008924616

Cover Illustrations: (Background) Prion disease pathology: brain sections showing spongiform neurodegeneration (Fig. 6, Chap. 14; see complete figure, caption, and discussion on p. 223). (Foreground) Epifluorescence microscopy of PrP fibrils stained with ThT (Fig. 4, Chap. 8; see complete figure and caption on p. 112 and discussion on p. 111).

Printed on acid-free paper

9 8 7 6 5 4 3 2 1

springer.com

Preface

Prion Protein Protocols brings together a collection of current protocols in the field of mammalian prion disease research. Since identification of the prion protein gene some 20 years ago, what were once thought to be rare, neurodegenerative diseases of humans and animals have become a major research area.

The major interest in this field results from the epidemic of bovine spongiform encephalopathy and the subsequent human prion disease, variant Creutzfeldt–Jakob disease, which still poses an unknown risk to human health. The "unknowns" in the field stem from the enigmatic nature of infectious prions, the infectious agent that can transmit these diseases between individuals. Although much information has been gained over the past two decades about the molecular nature of prion proteins, only very recently have several research groups begun to cast light on how to turn the normal prion protein into its aberrant, infectious form.

Prion Protein Protocols brings together a collection of protocols from 13 different laboratories in five countries covering basic science and diagnostic areas of prion research. Together, the chapters provide an up-to-date collection of current methods in this unique area of neuroscience. The notes section at the end of each methods chapter provides useful insight into the experimental techniques, and they are no doubt a benefit to researchers wanting to use these technologies.

With the novel nature of the infectious agent and the biosafety issues surrounding prion research come several innovative approaches to studying the disease process. A series of chapters on the cell biology of prions outlines techniques and approaches for studying prion infection in cultured cells, and in doing so they highlight how these systems can be used as a rapid bioassay. Much research also has centered on identifying how the normal form of the prion protein can misfold into the abnormal, infectious form; this is the basis for several chapters in this volume. The methods chapters are separated by review chapters that describe some of the historical background and provide useful reference to other work in these areas.

I thank the contributing authors for their efforts in writing the chapters and John Walker, Series Editor, for his patience and helpful advice in producing this volume.

Andrew F. Hill

Contents

Contributors

Kevin J. Barnham
Department of Pathology and Mental Health Research Institute of Victoria,
Bio21 Molecular Science and Biotechnology Institute, University of Melbourne,
Melbourne, Australia

Ilia V. Baskakov
Medical Biotechnology Center, University of Maryland Biotechnology Institute,
Baltimore, MD

Jonathan A. Beck
MRC Prion Unit, University College London Institute of Neurology, London, UK

Katarina Bedecs
Department of Biochemistry and Biophysics, Stockholm University, Sweden

Sebastian Brandner
MRC Prion Unit, University College London Institute of Neurology, London, UK

Leonid R. Breydo
Medical Biotechnology Center, University of Maryland Biotechnology Institute,
Baltimore, MD

David R. Brown
Department of Biochemistry, Bath University, Bath, UK

Roberto Cappai
Department of Pathology and Mental Health Research Institute of Victoria,
Bio21 Molecular Science and Biotechnology Institute, University of Melbourne,
Melbourne, Australia

Giuseppe D. Ciccotosto
Department of Pathology and Mental Health Research Institute of Victoria,
Bio21 Molecular Science and Biotechnology Institute, University of Melbourne,
Melbourne, Australia

John Collinge
MRC Prion Unit, University College London Institute of Neurology, London, UK

Steven J. Collins
Australian National Creutzfeldt–Jakob Disease Registry, Department of
Pathology, University of Melbourne, Melbourne, Australia

Nathan R. Deleault
Departments of Biochemistry and Medicine, Dartmouth Medical School,
Hanover, NH

Cheryl A. Demczyk
Department of Infectology, Scripps Florida, Jupiter, FL

Simon C. Drew
Department of Pathology and Mental Health Research Institute of Victoria,
Bio21 Molecular Science and Biotechnology Institute, University of Melbourne,
Melbourne, Australia
School of Physics, Monash University, Melbourne, Australia

Hanna Gyllberg
Department of Biochemistry and Biophysics, Stockholm University, Sweden

Cathryn L. Haigh
Department of Pathology and Mental Health Research Institute of Victoria,
University of Melbourne, Melbourne, Australia

Sen Han
Department of Biochemistry and Molecular Biology and Department of
Pathology, Bio21 Molecular Science and Biotechnology Institute, University of
Melbourne, Melbourne, Australia

Andrew F. Hill
Department of Biochemistry and Molecular Biology, Bio21 Molecular Science
and Biotechnology Institute, Department of Pathology and Mental Health
Research Institute of Victoria, University of Melbourne, Melbourne, Australia

Susan Joiner
MRC Prion Unit, University College London Institute of Neurology, London, UK

Peter-Christian Klohn
MRC Prion Unit, Institute of Neurology, London, UK

Genevieve M. Klug
Australian National Creutzfeldt-Jakob Disease Registry, Department of Pathology,
University of Melbourne, Melbourne, Australia

Victoria A. Lawson
Department of Pathology and Mental Health Research Institute of Victoria,
University of Melbourne, Melbourne, Australia

Victoria Lewis
Australian National Creutzfeldt-Jakob Disease Registry, Department of Pathology,
University of Melbourne, Melbourne, Australia

Jacqueline M. Linehan
MRC Prion Unit, University College London Institute of Neurology, London, UK

Kajsa Löfgren
Department of Biochemistry and Biophysics, Stockholm University, Stockholm,
Sweden

Sukhvir P. Mahal
Department of Infectology, Scripps Florida, Jupiter, FL

Natallia Makarava
Medical Biotechnology Center, University of Maryland Biotechnology Institute,
Baltimore, MD

Simon Mead
MRC Prion Unit, University College London Institute of Neurology, London, UK

Caroline Powell
MRC Prion Unit, University College London Institute of Neurology, London, UK

Judy R. Rees
Department of Community and Family Medicine, Dartmouth Medical School,
Hanover, NH

Emery W. Smith, Jr.
Department of Infectology, Scripps Florida, Jupiter, FL

Surachai Supattapone
Departments of Biochemistry and Medicine[1], Dartmouth Medical School,
Hanover, NH

Glenn C. Telling
Department of Microbiology, Immunology and Molecular Genetics
Department of Neurology
Sanders Brown Center on Aging, University of Kentucky, Lexington, KY

Carina Treiber
Institut für Chemie und Biochemie, Freie Universität Berlin, Berlin, Germany

Laura J. Vella
Department of Biochemistry and Molecular Biology and Department of
Pathology, Bio21 Molecular Science and Biotechnology Institute, University of
Melbourne, Melbourne, Australia

Jonathan D. F. Wadsworth
MRC Prion Unit, University College London Institute of Neurology, London, UK

Charles Weissmann
Department of Infectology, Scripps Florida, Jupiter, FL

Anthony R. White
Department of Pathology and Mental Health Research Institute of Victoria,
University of Melbourne, Melbourne, Australia

Chapter 1
Cell Culture Models to Unravel Prion Protein Function and Aberrancies in Prion Diseases

Katarina Bedecs

Summary From an early stage of prion research, tissue cultures that could support and propagate the scrapie agent were sought after. The earliest attempts were explants from brains of infected mice, and their growth and morphological characteristics were compared with those from uninfected mice *(1)*. Using the explant technique, several investigators reported increased cell growth in cultures established from scrapie-sick brain compared with cultures from normal mice *(1, 2)*. These are odd findings in the light of the massive neuronal cell death known to occur in scrapie-infected brains; however, the cell types responsible for the increased cell growth in the scrapie-explants most probably were not neuronal. The first successful cell culture established in this way, in which the scrapie agent was serially and continuously passaged beyond the initial explant, was in the scrapie mouse brain culture *(3)*, which is still used today *(4, 5)*. This chapter describes the generation and use of chronically prion-infected cell lines as cell culture models of prion diseases. These cell lines have been crucial for the current understanding of the cell biology of both the normal (PrP^C) and the pathogenic isoform (PrP^{Sc}) of the prion protein. They also have been useful in the development of antiprion drugs, prospectively used for therapy of prion diseases, and they offer an alternative approach for transmission/infectivity assays normally performed by mouse bioassay. Cell culture models also have been used to study prion-induced cytopathological changes, which could explain the typical spongiform neurodegeneration in prion diseases.

Keywords Cell culture; neuroblastoma; neurodegeneration; prion; scrapie.

1. Generation of Scrapie-infected Cells

Initial attempts to create scrapie-propagating cell cultures were performed by cultivating explants from scrapie-infected brains *(1, 2, 6)*. However, only the scrapie mouse brain (SMB) cells remain infectious after years of in vitro passage *(3–5)*. The apparent disadvantage of generating scrapie-infected cell cultures by this technique is the lack of uninfected controls. Therefore, alternative methods were developed.

From: *Prion Protein Protocols.*
Methods in Molecular Biology, Vol. 459.
Edited by: A. F. Hill © Humana Press, Totowa, NJ

One such method was to intravenously inoculate splenotropic tumor cell lines derived from, for example, macrophage, B-, or T-lymphocytes and erythroid lineages, into scrapie-infected mice (which after a few weeks after infection produce high titers of scrapie in the spleen). The presumably scrapie-infected tumors were subsequently explanted, and their in vitro passaging was resumed. Sadly, none of the cultures established in this way were infected *(7)*. After these primary attempts, cells from several sources and a variety of experimental approaches have been used to establish scrapie-infected cultures. The most straightforward approach, that is, the exposure of cell monolayer or cell suspensions to homogenates of scrapie-infected brains or partly or highly purified preparations, proved to be successful to many if not all cell types *(7, 8)*. Many of the currently used chronically prion-infected cell cultures have been generated this way (**Table 1**).

This chapter focuses on the N2a and GT1 cells because the pathogenic isoform of the prion protein (PrP^{Sc}) accumulates mainly in neurons, and these are the targets for prion-induced neurodegeneration. However, non-neuronal cells, for example, the recently developed Schwann cell cultures, may be useful to study the peripheral steps of prion invasion *(20, 21)*.

A striking feature of the cell lines shown in **Table 1** is that most of the cell lines susceptible for persistent prion infection are of non-neuronal origin. Among the murine neuronal cell lines capable of continuously replicating prions are the N2a, N1E-115, and C-1300 neuroblastoma cell lines. Interestingly, all these cell lines have a common origin, derived from a spontaneous tumor arising in A/J mice, a *Prnpa/a* mouse strain *(32)*, but with different passage histories. This particular cell

Table 1 Cell Lines Supporting Prion Infection

Cell type[a]	Species	Tissue or cell type of origin	Prion isolate	Comments	References
Neuronal cell types					
N2a	Mouse	Neuroblastoma	Chandler		*(7–9)*
C-1300	Mouse	Neuroblastoma	Chandler		*(7, 9)*
N1E-115	Mouse	Neuroblastoma	Chandler, C506		*(7, 10, 11)*
N2a #58	Mouse	Neuroblastoma	Chandler, FU	N2a 6x overexpressing PrP^C	*(12, 13)*
SHSY-5Y	Human	Neuroblastoma	CJD		*(14)*
GT1-1	Mouse	Hypothalamic neuronal cell subclone 1	Chandler, RML, FU	Large T-immortalized cells	*(13, 15)*
GT1-7	Mouse	Hypothalamic neuronal cell subclone 7	Chandler, 139A, 22L, FU, SY	Large T-immortalized cells	*(12, 13, 15, 16)*
PC12	Rat	Pheochromo-cytoma	139A/ME7	Neuronal differentiation in presence of NGF	*(17, 18)*
SN56	Mouse	Hybrid septal neuron/ neuroblastoma	Chandler/ ME7/22L		*(19)*

(continued)

Table 1 (continued)

Cell type[a]	Species	Tissue or cell type of origin	Prion isolate	Comments	References
Non-neuronal cell types					
MSC-80	Mouse	Schwann cell	Chandler		*(20)*
MovS6/S2	Tg mouse	Schwann cell-like from dorsal root ganglia	PG127	Ovine PrP in Tg mouse	*(21)*
HaB	Hamster	Non-neuronal hamster brain cell	Sc237 rods	Spontaneously immortalized cells	*(22)*
Glial cell	Rat	Glial cells from rat trigeminal ganglion	Chandler	Ethylnitrosourea-induced ganglion tumour	*(23)*
SMB	Mouse	Mesodermal cells from scrapie-infected mouse brain (Chandler)		Still propagating scrapie	*(3, 24)*
SMB-PS	Mouse		Chandler, 22F	Pentosan sulfate-cured SMB	*(5)*
L-fibroblast	Mouse	Subclone of L929 fibroblast cell	Chandler		*(25)*
L23	Mouse	Subclone of L929 fibroblast cell	Compton ME7		*(26)*
L929	Mouse	Fibroblast cell	RML, 22L, ME7		*(27)*
NIH/3T3	Mouse	Fibroblast cell	22L		*(27)*
NS1	Mouse	Spleen cell from scrapie-sick mouse fused with NS1 cell	Chandler		*(28)*
Rov	Rabbit	Kidney epithelial cell	PG127, LA404	RK-13–expressing ovine PrP^C	*(29, 30)*
C2C12	Mouse	Myoblasts	22L	In coculture with 22L-infected N2a.	*(31)*

[a]FU, mouse-adapted Fukuoka-1 (familial GSS); SY. mouse-adapted sporadic CJD.

lineage, therefore, seems to be especially susceptible to scrapie infection when infected with mouse–adapted prion isolates. This is clearly apparent in view of the large number of neuronal or neural cell lines that could not be infected *(7, 8, 10, 28, 33)*. Except for the C-1300-derived clones, only one other murine neuronal cell line is easily infected. That is the GT1 cell line, originating from hypothalamic neurons and immortalized by genetically targeted tumorigenesis in transgenic mice *(34)*. The GT1 cells are highly differentiated gonadotropin-releasing hormone neurons, and they are in contrast to the C-1300 cells, susceptible to prions other than the Chandler/RML isolates *(15)*. The GT1 cells have shown to replicate both the scrapie-derived

139A and 22L prions, as well as familial GSS- and sporadic CJD-derived FU and SY prions, respectively, whereas only transfected N2a cells (N2a#58) overexpressing the normal isoform of the prion protein (PrP^C) are possible to infect with isolates other than Chandler/RM *(12, 13)*.

GT1 cells offer several important advantages over the use of N2a cells, especially for studying cytopathological effects provoked by prion infection. Initially, they express approx 8 times higher levels of endogenous PrP^C *(12)*, and they are much more susceptible to prion infection than N2a cells (in our lab every attempt to RML-infect GT1 cells has been successful), providing a simple means to produce several independently prion-infected ScGT1 cell lines.

In contrast, in typical N2a cultures exposed to prions, <2% of the cells become infected, and only low levels of prions are produced. Also, these cultures frequently loose infectivity within 10–15 passage *(8, 9, 35)*. To obtain persistently prion-infected cultures that produce sufficient quantities of PrP^{Sc}, the prion-infected N2a cultures (ScN2a) must be subcloned *(8, 36)*. By this method, ScN2a lines have been obtained, in which 80–90% of the cells are infected. These ScN2a subclones have been very useful for studying the cell biology of prion replication, that is, to determine the subcellular location of PrP^{Sc} and the metabolism and kinetics of PrP^{Sc} formation *(22, 37–40)*. ScN2a subclones are also suitable in the search for inhibitors of PrP^{Sc} formation and in the development of antiprion therapeutic agents *(41–43)*. However, for studies aiming at studying potential neurotoxic changes induced by prion infection, these subclones are less suitable, because in these studies the ScN2a cells are subcloned from a population of cells to which they are then compared. The observed differences between ScN2a and the uninfected N2a population might just represent cloning artifacts or be due to a selection artifact during the prion infection. An elegant way to avoid this possibility was to derive highly susceptible N2a sublines (by subcloning the N2a population before prion infection) from which prion-infected cultures were generated, without further subcloning *(35)*. These infected sublines can be compared with the corresponding uninfected sublines, without interference from potential cloning artifacts.

The GT1 cells, however, do not need to be subcloned after prion infection, because a much higher proportion of the cells become infected *(12, 15)*; therefore, these cells represent an excellent culture model, avoiding clonal differences. In addition, GT1 cells are the only CNS-derived neuronal cells susceptible to prion infection today.

2. Detection of PrP^{Sc} in Infected Cells: Definition of Prion Infection in Cell Cultures

Although not fully understood, the infectious agent is largely, if not exclusively an abnormal form of the host's own PrP^C (reviewed in **ref. *44***). The pathogen-associated form, PrP^{Sc}, differs from the normal form in its biochemical and biophysical properties, including protease resistance, solubility in nondenaturing detergents, and resistance to phosphatidylinositol-specific phospholipase C *(45)*.

The most straightforward approach to detect PrP^{Sc} would be to use a PrP^{Sc}-specific antibody. Despite years of intense effort, only two PrP^{Sc}-specific antibodies have been described—a 1997 report that has neither been confirmed nor extended in the literature *(46)* and a recent report by the Cashman group *(47)*.

Instead, infectivity has traditionally been linked to the presence of a proteinase K (PK)-resistant core of PrP^{Sc}, PrP$^{27\text{-}30}$ *(45)*, which can be detected by Western blot analysis using C-terminal anti-PrP antibodies. However, in some experimental set-ups, protease-resistant PrP could not be found in samples that contained prion infectivity *(48–50)*. Conversely, not all PK-resistant PrP species are associated with prion infectivity *(51, 52)*. Thus, it is important to recognize that the presence of PK-resistant PrP species is a mere surrogate marker for prion infection.

However, in most of the cell cultures described in **Table 1**, a correlation between PK-resistant PrP and prion infectivity was confirmed by inoculating mice with cell culture extracts (after sufficient passages to rule out remaining infectivity from the initial inoculum). Several techniques with different sensitivities to detect PrP^{Sc} in prion-infected cultures have been developed (**Fig. 1**).

Besides these mainly nonquantitative techniques based on the detection of PrP^{Sc}, a cell-based bioassay, named scrapie cell (SC) assay has been developed by Charles Weissmann's group *(53)*. The SC assay is quantitative, about as sensitive as the mouse bioassay, 10 times faster, >2 orders of magnitude less expensive, and suitable for robotization. This assay is described in detail in Chapter 4.

With the generation of more specific PrP antibodies, the standard PK-digestion-Western blot assay is progressively replaced by quicker and more sensitive techniques. Whereas Western blot could detect PrP^{Sc} when 10% of the cells in a 6-cm-diameter dish were infected, a sensitive cell blot technique could detect PrP^{Sc} when only 1% of the cells in a single well of a 24-well plate were infected, corresponding to a 150-fold increase in sensitivity *(35)*. Another method is the filter-retention assay (slot blot),

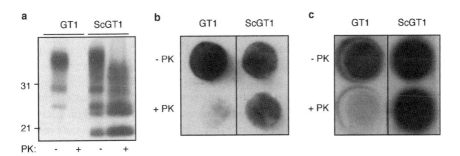

Fig. 1 Different methods to detect PrP^{Sc} in prion-infected cells. (**A**) Western blot. Lysates of GT1 or ScGT1 are treated with PK and analyzed by SDS-PAGE and Western blotting with anti-PrP antibody. PrP$^{27\text{-}30}$, the resistant core of PrP^{Sc}, is detected only in infected cells. (**B**) Cell blot. GT1 and ScGT1 cells grown on coverslips are blotted directly on to a nitrocellulose membrane, and PrP^{Sc} is detected, after PK digestion and denaturation with guanidine isothiocyanate of the membrane, by immunoblotting with anti-PrP antibodies. (**C**) Dot blot. Cell lysates of GT1 and ScGT1 cells are filtered through a nitrocellulose membrane with a dot blot device, and PrP^{Sc} is detected as described for B

which takes into account two diagnostic criteria in combination: the protease resistance and the presence of detergent-insoluble aggregates of PrP^{Sc} *(54)*. This assay, and classical enzyme-linked immunosorbent assay *(55)*, are well-suited assays for high-throughput screenings of therapeutic compounds. With none of these methods can the percentage of cells producing PrP^{Sc} be determined. However, a postembedded method has been developed that enables the detection of a single infected cell *(29)*.

3. Cell Biology and Putative Functions of PrP^C

The development of cell cultures has been fundamental for the understanding of the cell biology of PrP^C, including its intracellular trafficking, localization, and interaction with other proteins/molecules. Several techniques, including immunofluorescence, metabolic labeling, cell surface biotinylation, Western blot, immunoprecipitation, and crosslinking have been used to determine the trafficking and localization of PrP^C. PrP^C is a glycosylphosphatidylinositol-anchored glycoprotein that is generally found at the cell membrane associated with cholesterol and glycolipid-rich microdomains, called lipid rafts or detergent-resistant microdomains *(39, 40)*. Once on the cell surface, PrP^C is endocytosed, cleaved in its central domain, and then recycled to the surface *(56)*.

In the search of PrP^C function, several putative binding partners have been identified by using the two-hybrid system screened against cDNA libraries from HeLa cells or mouse brain. Among them, the antiapoptotic Bcl-2 *(57)*, the chaperone Hsp6 *(58)*, the 37-kDa laminin receptor precursor (LRP *(59)*), and the adaptor protein Grb2, Synapsin1, and an unknown protein, Pint1, were identified *(60)*. Because PrP^C is predominantly expressed as a glycosylphosphatidylinositol (GPI)-anchored plasma membrane protein, an interaction with many of these proteins may present a logistic problem. In addition, the biological significance of many of these interactions remains unclear, and confirmation in functional assays has yet to be established.

However, two proteins have been singled out from these studies as putative binding partner: the 37-kDa LRP and its mature 67-kDa form, termed the high-affinity laminin receptor L *(59, 61)*. The LRP/LR was suggested to act as a putative PrP receptor because it was shown, by immunofluorescence studies, to colocalize with PrP^C on the surface of both N2a and baby hamster kidney (BHK) cells. Cell binding assays with exogenously applied PrP on these cell cultures revealed an LRP/LR-dependent binding and internalization of PrP^C, which was inhibited in the presence of an anti-LRP antibody *(61)*. Furthermore, the LRP/LR was also shown to be required for PrP^{Sc} propagation in ScN2a and ScGT1 cells, because expression of an antisense LRP RNA-expression plasmid, transfection with small interfering RNAs specific for the LRP mRNA, or incubation with an anti-LRP/LR antibody inhibited the accumulation of PrP^{Sc} in these cells. These findings suggest that the LRP/LR may not only act as a cell surface receptor for PrP^C, possibly involved in the endocytic pathway of PrP^C, but also might represent the portal of entry for PrP^{Sc} in prion infection *(62)*. Another report also suggests PrP^C to be a ligand to a yet unknown heterophilic receptor mediating neuronal recognition, as shown by increased neurite outgrowth from primary neurons on a PrP^C-coated substratum, compared with non–PrP-coated dishes (63).

Alternative methods to find interacting partners are coimmunoprecipitation or chemical crosslinking, to demonstrate either a direct physical interaction or at least colocalization. PrP^C could be immunoprecipitated with antibodies directed against stress-inducible protein (STI) from PrP^C-transfected human embryonic kidney (HEK) 293T cell (64), and Grb2, Synapsin1, and Pint1 were coimmunoprecipitated from BHK cells, showing that PrP^C can specifically interact with these proteins not only in a yeast model but also in mammalian cells (60). Coimmunoprecipitation from hamster brain extracts, in the presence of detergents that either preserve or completely dissociate lipid rafts, showed that PrP^C interacts strongly with neuronal nitric-oxide synthase and dystroglycan, a transmembrane protein that is the core of the dystrophin-glycoprotein complex (65). Binding of PrP^C to neural cell adhesion molecules (N-CAMs) was demonstrated in N2a and ScN2a cells by use of mild formaldehyde crosslinking, followed by sodium dodecyl sulfate-polyacrylamide gel electrophoresis (SDS-PAGE) and HPLC/mass spectrometry. This interaction was further shown to occur in raft domain (66).

Many studies indicate that heparan sulfate (HS) proteoglycans play a role in the life cycle of PrP^C and possibly also in the formation of PrP^{Sc} (reviewed in **ref.** 67). First, HS interacts with PrP^C, an interaction possibly involving the N-terminal octarepeat region of PrP (68, 69). Second, a variety of sulfated glycans, such as HS mimetic (68, 70), pentosan polysulfate, and dextran sulfate (71), reduce the formation of PrP^{Sc} in ScN2a and ScGT1 cultures, and in some cases prolong the incubation time of experimental prion disease (72). HS also was shown to accumulate in cerebral prion amyloid plaques (73), and it was associated with the more diffuse PrP^{Sc} deposits that occur in early stages of prion disease (74). Functionally, HS and sulfated cell-surface glycans seem to play a role in PrP^C internalization, because treatment of N2a cells with sulfated glycans stimulates endocytosis of PrP^C (75). This stimulation may account for the antiprion effect of sulfated glycans. Alternatively, these findings suggest that exogenously applied sulfated glycans inhibit PrP^{Sc} formation by competing with the binding of PrP^C, PrP^{Sc}, or both to a putative cellular HS proteoglycan. The latter scenario was recently corroborated by Taraboulos' group, showing the involvement of a cellular HS proteoglycan (68, 76).

4. PrP Knockout Mice

Several lines of mice devoid of PrP^C have been generated by homologous recombination in embryonic stem cells, by using either of two strategies. In one strategy, the disruptive modifications are restricted to the open reading frame, generating the Np (77) and Zrch (78) lines. Mice homozygous for the inactivated gene (Prnp–/–) develop normally, show no striking pathology, and they are resistant to prion infection. The other strategy involves deletion of not only the reading frame but also its flanking regions. The three lines generated this way; Ngs *(79)*, Rcm *(80)*, and Zrch *(81)*, also develop normally, but they exhibit severe ataxia and Purkinje cell loss later in life *(82)*. In these mice, Purkinje cell loss and ataxia were shown to be due to ectopic expression of an unknown protein, doppel, which is not expressed in the Npu and Zrch1 mice *(82)*. Doppel (German for double) or Dpl (downstream of the Prnp locus) was expressed because of an intergenic splicing event that placed the Dpl gene under the control of the Prnp promoter. Thus, ectopic expression of Dpl in the absence of PrP^C, rather than absence of PrP^C itself causes Purkinje cell loss.

To circumvent the problems involved in the interpretation of PrP^C-null mice, two lines of conditional knockout mice have been generated, with an intact PrP^C expression during embryogenesis, which can be knocked down in the adult mouse. Unfortunately, these mice did not reveal the function of PrP^C, because no obvious effects upon neuronal viability, neuropathological abnormalities, or neurological status were observed *(83)*.

5. Involvement of PrP^C in Copper Metabolism and Oxidative Stress

PrP^C is a metal ion-binding protein, binding copper ions (Cu^{2+}) with high affinity, and also nickel, zinc, and manganese cations, but with lower affinities. Full-length recombinant PrP^C can potentially bind up to five copper ions, four in the highly conserved N-terminal octarepeat region PHGGGWGQ *(84, 85)*, with a fifth Cu^{2+} binding site around residues His-96 and His-111 *(85)*. Several lines of evidence suggest a functional role of PrP^C in cellular copper metabolism and maintenance of the proper oxidative balance, possibly through a regulation of intracellular copper transport *(86–89)*.

6. Role of PrP^C in Cell Survival and Differentiation

Several lines of evidence suggest that PrP^C may play an important role in neuronal survival, differentiation, or both, possibly acting like a cell surface receptor itself. A putative role of PrP^C in neuronal differentiation was demonstrated by laminin-induced neuritogenesis of primary neurons from wild-type but not PrP-null (Zrch1) mice *(90)*. This finding was confirmed in the PC12 cell model, showing that antibodies against PrP^C inhibit cell adhesion to laminin-coated dishes and laser-induced ablation of PrP^C inhibits laminin-induced differentiation and promotes retraction of preformed neurite *(91)*. Additional results supporting a role of PrP^C in cell–cell interaction and differentiation are that the PrP^C expression in a rat neuroblastoma cell B104 was tightly regulated both as a function of cell density and during neuro-nal differentiation *(92)*.

A neuroprotective function of PrP^C has been suggested as PrP-null (Ngsk) neu-ronal cell lines are more vulnerable to serum deprivation than their PrP^C-expressing counterpart *(93)*, and PrP^C could protect human primary neurons against Bax-medi-ated apoptosis, with the same neuroprotective potency as Bcl-2 *(94)*. Interesting, in the light of the finding that PrP^C can bind Bcl-2 *(57)*. Moreover, antibody-mediated ligation of PrP^C or binding of a PrP^C-binding peptide protects retinal neuroblastic layer cells from anisomycin-induced apoptosis from wild-type, but not PrP^C-null mice, in a cAMP/protein kinase A-dependent manner *(95)*. In addition, PrP^C was shown to be dramatically upregulated in tumor necrosis factor (TNF)-α–resistant human breast carcinoma MCF7 subclones, and overexpression of PrP^C in normally TNF-α-sensitive MCF7 cells protects them against TNF-α–induced apoptosis *(96)*. This antiapoptotic function could be due to a PrP^C-mediated decrease in the expres-sion of several proapoptotic proteins, because PrP^C expression in a PrP^C-null neu-ronal cell line decreased the levels of p53, Bax, and caspase-3, and it increased Bcl-2 level *(97)*. These data are contradicted by others, who suggest a proapoptotic function of PrP^C. These authors reported an increase in the expression and activity of p53 and caspase-3 when expressing PrP^C in TSM1 and HEK 239 cell *(98)* or in primary cultures from PrP^C-null mice *(98)*.

These apparently paradoxical findings might be explained by the use of neuronal cells derived from PrP^C-null mice generated by the two different strategies (described in the PrP^C knockout mice section). Thus, Kim et al. *(97)* used the Ngsk mouse, which besides ablated PrP^C expression also express the doppel protein, whereas Paitel et al. *(98)* derived their cells from the Zrch1 mouse, which does not express doppel. However, the expression of doppel in these cells or the influence of doppel on the measured parameters was not addressed in these articles. These con-tradictory findings suggest that the interpretation of results obtained from PrP-null cell lines may depend on which type of knockout mouse they were derived from and could gain being revisited.

7. Signal Transduction by PrP^C

Because PrP^C may act as a cell surface receptor for a still hypothetical extracel-lular ligand, various studies aiming at unraveling its signaling potential has been performed. Because the ligand is still unknown, one way to activate or stimulate PrP^C was to crosslink PrP^C by use of specific anti-PrP^C antibodies, directed against PrP(142–160). Using this technique, it was demonstrated that PrP^C can function in a signal transduction cascade upstream of the protein tyro-sine kinase Fyn. A neuroectodermal progenitor cell line (1C11) was used, and a PrP^C-dependent Fyn activation was observed when the 1C11 cells had been differentiated to their serotonergic or noradrenergic progeny. Because PrP^C and Fyn are bound to opposing faces of the membrane, a transmembrane factor that could function as a link between PrP^C and Fyn was searched for. Caveolin-1 was presented as a candidate based on its coimmunoprecipitative behavior and the finding that cellular bombardment with caveolin-1–directed antibodies abolished PrP^C-dependent Fyn activation *(99, 100)*. Interestingly, crosslinking of PrP^C in vivo with antibodies directed against PrP(95–105) was found to trigger rapid and extensive apoptosis in hippocampal and cerebellar neurons, but not an antibody directed against PrP(133–157) *(101)*, the same epitope recognized by the anti-PrP antibody shown to induce Fyn activation in 1C11 cells *(99)*.

In a follow-up study, the same authors showed that after ligation of PrP^C, NADPH oxidase, a major cellular generator of reactive oxygen species (ROS) and extracellular-regulated kinase (ERK)1/2 members of the mitogen-activated protein kinase (MAPK) family, are targets of the caveolin-dependent Fyn activation, but only in the fully differentiated serotonergic or adrenergic 1C1 *(102)*. These findings are attractive because of the potential role of PrP^C in neuronal survival, differentiation, or both, because both ROS and ERK1/2 are important mediators in these cellular processes. They also may clarify the potential link between PrP^C and the cellular redox state. Several other studies also have proposed a link between PrP^C and Fyn or Src, the prototype member of the Src-family kinases. In enterocytes, PrP^C and Src were shown to colocalize in raft domains in cell–cell junctional complexes, and a direct physical interaction with Src was also demonstrated by coimmunoprecipitation of PrP^C *(103)*. In the human T-cell line CEM, PrP^C colocalizes with and coimmunoprecipitates Fyn, and it also was shown to interact with the tyrosine kinase ZAP70 after T-cell activation, suggesting that PrP^C is a component of the signaling complex involved in T-cell activation *(104)*. An interesting finding in this context is that STI571 (Gleevec® (imatinib mesylate), Novartis Pharmaceuticals, East Hanover, NJ, a potent inhibitor of the BCR-Abl tyrosine kinase), potently inhibits PrP^{Sc} replication in ScN2a, ScGT1 and SMB cells, via lysosomal degradation of existing PrP^{Sc} *(105)*.

Because some neuronal cell lines, such as the GT1- *(102)* and N2 *(39)*, do not express caveolin, an alternative signaling route from PrP^C to Fyn is suggested. One such candidate could be N-CAM, which was shown to associate with PrP^C *(66)*. This scenario is supported by data showing a direct interaction of N-CAM with Fyn and the selective inhibition of N-CAM–dependent neurite outgrowth in neurons from Fyn-null mice *(106)*.

Alternatively, no transmembrane intermediate may be necessary at all, suggested by previous studies showing that antibody ligation of the glycosphingolipids in rafts, and that of GPI-anchored proteins, induces a transient increase in the tyrosine phosphorylation of several substrates. This model thus proposes that glycosphingolipid crosslinking may induce a redistribution and clustering of signaling components in rafts, including Src-family kinases, on the opposite cytoplasmic leaflet, resulting in the activation of these kinases (reviewed in **ref. *107***).

Consistent with a role of PrP^C in signal transduction is the finding that PrP^C interacts with Grb2, as shown by coimmunoprecipitation from PrP^C- and Grb2-transfected

HEK 293T cell *(60)*. This interaction might be the result of their colocalization in rafts, because both of these proteins, together with other signaling intermediates in the MAPK pathway, such as Shc, Ras, phosphatidylinositol 3-kinase, and ERK1/2, are concentrated to these *(108)*. Together, these findings suggest that PrP^C has a significant role in signal transduction via lipid rafts, and the next question is how signal transduction may be affected by PrP^{Sc} accumulation in general and in rafts in particular.

8. Alterations in Cell Biology and Biochemistry in Prion-infected Cells: A Long List of Biochemical Abnormalities

One of the major objectives in the development of prion-infected cell cultures was to look for morphological and cytopathological manifestations of the prion infection. However, a striking features was the lack of any obvious signs of cell death. Moreover, several studies described an increased cell growth in cultures established from scrapie-sick brain compared with cultures from uninfected mice *(1, 2, 6)*. Increased cell growth and transformation also were reported when N1E-115 was inoculated with brain homogenate from C506-infected mice *(10)*. Unfortunately, it is not clear whether the changes described were necessarily due only to the scrapie agent rather than to other factors present in the brain homogenate.

Although no gross morphological differences have been observed, alterations in the expression, function, or both of several proteins in scrapie-infected cells have been described, possibly corresponding to the preclinical and subacute manifestations in prion diseases.

In ScN2a and ScHab cells, bradykinin- and platelet-derived growth factor-induced calcium responses were significantly reduced, compared with uninfected cell *(109)*. Although the number of ^{125}I-bradykinin binding sites was increased fourfold in ScN2a, their binding affinity was reduced tenfold, probably explaining the decreased Ca^{2+} response *(110)*. These changes were further ascribed to a significant reduction of plasma membrane fluidity *(110)*. The authors hypothesize that the conversion of PrP^C to PrP^{Sc} and the subsequent shunting to secondary lysosomes may alter protein and lipid trafficking pathways sufficiently to change the composition and properties of the plasma membrane, resulting in abnormal receptor-mediated functions.

We have speculated that prion infection may alter the expression, processing, function, or a combination, of neurotrophic receptors and thereby contribute to neurodegeneration in prion diseases by weakening the trophic support in prion-infected neurons. Indeed, our results show that scrapie infection induces a two- and fourfold increase in insulin receptor (IR) *(111)* and insulin-like growth factor-1 receptor (IGF-1R) *(11)* protein levels, respectively, in two independently scrapie-infected cells lines: ScN2a and ScN1E-115. However, despite the increased IR/IGF-1R expression in ScN2a, receptor binding studies revealed an important decrease in ^{125}I-insulin and ^{125}I-IGF-1 binding sites compared with the amount of immunoreactive receptors. In the IGF-1R, the absence of increased binding sites was due to a sevenfold decrease in IGF-1R binding affinity in ScN2a compared with N2a cells *(11)*. However, binding studies revealed no change in IR binding affinity, rather indicating a complete functional loss of a subpopulation of I *(111)*. In addition, ScN2a showed no significant difference in cell proliferation in the presence of insulin or IGF-1, as the only mitogen, despite the increased receptor expression, probably explained by the decrease in IR/IGF-1R binding sites.

Further studies revealed that the apparent loss of insulin binding sites was due to an increased formation of IR/IGF-1R hybrid receptors, with high affinity to IGF-1 but a 10- to 20-fold lower affinity to insulin than the homotypic I *(112)*. Moreover, the IR-α- and -β subunits are aberrantly processed with apparent molecular weights of 128 and 85 kDa in ScN2a, compared with 136 and 95 kDa in uninfected N2a cell *(111, 112)*, and the reason was shown to be due to altered glycosylation, by combined enzymatic or chemical deglycosylation of the I *(111, 112)*. In addition to these differences in IR properties, the basal ERK2 activity was significantly elevated and the insulin stimulated-ERK2 phosphorylation was subsequently decreased in ScN2a. This is interesting because antibody-mediated ligation of PrP^C induces activation of NADPH oxidase and ERK1/2 *(102)*. Thus, it can be speculated that the propagation and presence of PrP^{Sc} in raft domains may crosslink and oligomerize colocalized PrP^C, resulting in an uncontrolled NADPH oxidase and ERK1/2 activation.

The observed changes in IR/IGF-1R expression and function in ScN2a and ScN1E-115 cells, suggest that although these receptors are expressed, their folding, trafficking, or processing is disturbed by scrapie infection, resulting in decreased function, decreased levels of functional receptors, or both, which may contribute to neuronal cell death in prion diseases. Altered expression, processing, localization, and function have previously been described for several proteins in scrapie-infected cells and brain *(109, 110, 113–115)*.

The only prion-infected cell line exhibiting morphological signs of neurodegeneration and vacuolation is the ScGT1 cell line *(15)*. In scrapie-infected GT1 cells, stably transfected with trkA (the high-affinity nerve growth factor [NGF] receptor)-ScGT1-TrkA, NGF treatment increased the viability and reduced the morphological signs of apoptosis. ScGT1 cells also display a higher sensitivity to induced oxidative stress than GT1 cells *(16)*, together with an increased lipid peroxidation and reduction in the activities of the glutathione-dependent and superoxide dismutase antioxidant systems. These findings are indeed very important, because ScGT1 cells reproduce some of the major pathological changes found in prion-infected brains; therefore, they represent a good system for studying the molecular mechanisms underlying prion-induced neurodegeneration. Using the same cell line, we have recently shown that scrapie infection induces an important increase in the level of active Src kinase, together with a Src-dependent increase in protein tyrosine phosphorylation of several high-molecular-weight proteins *(116)*. The same result also was obtained in ScN2a cells *(116)*. Immunohistochemical studies of brains of scrapie-infected mice further confirm these results *(117)*. These findings suggest that an abnormal Src-kinase activation and subsequent protein tyrosine phosphorylation may be key elements in the neuropathology of the prion diseases.

9. Concluding Remarks

Although the establishment of scrapie-infected cell lines or cell lines expressing mutant PrP^C linked to hereditary prion diseases has been crucial for the understanding of the biogenesis and metabolism of PrP^{Sc}, the weak point in using scrapie-infected cell cultures to study molecular mechanisms underlying prion-induced neurodegeneration is the lack of apparent neurotoxicity in these cultures.

The discrepancy between in vivo and in vitro PrP^{Sc} neurotoxicity might be due to the transformed phenotypes of the available cell culture models today, masking the PrP^{Sc} neurotoxicity that may manifest only in finally differentiated cells, resembling more the postmitotic character of the adult CNS. This may explain why ScGT1 cells, which have a more differentiated neuronal phenotype than ScN2a and ScN1E-115 cells, do display signs of neurodegeneration, whereas ScN2a and ScN1E-115 cells do not.

An alternative strategy to the use of scrapie-infected immortalized cell lines has been to expose primary neuronal cultures to purified PrP^{Sc}; however, the acute toxic effect triggered by high doses of exogenous PrP^{Sc} probably differs from the delayed neurodegeneration in neurons propagating prions *(118, 119)*. To adress this issue, Cronier et al. *(120)* developed a cell system based on primary cerebellar cultures established from transgenic mice expressing ovine PrP and then exposed the cultures to sheep scrapie. Interestingly, those cultures were found to accumulate *de novo* PrP^{Sc} and infectivity, as assessed by mouse bioassay. Moreover, contrasting with data obtained in chronically infected cell lines, late-occurring apoptosis was consistently demonstrated in the infected primary neuronal cultures. This approach thus offers promise to study the mechanisms involved in prion-triggered neurodegeneration at a cellular level.

Alternatively, the formation, accumulation, or both of PrP^{Sc} are not neurotoxic per se but a "neuro-stressant," and they require an interaction with microglia/astrocytes to be neurotoxic. Although the cell types involved in scrapie pathology have not been completely identified, the emerging picture indicates a glial-derived inflammatory component in prion pathogenesis *(121)*. Only circumstantial evidence suggests that PrP^{Sc} is neurotoxic per se because its accumulation in scrapie-infected brains correlates to areas of vacuolation and astrogliosis. Studies addressing the spatial and temporal relationships among PrP^{Sc} accumulation, glial activation, neuronal vacuolation, and apoptosis indicate that microglia and astroglia activation precedes that of spongiform degeneration *(122)*. Together, this indicates an indirect glial neurotoxic effect, possibly mediated by micro- and astroglial cytokines and ROS in scrapie pathogenesis—an inflammatory "vicious cycle" also suggested as causative/contributory in other chronic neurodegenerative diseases, such as Alzheimer's and amyotrophic lateral sclerosis (**Fig. 2**).

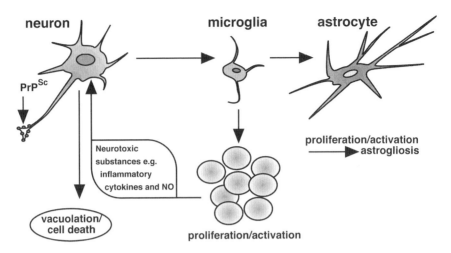

Fig. 2 Involvement of an inflammatory loop in prion-induced neurodegeneration. I hypothesize that the formation and accumulation of PrP^{Sc} in combination with an inflammatory glial response has a synergistic deteriorating effect on neurons, together adding up to a fatal threshold level of cellular stress resulting in neuronal apoptosis. Therefore, a truer cell culture model of prion-induced neurodegeneration might require the creation of a "mini-brain" in the culture dish, constituting besides neuronal cells also glial cells

References

1. Field, EJ, Windsor, GD (1965) Cultural characters of scrapie mouse brain. Res Vet Sci; 35:130–132.
2. Caspary, EA, Bell, TM (1971) Growth potential of scrapie mouse brain in vitro. Nature; 229:269–270.
3. Clarke, MC, Haig, DA (1970) Evidence for the multiplication of scrapie agent in cell culture. Nature; 225:100–101.
4. Bate, C, Salmona, M, Diomede, L, Williams, A (2004) Squalestatin cures prion-infected neurons and protects against prion neurotoxicity. J Biol Chem; 279:14983–14990.
5. Birkett, CR, Hennion, RM, Bembridge, DA, Clarke, MC, Chree, A, Bruce, ME, Bostock, CJ (2001) Scrapie strains maintain biological phenotypes on propagation in a cell line in culture. EMBO J; 20:3351–3358.
6. Haig, DA, Pattison, IH (1967) In vitro growth of pieces of brain from scrapie-affected mice. J Pathol Bacteriol; 93:724–727.
7. Race, R (1991) The scrapie agent in vitro. Curr Top Microbiol Immunol; 172:181–193.
8. Butler, DA, Scott, MR, Bockman, JM, Borchelt, DR, Taraboulos, A, Hsiao, KK, Kingsbury, DT, Prusiner, SB (1988) Scrapie-infected murine neuroblastoma cells produce protease-resistant prion proteins. J Virol; 62:1558–1564.
9. Race, RE, Fadness, LH, Chesebro, B (1987) Characterization of scrapie infection in mouse neuroblastoma cells. J Gen Virol; 68:1391–1399.
10. Markovits, P, Dautheville, C, Dormont, D, Dianoux, L, Latarjet, R (1983) In vitro propagation of the scrapie agent. I. Transformation of mouse glia and neuroblastoma cells after infection with the mouse-adapted scrapie strain c-506. Acta Neuropathol (Berl); 60:75–80.

11. Östlund, P, Lindegren, H, Pettersson, C, Bedecs, K (2001) Up-regulation of functionally impaired insulin-like growth factor-1 receptor in scrapie-infected neuroblastoma cells. J Biol Chem; 276:36110–36115.
12. Nishida, N, Harris, DA, Vilette, D, Laude, H, Frobert, Y, Grassi, J, Casanova, D, Milhavet, O, Lehmann, S (2000) Successful transmission of three mouse-adapted scrapie strains to murine neuroblastoma cell lines overexpressing wild-type mouse prion protein. J Virol; 74:320–325.
13. Arjona, A, Simarro, L, Islinger, F, Nishida, N, Manuelidis, L (2004) Two Creutzfeldt-Jakob disease agents reproduce prion protein-independent identities in cell cultures. Proc Natl Acad Sci U S A; 101:8768–8773.
14. Ladogana, A, Liu, Q, Xi, YG, Pocchiari, M (1995) Proteinase-resistant protein in human neuroblastoma cells infected with brain material from Creutzfeldt-Jakob patient. Lancet; 345:594–595.
15. Schatzl, HM, Laszlo, L, Holtzman, DM, Tatzelt, J, DeArmond, SJ, Weiner, RI, Mobley, WC, Prusiner, SB (1997) A hypothalamic neuronal cell line persistently infected with scrapie prions exhibits apoptosis. J Virol; 71:8821–8831.
16. Milhavet, O, McMahon, HE, Rachidi, W, Nishida, N, Katamine, S, Mange, A, Arlotto, M, Casanova, D, Riondel, J, Favier, A, Lehmann, S (2000) Prion infection impairs the cellular response to oxidative stress. Proc Natl Acad Sci U S A; 97:13937–13942.
17. Rubenstein, R, Carp, RI, Callahan, SM (1984) In vitro replication of scrapie agent in a neuronal model: infection of PC12 cells. J Gen Virol; 65:2191–2198.
18. Rubenstein, R, Deng, H, Race, RE, Ju, W, Scalici, CL, Papini, MC, Kascsak, RJ, Carp, RI (1992) Demonstration of scrapie strain diversity in infected PC12 cells. J Gen Virol; 73:3027–3031.
19. Baron, GS, Magalhaes, AC, Prado, MA, Caughey, B (2006) Mouse-adapted scrapie infection of SN56 cells: greater efficiency with microsome-associated versus purified PrP-res. J Virol; 80:2106–2117.
20. Follet, J, Lemaire-Vieille, C, Blanquet-Grossard, F, Podevin-Dimster, V, Lehmann, S, Chauvin, JP, Decavel, JP, Varea, R, Grassi, J, Fontes, M, Cesbron, JY (2002) PrP expression and replication by Schwann cells: implications in prion spreading. J Virol; 76:2434–2439.
21. Archer, F, Bachelin, C, Andreoletti, O, Besnard, N, Perrot, G, Langevin, C, Le Dur, A, Vilette, D, Baron-Van Evercooren, A, Vilotte, JL, Laude, H (2004) Cultured peripheral neuroglial cells are highly permissive to sheep prion infection. J Virol; 78:482–490.
22. Taraboulos, A, Serban, D, Prusiner, SB (1990) Scrapie prion proteins accumulate in the cytoplasm of persistently infected cultured cells. J Cell Biol; 110:2117–2132.
23. Roikhel, VM, Fokina, GI, Lisak, VM, Kondakova, LI, Korolev, MB, Pogodina, VV (1987) Persistence of the scrapie agent in glial cells from rat Gasserian ganglion. Acta Virol; 31:36–42.
24. Haig, DA, Clarke, MC (1971) Multiplication of the scrapie agent. Nature; 234:106–107.
25. Clarke, MC, Millson, GC (1976) Infection of a cell line of mouse L fibroblasts with scrapie agent. Nature; 261:144–145.
26. Cherednichenko Yu, N, Mikhailova, GR, Rajcani, J, Zhdanov, VM (1985) In vitro studies with the scrapie agent. Acta Virol; 29:285–293.
27. Vorberg, I, Raines, A, Story, B, Priola, SA (2004) Susceptibility of common fibroblast cell lines to transmissible spongiform encephalopathy agents. J Infect Dis; 189:431–439.
28. Elleman, CJ (1984) Attempts to establish the scrapie agent in cell lines. Vet Res Commun; 8:309–316.
29. Vilette, D, Andreoletti, O, Archer, F, Madelaine, MF, Vilotte, JL, Lehmann, S, Laude, H (2001) Ex vivo propagation of infectious sheep scrapie agent in heterologous epithelial cells expressing ovine prion protein. Proc Natl Acad Sci U S A; 98:4055–4059.
30. Sabuncu, E, Petit, S, Le Dur, A, Lan Lai, T, Vilotte, JL, Laude, H, Vilette, D (2003) PrP polymorphisms tightly control sheep prion replication in cultured cells. J Virol; 77:2696–2700.
31. Dlakic, WM, Grigg, E, Bessen, RA (2007) Prion infection of muscle cells in vitro. J Virol; 81:4615–4624.

32. Klebe, RJ, Ruddle, FH (1969) Neuroblastoma: cell culture analysis of a differentiating stem cell system. J Cell Biol; 43:69A.
33. Chesebro, B, Wehrly, K, Caughey, B, Nishio, J, Ernst, D, Race, R (1993) Foreign PrP expression and scrapie infection in tissue culture cell lines. Dev Biol Stand; 80:131–140.
34. Mellon, PL, Windle, JJ, Goldsmith, PC, Padula, CA, Roberts, JL, Weiner, RI (1990) Immortalization of hypothalamic GnRH neurons by genetically targeted tumorigenesis. Neuron; 5:1–10.
35. Bosque, PJ, Prusiner, SB (2000) Cultured cell sublines highly susceptible to prion infection. (2000) J Virol; 74:4377–4386.
36. Race, RE, Caughey, B, Graham, K, Ernst, D, Chesebro, B (1988) Analyses of frequency of infection, specific infectivity, and prion protein biosynthesis in scrapie-infected neuroblastoma cell clones. J Virol; 62:2845–2849.
37. Borchelt, DR, Scott, M, Taraboulos, A, Stahl, N, Prusiner, SB (1990) Scrapie and cellular prion proteins differ in their kinetics of synthesis and topology in cultured cells. J Cell Biol; 110:743–752.
38. Caughey, B, Raymond, GJ (1991) The scrapie-associated form of PrP is made from a cell surface precursor that is both protease- and phospholipase-sensitive. J Biol Chem; 266:18217–18223.
39. Gorodinsky, A, Harris, DA (1995) Glycolipid-anchored proteins in neuroblastoma cells form detergent-resistant complexes without caveolin. J Cell Biol; 129:619–627.
40. Taraboulos, A, Scott, M, Semenov, A, Avrahami, D, Laszlo, L, Prusiner, SB, Avraham, D (1995) Cholesterol depletion and modification of COOH-terminal targeting sequence of the prion protein inhibit formation of the scrapie isoform. J Cell Biol; 129:121–132.
41. Caughey, B, Race, RE (1992) Potent inhibition of scrapie-associated PrP accumulation by congo red. J Neurochem; 59:768–771.
42. Winklhofer, KF, Tatzelt, J (2000) Cationic lipopolyamines induce degradation of PrPSc in scrapie-infected mouse neuroblastoma cells. Biol Chem; 381:463–469.
43. Supattapone, S, Nishina, K, Rees, JR (2002) Pharmacological approaches to prion research. Biochem Pharmacol; 63:1383–1388.
44. Prusiner, SB (1998) Prions. Proc Natl Acad Sci U S A; 95:13363–13383.
45. Bolton, DC, McKinley, MP, Prusiner, SB (1984) Molecular characteristics of the major scrapie prion protein. Biochemistry; 23:5898–5906.
46. Korth, C, Streit, P, Oesch, B (1999) Monoclonal antibodies specific for the native, disease-associated isoform of the prion protein. Methods Enzymol; 309:106–122.
47. Paramithiotis, E, Pinard, M, Lawton, T, LaBoissiere, S, Leathers, VL, Zou, WQ, Estey, LA, Lamontagne, J, Lehto, MT, Kondejewski, LH, Francoeur, GP, Papadopoulos, M, Haghighat, A, Spatz, SJ, Head, M, Will, R, Ironside, J, O'Rourke, K, Tonelli, Q, Ledebur, HC, Chakrabartty, A, Cashman, NR (2003) A prion protein epitope selective for the pathologically misfolded conformation. Nat Med; 9:893–899.
48. Manuelidis, L, Sklaviadis, T, Manuelidis, EE (1987) Evidence suggesting that PrP is not the infectious agent in Creutzfeldt-Jakob disease. EMBO J; 6:341–347.
49. Hsiao, KK, Groth, D, Scott, M, Yang, SL, Serban, H, Rapp, D, Foster, D, Torchia, M, Dearmond, SJ, Prusiner, SB (1994) Serial transmission in rodents of neurodegeneration from transgenic mice expressing mutant prion protein. Proc Natl Acad Sci U S A; 91:9126–9130.
50. Lasmezas, CI, Deslys, JP, Robain, O, Jaegly, A, Beringue, V, Peyrin, JM, Fournier, JG, Hauw, JJ, Rossier, J, Dormont, D (1997) Transmission of the BSE agent to mice in the absence of detectable abnormal prion protein. Science; 275:402–405.
51. Hill, AF, Antoniou, M, Collinge, J (1999) Protease-resistant prion protein produced in vitro lacks detectable infectivity. J Gen Virol; 80:11–14.
52. Shaked, GM, Fridlander, G, Meiner, Z, Taraboulos, A, Gabizon, R (1999) Protease-resistant and detergent-insoluble prion protein is not necessarily associated with prion infectivity. J Biol Chem; 274:17981–17986.
53. Klöhn, PC, Stoltze, L, Flechsig, E, Enari, M, Weissmann, C (2003) A quantitative, highly sensitive cell-based infectivity assay for mouse scrapie prions. Proc Natl Acad Sci U S A; 100:11666–11671.

54. Winklhofer, KF, Hartl, FU, Tatzelt, J (2001) A sensitive filter retention assay for the detection of PrP(Sc) and the screening of anti-prion compounds. FEBS Lett; 503:41–45.
55. Serban, D, Taraboulos, A, DeArmond, SJ, Prusiner, SB (1990) Rapid detection of Creutzfeldt-Jakob disease and scrapie prion proteins. Neurology; 40:110–117.
56. Caughey, B, Raymond, GJ, Ernst, D, Race, RE (1991) N-terminal truncation of the scrapie-associated form of PrP by lysosomal protease(s): implications regarding the site of conversion of PrP to the protease-resistant state. J Virol; 65:6597–6603.
57. Kurschner, C, Morgan, JI (1995) The cellular prion protein (PrP) selectively binds to Bcl-2 in the yeast two-hybrid system. Brain Res Mol Brain Res; 30:165–168.
58. Edenhofer, F, Rieger, R, Famulok, M, Wendler, W, Weiss, S, Winnacker, EL (1996) Prion protein PrP^C interacts with molecular chaperones of the Hsp60 family. J Virol; 70:4724–4728.
59. Rieger, R, Edenhofer, F, Lasmezas, CI, Weiss, S (1997) The human 37-kDa laminin receptor precursor interacts with the prion protein in eukaryotic cells. Nat Med; 3:1383–1388.
60. Spielhaupter, C, Schatzl, HM (2001) PrP^C directly interacts with proteins involved in signaling pathways. J Biol Chem; 276:44604–44612.
61. Gauczynski, S, Peyrin, JM, Haik, S, Leucht, C, Hundt, C, Rieger, R, Krasemann, S, Deslys, JP, Dormont, D, Lasmezas, CI, Weiss, S (2001) The 37-kDa/67-kDa laminin receptor acts as the cell-surface receptor for the cellular prion protein. EMBO J; 20:5863–5875.
62. Leucht, C, Simoneau, S, Rey, C, Vana, K, Rieger, R, Lasmezas, CI, Weiss, S (2003) The 37 kDa/67 kDa laminin receptor is required for PrP(Sc) propagation in scrapie-infected neuronal cells. EMBO Rep; 4:290–295.
63. Chen, S, Mange, A, Dong, L, Lehmann, S, Schachner, M (2003) Prion protein as trans-interacting partner for neurons is involved in neurite outgrowth and neuronal survival. Mol Cell Neurosci; 22:227–233.
64. Zanata, SM, Lopes, MH, Mercadante, AF, Hajj, GN, Chiarini, LB, Nomizo, R, Freitas, AR, Cabral, AL, Lee, KS, Juliano, MA, de Oliveira, E, Jachieri, SG, Burlingame, A, Huang, L, Linden, R, Brentani, RR, Martins, VR (2002) Stress-inducible protein 1 is a cell surface ligand for cellular prion that triggers neuroprotection. EMBO J; 21:3307–3316.
65. Keshet, GI, Bar-Peled, O, Yaffe, D, Nudel, U, Gabizon, R (2000) The cellular prion protein colocalizes with the dystroglycan complex in the brain. J Neurochem; 75:1889–1897.
66. Schmitt-Ulms, G, Legname, G, Baldwin, MA, Ball, HL, Bradon, N, Bosque, PJ, Crossin, KL, Edelman, GM, DeArmond, SJ, Cohen, FE, Prusiner, SB (2001) Binding of neural cell adhesion molecules (N-CAMs) to the cellular prion protein. J Mol Biol; 314:1209–1225.
67. Diaz-Nido, J, Wandosell, F, Avila, J (2002) Glycosaminoglycans and beta-amyloid, prion and tau peptides in neurodegenerative diseases. Peptides; 23:1323–1332.
68. Gabizon, R, Meiner, Z, Halimi, M, Ben-Sasson, SA (1993) Heparin-like molecules bind differentially to prion-proteins and change their intracellular metabolic fate. J Cell Physiol; 157:319–325.
69. Caughey, B, Brown, K, Raymond, GJ, Katzenstein, GE, Thresher, W (1994) Binding of the protease-sensitive form of PrP (prion protein) to sulfated glycosaminoglycan and congo red. J Virol; 68:2135–2141.
70. Schonberger, O, Horonchik, L, Gabizon, R, Papy-Garcia, D, Barritault, D, Taraboulos, A (2003) Novel heparan mimetics potently inhibit the scrapie prion protein and its endocytosis. Biochem Biophys Res Commun; 312:473–479.
71. Caughey, B, Raymond, GJ (1993) Sulfated polyanion inhibition of scrapie-associated PrP accumulation in cultured cells. J Virol; 67:643–650.
72. Ehlers, B, Diringer, H (1984) Dextran sulphate 500 delays and prevents mouse scrapie by impairment of agent replication in spleen. J Gen Virol; 65:1325–1330.
73. Snow, AD, Kisilevsky, R, Willmer, J, Prusiner, SB, DeArmond, SJ (1989) Sulfated glycosaminoglycans in amyloid plaques of prion diseases. Acta Neuropathol (Berl); 77:337–342.
74. McBride, PA, Wilson, MI, Eikelenboom, P, Tunstall, A, Bruce, ME (1998) Heparan sulfate proteoglycan is associated with amyloid plaques and neuroanatomically targeted PrP pathology throughout the incubation period of scrapie-infected mice. Exp Neurol; 149:447–454.

75. Shyng, SL, Lehmann, S, Moulder, KL, Harris, DA (1995) Sulfated glycans stimulate endocytosis of the cellular isoform of the prion protein, PrPC, in cultured cells. J Biol Chem; 270:30221–30229.

76. Ben-Zaken, O, Tzaban, S, Tal, Y, Horonchik, L, Esko, JD, Vlodavsky, I, Taraboulos, A (2003) Cellular heparan sulfate participates in the metabolism of prions. J Biol Chem; 278:40041–40049.

77. Bueler, H, Fischer, M, Lang, Y, Bluethmann, H, Lipp, HP, DeArmond, SJ, Prusiner, SB, Aguet, M, Weissmann, C (1992) Normal development and behaviour of mice lacking the neuronal cell-surface PrP protein. Nature; 356:577–582.

78. Manson, JC, Clarke, AR, Hooper, ML, Aitchison, L, McConnell, I, Hope, J (1994) 129/Ola mice carrying a null mutation in PrP that abolishes mRNA production are developmentally normal. Mol Neurobiol; 8:121–127.

79. Sakaguchi, S, Katamine, S, Nishida, N, Moriuchi, R, Shigematsu, K, Sugimoto, T, Nakatani, A, Kataoka, Y, Houtani, T, Shirabe, S, Okada, H, Hasegawa, S, Miyamoto, T, Noda, T (1996) Loss of cerebellar Purkinje cells in aged mice homozygous for a disrupted PrP gene. Nature; 380:528–531.

80. Moore, R Gene targeting studies at the mouse prion protein locus. (1997) PhD thesis, University of Edinburgh, Edinburgh, Scotland.

81. Weissmann, C, Aguzzi, A (1999) Perspectives: neurobiology. PrP's double causes trouble. Science; 286:914–915.

82. Moore, RC, Lee, IY, Silverman, GL, Harrison, PM, Strome, R, Heinrich, C, Karunaratne, A, Pasternak, SH, Chishti, MA, Liang, Y, Mastrangelo, P, Wang, K, Smit, AF, Katamine, S, Carlson, GA, Cohen, FE, Prusiner, SB, Melton, DW, Tremblay, P, Hood, LE, Westaway, D (1999) Ataxia in prion protein (PrP)-deficient mice is associated with upregulation of the novel PrP-like protein doppel. J Mol Biol; 292:797–817.

83. Tremblay, P, Meiner, Z, Galou, M, Heinrich, C, Petromilli, C, Lisse, T, Cayetano, J, Torchia, M, Mobley, W, Bujard, H, DeArmond, SJ, Prusiner, SB (1998) Doxycycline control of prion protein transgene expression modulates prion disease in mice. Proc Natl Acad Sci U S A; 95:12580–12585.

84. Viles, JH, Cohen, FE, Prusiner, SB, Goodin, DB, Wright, PE, Dyson, HJ (1999) Copper binding to the prion protein: structural implications of four identical cooperative binding sites. Proc Natl Acad Sci U S A; 96:2042–2047.

85. Burns, CS, Aronoff-Spencer, E, Legname, G, Prusiner, SB, Antholine, WE, Gerfen, GJ, Peisach, J, Millhauser, GL (2003) Copper coordination in the full-length, recombinant prion protein. Biochemistry; 42:6794–6803.

86. Brown, DR, Qin, K, Herms, JW, Madlung, A, Manson, J, Strome, R, Fraser, PE, Kruck, T, von Bohlen, A, Schulz-Schaeffer, W, Giese, A, Westaway, D, Kretzschmar, H (1997) The cellular prion protein binds copper in vivo. Nature; 390:684–687.

87. Brown, DR, Schulz-Schaeffer, WJ, Schmidt, B, Kretzschmar, HA (1997) Prion protein-deficient cells show altered response to oxidative stress due to decreased SOD-1 activity. Exp Neurol; 146:104–112.

88. Pauly, PC, Harris, DA (1998) Copper stimulates endocytosis of the prion protein. J Biol Chem; 273:33107–33110.

89. Klamt, F, Dal-Pizzol, F, Conte da Frota, MJ, Walz, R, Andrades, ME, da Silva, EG, Brentani, RR, Izquierdo, I, Fonseca Moreira, JC (2001) Imbalance of antioxidant defense in mice lacking cellular prion protein. Free Radic Biol Med; 30:1137–1144.

90. Graner, E, Mercadante, AF, Zanata, SM, Forlenza, OV, Cabral, AL, Veiga, SS, Juliano, MA, Roesler, R, Walz, R, Minetti, A, Izquierdo, I, Martins, VR, Brentani, RR (2000) Cellular prion protein binds laminin and mediates neuritogenesis. Brain Res Mol Brain Res; 76:85–92.

91. Graner, E, Mercadante, AF, Zanata, SM, Martins, VR, Jay, DG, Brentani, RR (2000) Laminin-induced PC-12 cell differentiation is inhibited following laser inactivation of cellular prion protein. FEBS Lett; 482:257–260.

92. Monnet, C, Marthiens, V, Enslen, H, Frobert, Y, Sobel, A, Mege, RM (2003) Heterogeneity and regulation of cellular prion protein glycoforms in neuronal cell lines. Eur J Neurosci; 18:542–548.

93. Kuwahara, C, Takeuchi, AM, Nishimura, T, Haraguchi, K, Kubosaki, A, Matsumoto, Y, Saeki, K, Yokoyama, T, Itohara, S, Onodera, T (1999) Prions prevent neuronal cell-line death. Nature; 400:225–226.

94. Bounhar, Y, Zhang, Y, Goodyer, CG, LeBlanc, A (2001) Prion protein protects human neurons against Bax-mediated apoptosis. J Biol Chem; 276:39145–39149.

95. Chiarini, LB, Freitas, AR, Zanata, SM, Brentani, RR, Martins, VR, Linden, R (2002) Cellular prion protein transduces neuroprotective signals. EMBO J; 21:3317–3326.

96. Diarra-Mehrpour, M, Arrabal, S, Jalil, A, Pinson, X, Gaudin, C, Pietu, G, Pitaval, A, Ripoche, H, Eloit, M, Dormont, D, Chouaib, S (2004) Prion protein prevents human breast carcinoma cell line from tumor necrosis factor alpha-induced cell death. Cancer Res; 64:719–727.

97. Kim, BH, Lee, HG, Choi, JK, Kim, JI, Choi, EK, Carp, RI, Kim, YS (2004) The cellular prion protein (PrPC) prevents apoptotic neuronal cell death and mitochondrial dysfunction induced by serum deprivation. Brain Res Mol Brain Res; 124:40–50.

98. Paitel, E, Sunyach, C, Alves da Costa, C, Bourdon, JC, Vincent, B, Checler, F (2004) Primary cultured neurons devoid of cellular prion display lower responsiveness to staurosporine through the control of p53 at both transcriptional and post-transcriptional levels. J Biol Chem; 279:612–618.

99. Mouillet-Richard, S, Ermonval, M, Chebassier, C, Laplanche, JL, Lehmann, S, Launay, JM, Kellermann, O (2000) Signal transduction through prion protein. Science; 289:1925–1928.

100. Toni, M, Spisni, E, Griffoni, C, Santi, S, Riccio, M, Lenaz, P, Tomasi, V (2006) Cellular prion protein and caveolin-1 interaction in a neuronal cell line precedes fyn/erk 1/2 signal transduction. J Biomed Biotechnol; 69469.

101. Solforosi, L, Criado, JR, McGavern, DB, Wirz, S, Sanchez-Alavez, M, Sugama, S, DeGiorgio, LA, Volpe, BT, Wiseman, E, Abalos, G, Masliah, E, Gilden, D, Oldstone, MB, Conti, B, Williamson, RA (2004) Cross-linking cellular prion protein triggers neuronal apoptosis in vivo. Science; 303:1514–1516.

102. Schneider, B, Mutel, V, Pietri, M, Ermonval, M, Mouillet-Richard, S, Kellermann, O (2003) NADPH oxidase and extracellular regulated kinases 1/2 are targets of prion protein signaling in neuronal and nonneuronal cells. Proc Natl Acad Sci U S A; 100:13326–13331.

103. Morel, E, Fouquet, S, Chateau, D, Yvernault, L, Frobert, Y, Pincon-Raymond, M, Chambaz, J, Pillot, T, Rousset, M (2004) The cellular prion protein PrPc is expressed in human enterocytes in cell-cell junctional domains. J Biol Chem; 279:1499–1505.

104. Mattei, V, Garofalo, T, Misasi, R, Circella, A, Manganelli, V, Lucania, G, Pavan, A, Sorice, M (2004) Prion protein is a component of the multimolecular signaling complex involved in T cell activation. FEBS Lett; 560:14–18.

105. Ertmer, A, Gilch, S, Yun, SW, Flechsig, E, Klebl, B, Stein-Gerlach, M, Klein, MA, Schatzl, HM (2004) The tyrosine kinase inhibitor STI571 induces cellular clearance of PrPSc in prion-infected cells. J Biol Chem; 279:41918–41927.

106. Beggs, HE, Soriano, P, Maness, PF (1994) NCAM-dependent neurite outgrowth is inhibited in neurons from Fyn-minus mice. J Cell Biol; 127:825–833.

107. Kasahara, K, Sanai, Y (1999) Possible roles of glycosphingolipids in lipid rafts. Biophys Chem; 82:121–127.

108. Wu, CB, Butz, S, Ying, YS, Anderson, RGW (1997) Tyrosine kinase receptors concentrated in caveolae-like domains from neuronal plasma membrane. J Biol Chem; 272:3554–3559.

109. DeArmond, SJ, Kristensson, K, Bowler, RP (1992) PrPSc causes nerve cell death and stimulates astrocyte proliferation: a paradox. Prog Brain Res; 94:437–446.

110. Wong, K, Qiu, Y, Hyun, W, Nixon, R, VanCleff, J, Sanchez-Salazar, J, Prusiner, SB, DeArmond, SJ (1996) Decreased receptor-mediated calcium response in prion-infected cells correlates with decreased membrane fluidity and IP3 release. Neurology; 47:741–750.

111. Östlund, P, Lindegren, H, Pettersson, C, Bedecs, K Altered insulin receptor processing and function in scrapie-infected neuroblastoma cell lines. Brain Res Mol Brain Res 2001;97:161–170.

112. Nielsen, D, Gyllberg, H, Östlund, P, Bergman, T, Bedecs, K (2004) Increased levels of insulin and insulin-like growth factor-1 hybrid receptors and decreased glycosylation of the insulin

receptor alpha- and beta-subunits in scrapie-infected neuroblastoma N2a cells. Biochem J; 380:571–579.

113. Diez, M, Koistinaho, J, Dearmond, SJ, Groth, D, Prusiner, SB, Hokfelt, T (1997) Marked decrease of neuropeptide Y Y2 receptor binding sites in the hippocampus in murine prion disease. Proc Natl Acad Sci U S A; 94:13267–13272.

114. Tatzelt, J, Zuo, J, Voellmy, R, Scott, M, Hartl, U, Prusiner, SB, Welch, WJ (1995) Scrapie prions selectively modify the stress response in neuroblastoma cells. Proc Natl Acad Sci U S A; 92:2944–2948.

115. Ovadia, H, Rosenmann, H, Shezen, E, Halimi, M, Ofran, I, Gabizon, R (1996) Effect of scrapie infection on the activity of neuronal nitric-oxide synthase in brain and neuroblastoma cells. J Biol Chem; 271:16856–16861.

116. Gyllberg, H, Lofgren, K, Lindegren, H, Bedecs, K (2006) Increased Src kinase level results in increased protein tyrosine phosphorylation in scrapie-infected neuronal cell lines. FEBS Lett; 580:2603–2608.

117. Nixon, RR (2005) Prion-associated increases in Src-family kinases. J Biol Chem; 280:2455–2462.

118. Giese, A, Brown, DR, Groschup, MH, Feldmann, C, Haist, I, Kretzschmar, HA (1998) Role of microglia in neuronal cell death in prion disease. Brain Pathol; 8:449–457.

119. Muller, WE, Ushijima, H, Schroder, HC, Forrest, JM, Schatton, WF, Rytik, PG, Heffner-Lauc, M (1993) Cytoprotective effect of NMDA receptor antagonists on prion protein (PrionSc)-induced toxicity in rat cortical cell cultures. Eur J Pharmacol; 246:261–267.

120. Cronier, S, Laude, H, Peyrin, JM (2004) Prions can infect primary cultured neurons and astrocytes and promote neuronal cell death. Proc Natl Acad Sci U S A; 101:12271–12276.

121. Ju, WK, Park, KJ, Choi, EK, Kim, J, Carp, RI, Wisniewski, HM, Kim, YS (1998) Expression of inducible nitric oxide synthase in the brains of scrapie-infected mice. J Neurovirol; 4:445–450.

122. Williams, A, Van Dam, AM, Ritchie, D, Eikelenboom, P, Fraser, H (1997) Immunocytochemical appearance of cytokines, prostaglandin E2 and lipocortin-1 in the CNS during the incubation period of murine scrapie correlates with progressive PrP accumulations. Brain Res; 754:171–180.

Chapter 2
Investigation of PrP^C Metabolism and Function in Live Cells

Methods for Studying Individual Cells and Cell Populations

Cathryn L. Haigh and David R. Brown

Summary Prion protein (PrP)^c expression levels and protein localization are known to be affected by factors such as metal ions and oxidative stress. By the development of a green fluorescent protein (GFP)-PrP^c fusion protein, the movement of PrP can be followed in real time. Furthermore, alterations in cellular metabolism can be detected while cells are still viable. The internalization response of PrP to 20 µM manganese (Mn) in divalent metal ion-depleted media is used to demonstrate the movement of GFP-tagged proteins in live cells and real time. A live cell microtiter plate assay shows that PrP null cells are less capable of dealing with Mn-induced oxidative stress. In addition, this chapter outlines several complementary techniques for studying live cells and GFP fusion proteins.

Keywords Confocal imaging; colocalization; dichlorodihydrofluorescein diacetate (DCFDA); native polyacrylamide gel electrophoresis (PAGE); reactive oxygen species (ROS); trafficking.

1. Introduction

Prion protein (PrP)^c is a cell surface protein, bound to the outer leaflet of the cell membrane by a glycosylphosphatidylinositol (GPI) anchor. PrP^c is trafficked through the cell membrane via the exocytic pathway *(1, 2)*. The octameric repeat domain of the PrP^c N terminus is able to bind four copper (Cu) molecules, with a fifth copper binding site located adjacent, and C-terminal, to the repeat domain *(3–6)*. Previous studies have shown that PrP^c is internalized from the cell membrane in response to copper *(7, 8)*. The rate of this reaction is dependent both on copper concentration and on functional domains of the prion protein *(8–10)*.

From: *Prion Protein Protocols.*
Methods in Molecular Biology, Vol. 459.
Edited by: A. F. Hill © Humana Press, Totowa, NJ

Early changes in prion disease include altered metal ion concentrations in the brain and an associated inability to deal with oxidative stress insults *(11–13)*. Therefore, PrP^C may have a role as a protective molecule. The internalization of PrP^C in response to copper has prompted the theory that PrP^C acts as a copper chaperone, delivering the metal ion to enzymes within the cell such as superoxide dismutases (SODs), which mediate the cellular protective response against oxidative stress *(14)*. In addition, PrP^C has been shown to have SOD-like activity both in vitro and in vivo *(12, 15, 16)*, so it may itself impart a protective response at the cell membrane. The SOD-like response of PrP^C is dependent on copper binding to the octameric repeat region, and it requires at least two copper molecules to be bound *(15)*. The alteration in metal ion concentrations in the brain during disease may interfere with copper binding. Other metals, including manganese (Mn) and zinc (Zn), have been shown to bind to the octarepeats, albeit with lower affinities *(6, 17)*, and the binding of these metals may interfere with the normal function of PrP^C during disease progression.

Techniques looking at the activity of living cells are valuable tools providing information on how both individual cells and cell populations respond to stimuli under simulated physiological conditions. Cellular activity can be monitored in real time, giving accurate information on reaction rates and cellular metabolism. These techniques are described herein with examples of how they have been applied to studying the response of PrP^C null and wild-type cells to Mn.

2. Materials

2.1. Cell Culture and Transfection

1. F14 mouse PrP null and F21 mouse PrP wild-type neuroblastoma fusion cell lines *(17)*.
2. Dulbecco's modified Eagle's medium (DMEM) with 4500 mg/l glucose and minus L-glutamine (Sigma-Aldrich, Poole, Dorset, UK), supplemented with 10% (v/v) heat-inactivated fetal bovine serum (FBS) (Sigma-Aldrich), and 5 ml of 100× penicillin/streptomycin solution (Sigma-Aldrich).

   a. The FBS and penicillin/streptomycin solutions should be aliquoted appropriately so as to avoid repeated freeze-thawing.

3. FuGENE 6 transfection reagent (Roche Diagnostics, Indianapolis, IN).
4. Trypsin-EDTA solution (Sigma-Aldrich).
5. G-418 (Geneticin®, Invitrogen, Paisley, UK).

   a. A stock solution (100 mg/ml, w/v) should be prepared based on the potency of the antibiotic, this information is provided by the manufacturer.

b. To calculate the volume of water in which to dilute the antibiotic use the following formulas: potency (g/g) × weight (g) = active weight (g) and active weight (g)/desired final concentration (g/ml) = volume (ml).

c. Dilute all the provided G-418 powder in the volume of water calculated.

d. Rinse out the container in which it was provided, and ensure the powder is fully dissolved.

e. Filter sterilize the solution by using a syringe to pass the liquid through a 0.22-nm polyvinylidene difluoride (PVDF) filter, and aliquot appropriately.

f. Store at 4°C until use or at −20°C for longer term storage.

2.2. Live Cell Trafficking

1. Four-well borosilicate-chambered coverslips (LabTek, supplied by Fisher Scientific, Loughborough, Leicestershire, UK).
2. Opti-MEM® I reduced serum medium (1×), liquid (without phenol red) (Invitrogen).
3. Chelex-treated DMEM.

   a. Normal DMEM treated with chelex ion exchange resin to remove divalent ions.

   b. Weigh 1 g of Chelex-100 resin (Bio-Rad, Hercules, CA) into a sterile 50-ml tube.

   c. Transfer 20 ml of media into the tube and agitate for 1 h at room temperature.

   d. Allow resin to settle and separate from the media.

   e. Pipette media into the wells of a six-well plate (carefully so as not to collect the resin in the pipette).

   f. Allow the media to equilibrate under normal cell culture conditions until the pH balance is restored.

4. Cell dissociation buffer, enzyme free, Hanks'-based (Invitrogen).

2.3. Colocalization Studies

1. Four-well borosilicate-chambered coverslips (LabTek, supplied by Fisher Scientific).
2. High-quality anhydrous dimethyl sulfoxide (DMSO) (Sigma-Aldrich).
3. Cell dissociation buffer, enzyme free, Hanks'-based (Invitrogen).
4. Hanks' balanced salt solution (Sigma-Aldrich).
5. Lyso-Tracker® Red (Invitrogen).
6. Mito-Tracker Red (Invitrogen).

a. Probe should be dissolved in high-quality anhydrous DMSO to a final concentration of 1 mM.
b. Aliquot appropriately (avoid freeze-thaw cycles).
c. Store at −20°C.

7. N-((4-(4,4-Difluoro-5-(2-thienyl)-4-bora-3a,4a-diaza-s-indacene-3-yl)phenoxy) acetyl)sphingosine (BODIPY® TR ceramide) Golgi probe (Invitrogen).

a. Prepare as a bovine serum albumin (BSA)-complexed solution, as described in the product protocol.
b. Briefly, the 250 µg of the probe provided is dissolved in 354 µl of a 19:1 (v/v) chloroform:ethanol solution to produce a 1 mM stock solution of probe.
c. 50 µl of stock solution is aliquoted into a small glass test tube, dried under a stream of nitrogen, and left under vacuum for 1 h.
d. Dry probe is redissolved in 200 µl of absolute ethanol.
e. 10 ml of Hanks' buffer with 10 mM HEPES, pH 7.4, is prepared in a 50ml Falcon tube.
f. 3.4 mg of defatted BSA (Sigma-Aldrich) is added to the solution above.
g. The probe is added to the tube while agitating on a vortex mixer.
h. The resulting solution is 5 µM sphingolipid–5 µM defatted BSA solution.

8. pDsRed2-ER vector (Clontech BD Biosciences, Mountain View, CA).
9. Opti-MEM I reduced serum medium (1×), liquid (without phenol red) (Invitrogen).

2.4. 2′,7′-Dichlorodihydrofluorescein Diacetate (DCFDA) Assay for Reactive Oxygen Species

1. 96-well cell culture microtiter plate (Nunc, supplied by Fisher Scientific).
2. 5-(and-6)-Chloromethyl-2′,7′-dichlorodihydrofluorescein diacetate, acetyl ester (CM-H$_2$DCFDA) (Invitrogen).
3. Opti-MEM I reduced serum medium (1×), liquid (without phenol red) (Invitrogen).
4. Fluorescent microplate reader with light source and filter set suitable for detecting fluorescein isothiocyanate (FITC) (excitation, 488 nm; emission, 530 nm).
5. Phosphate-buffered saline tablets (Sigma-Aldrich).

a. Dissolve one tablet per 100 ml of deionized H$_2$O (dH$_2$O).
b. Autoclave to sterilize.
c. Store at room temperature.

6. High-quality anhydrous DMSO (Sigma-Aldrich).

2.5. Native PAGE

1. 30% (w/v) stabilized acrylamide solution, acrylamide/methylene bisacrylamide solution (37.5:1), with zero fluorescence (Protogel, supplied by Fisher, UK).
2. 1.5 M Tris-HCl, pH 8.8 (Merck, Darmstadt, Germany).
3. Ammonium persulfate (APS) (Sigma-Aldrich, UK): 10% (w/v) solution in water. Store powder away from moisture, and once in liquid form store at 4°C for immediate use or aliquot and store at −20°C.
4. N,N,N',N'-Tetramethylethylene-diamine (TEMED).
5. 4× native loading dye: 2.35 ml of dH_2O, 1.25 ml of 0.5 M Tris-HCl, pH 6.8, 6.0 ml of glycerol, 0.4 ml of 0.5% (w/v) bromophenol blue. Store at room temperature.
6. 10× native running buffer: 188 g of glycine, 30.2 g of Tris base, and dH_2O to 1 liter. Dilute 1 in 10 with distilled water for use. Store at room temperature.

3. Methods

Enhanced green fluorescent protein (EGFP) is the form of GFP that has become most widely available in cloning vectors. It is a red-shifted variant of GFP; this serves to make the excitation and emission spectra of the protein more compatible with the standard filter sets found on fluorescent microscopes. The mammalian expression vector pEGFP-C1 contains endonuclease recognition sites (for cleavage of the vector by the corresponding endonucleases) located C-terminal to the GFP gene sequence, which allows for the coding sequence of the gene of interest to be inserted C-terminal to the GFP signal. However, several sites are available N-terminal to the GFP sequence that also may be used for gene cloning. When designing a GFP fusion protein, care must be taken that the GFP tag does not interfere with functional domains of the protein or interrupt the open reading frame. PrP^C has a N-terminal signal sequence required to direct it into the endoplasmic reticulum (ER) and a C-terminal signal sequence that determines the attachment site of the GPI anchor, the GFP tag must not interfere with either of these sequences and so should not simply be attached at one end of the protein. In addition to these sequences, the C-terminal region of PrP^C is highly structured, and the N-terminal region contains the highly conserved octameric repeat motifs, which are also thought to be essential for protein function and so must not be altered by the presence of the GFP tag. Therefore, for the development of this fusion protein, it was determined most appropriate to insert the EGFP tag at amino acid 38 of the PrP^C coding sequence, because this amino acid is located after the N-terminal signal sequence but before the octameric repeat motifs *(17)*. This was achieved using both the N- and C-terminal restriction sites available in the expression vector. The resulting construct is termed GFP-PrP.

3.1. Transfection, Generation of Stable Cell Lines, and Routine Maintenance

1. Plate cells at 50% confluence in a six-well plate 24 h before transfection in normal culture medium. To do this, remove media from a T75 (75-cm^2 area) cell culture flask, wash once in Hanks' buffer, and then add 1 ml of 1× trypsin-EDTA solution and ensure complete coverage of the cell monolayer. Return cells to the incubator, and, depending on the adherence of the cell line, cells will lift off the base within 5–10 min. Resuspend cells in normal culture medium diluting appropriately to result in 50% coverage of the base of individual wells of a six-well plate when 2 ml of cell suspension is added per well.

2. Per transfection; pipette 100 μl of serum-free medium (DMEM) into a microcentrifuge tube.

3. For a 3:1 reagent:DNA reaction (*see* **Note 1**); pipette 3 μl of FuGENE 6 transfection reagent directly into the serum-free medium. Avoid contact with the sides of the tube, because this may inactivate the transfection reagent.

4. Tap gently to mix.

5. Add 1 ng of DNA directly into the serum-free medium–FuGENE 6 mixture.

6. Tap gently to mix.

7. Incubate for 15 min at room temperature (the reaction mixture is stable for up to 45 min).

8. Dropwise, add all the reaction mixture to the well and rock gently forward and backward followed by side to side to mix. Do not swirl; this will result in the mixture localizing to the center of the well.

9. Return the plate to the incubator for 24 h before starting antibiotic selection.

10. Add G-418 (*see* **Note 2**) at the concentration that corresponds to 90% of the lethal dose (LD$_{90}$) after a kill curve of 4 days (*see* **Note 3**). For the F14 cell line, this corresponds to 1 mg/ml (a 1 in 100 dilution of the 100 mg/ml stock of G-418).

11. Maintain cells in G-418 at LD$_{90}$ during normal culture.

3.2. Live Cell Trafficking

Changes in protein localization can be observed in cells expressing the GFP-PrP fusion construct in real time. Studies in live cells can be used to look at colocalization of proteins or alterations in movement of a protein of interest through the cell. These studies have been used to look at how increasing exogenous manganese concentrations alter the trafficking and localization of GFP-PrP.

1. Plate cells at approx 50% confluence in four-well chambered coverslips. To do this, remove the medium from a T75 flask of 85–95% confluent cells, wash once in Hanks' buffer, and resuspend cells using 1 ml of enzyme-free Hanks'-based cell dissociation buffer (*see* **Note 4**). Depending on the cell line used, cells will lift off the base of the flask after approx 10 min incubation at room temperature, the flask should be gently rocked every few minutes to ensure complete coverage

of the buffer. Cells should then be diluted suitably in 500 µl of culture medium per well to attain 50% confluence, and returned to the incubator allowing sufficient time for the cells to adhere to the coverslip (overnight should be suitable for most cell lines; *see* **Note 5**).

2. Before the start of the assay, transfer cells into fresh, prewarmed phenol red-free Opti-MEM (*see* **Note 6**). When performing experiments with Chelex-100–treated media, use media containing phenol red so pH balance can be ensured.

3. Locate cells of interest under a confocal laser scanning microscope and record the grid reference (most microscopes will do this digitally).

4. At the start of the assay, collect the cell image and optimize the picture; the setup and focal plane should then be maintained for the duration of the experiment.

5. Add the test and control reagents directly to the wells and the collect the cell images at appropriate times, 2 h after treatment was used for the copper internalization images shown in **Fig. 1**.

6. Throughout the assay, the cell environment should be maintained as close to ideal growth conditions as possible. This can be achieved using a CO_2 chamber

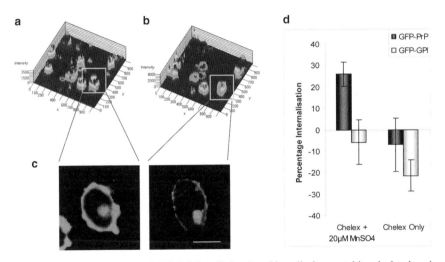

Fig. 1 MnSO$_4$ internalization of GFP-PrP in cells incubated in a divalent metal ion-depleted environment. Confocal images of a 146.2- ×146.2-µm field of GFP-PrP–expressing cells were observed using a ×63 magnification oil immersion lens and captured using the camera and digital imaging software for the Zeiss LS 550 confocal microscope. Images were the average of four scans with a 1.60-µs pixel time. All image settings including laser power, pinhole size, and detector gain were kept the same throughout the experiment. **A** and **B** represent the pixel intensity data plots produced by the Zeiss confocal imaging software on the entire microscope field. A is the image and data collected at time 0, and B is the same field captured after 2-h incubation in Chelex-pretreated media supplemented with 20 µM MnSO$_4$. **C** shows the reaction of an individual cell from this field. Bar = 10 µm. After the treatment, the membrane GFP-PrP intensity is visibly lower and the interior of the cell shows a more punctuate pattern of staining. The intensity data recorded in A and B was used to measure only the membrane intensity of GFP-PrP. The background fluorescence was subtracted, and the intensity change in response to the treatment was plotted as percentage of internalization, shown in **D**. D also includes the data acquired for a chelex-treated media only control and a GPI anchored GFP control to allow the PrP-specific effects to be detected. The plot is the average of 20 cells per condition from four different experiments. Error bars represent the SEM

and heat pad, and the cells should be out of the incubator for the shortest possible time when being transferred to the microscope.

7. Pixel intensity data is recorded by the confocal microscope software as part of the image file collected. This can used to quantify the location and intensity of protein signal depending on the aim of the assay.

3.3. Colocalization Studies

3.3.1. Preparation of Cells

For Lyso-Tracker, Mito-Tracker, and Golgi probes, plate cells at 50% confluence in four-well chambered coverslips as described in **step 1** of the trafficking studies protocol (*see* **Subheading 3.2.**), return to the incubator, and allow to adhere. For the endoplasmic reticulum (ER) probe, plate cells 2 days before assay at 30–50% confluence (adjust depending on the growth rate of the cells).

3.3.2. Lyso-Tracker Red

1. Dilute the Lyso-Tracker probe to a final concentration of 50 nM in prewarmed growth media.
2. Remove media from cells and replace with the fresh media containing probe. Incubate for 30 min under normal culture conditions.
3. Remove media containing probe and rinse cells once with Hanks' buffer.
4. Transfer cells into 500 µl of prewarmed phenol red-free Opti-MEM per well.
5. View under a confocal laser scanning microscope. The Lyso-Tracker probe can be visualized using optimal excitation (577-nm) and emission (590-nm) wavelengths.

3.3.3. Mito-Tracker Red

1. Dilute Mito-Tracker Red probe in prewarmed normal culture media to a final concentration of 100 nM.
2. Remove media from the cells and replace with the fresh media containing probe. Incubate for 30 min under normal culture conditions.
3. Remove media containing probe and rinse cells once with Hanks' buffer.
4. Transfer cells into 500 µl of prewarmed phenol red-free Opti-MEM per well.
5. View under a confocal laser scanning microscope. Visualize the probe using optimal excitation (588-nm) and emission (633-nm) wavelengths.

3.3.4. Sphingolipid Golgi Probe

1. Remove normal culture media from cells and wash once in Hanks' buffer.
2. Incubate cells in 200 µl of probe solution per well for 30 min at 4°C.

3. Wash cells three times in ice-cold Hanks' buffer.
4. Transfer cells into 500 μl of prewarmed normal culture media and incubate under normal culture conditions for 30 min.
5. Remove culture medium and replace with 500 μl of prewarmed phenol red-free Opti-MEM.
6. View under a confocal laser scanning microscope. Visualize probe using optimal excitation (589-nm) and emission (617-nm) wavelengths.

3.3.5. ER Probe

The pDsRed2-ER probe used is genetically encoded DsRed protein fused to the ER targeting sequence of calreticulin and the KDEL ER retention sequence, delivered in a plasmid vector. Colocalization studies were done using transient transfections.

1. For each well of the chambered coverslip to be transfected, pipette 20 μl of serum-free media into a microcentrifuge tube.
2. Add 0.6 μl of FuGENE6 transfection reagent directly into the serum-free medium.
3. Tap gently to mix.
4. Add 0.2 μg each of DNA encoding both the GFP-PrP and the DsRed2-ER constructs.
5. Tap gently to mix.
6. Incubate for 15 min at room temperature.
7. Add dropwise to each well.
8. Incubate for at least 24 hours before the start of the experiment to allow full expression of both proteins.
9. View under a confocal laser scanning microscope. The DsRed protein can be visualized using optimal excitation (558-nm) and emission (583-nm) wavelengths.

3.4. Detection of ROS in Live Cells

PrP^C has been proposed to have a role in cellular protection against oxidative stress. Direct measurement of ROS within live cells was made using a microplate assay using CM-H$_2$DCFDA, which is a chemically reduced, acetmethoxy ester of 2′,7′-dichlorofluorescein. This compound is not fluorescent until the acetate groups are removed by intracellular esterases and oxidation occurs within the cell. CM-H$_2$DCFDA is cell permeable until it is oxidized to its fluorescent product inside the cell. Oxidation may be induced by hydrogen peroxide (with the assistance of endogenous metal ions), organic hydroperoxides, nitric oxide, and peroxynitrite (*19*).

1. Plate cells so that they are 90–95% confluent and adhered to the plate at the time of the assay (preferably allow an overnight incubation before the start of the assay).
2. Prepare a 1 mM probe stock solution by diluting 50 μg (1 vial) into 86.54 μl of high-quality anhydrous DMSO.

3. Prepare the 5 µM working solution by diluting 5 µl of probe stock per 1 ml of phosphate-buffered saline (PBS).
4. Remove media from cells.
5. Add 50 µl of the 5 µM probe-PBS solution per well.
6. Incubate in the dark for 20 min at 37°C.
7. Remove probe solution.
8. Optional (*see* **Note 7**); wash 1 in PBS.
9. Add 100 µl of Opti-MEM to each well.
10. Read the time 0 fluorescence intensity by using fluorescent microplate reader set to FITC wavelengths excitation, 488 nm and emission, 530 nm. Readings should be taken from a minimum of three points per well to account for variation due to cell distribution.
11. Add control and test reagents to wells. Empty wells should be used as a background control and treated in the same way as test wells. Control and test conditions should be run at least in triplicate.
12. After reagent addition, plates should be kept under incubator conditions, and fluorescence intensity can be measured as often as required. The response of PrP^C null and wild-type cells to 100 µM $MnSO_4$ is shown in **Fig. 2**.

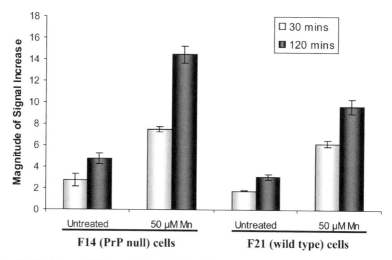

Fig. 2 $MnSO_4$ induced ROS production in PrP null and wild-type cells. The cells were treated for 2 h with 50 µM $MnSO_4$, and fluorescence readings were taken at time 0, 30 min, and 2 h. Untreated cells were followed simultaneously as controls for the production of ROS by normal cellular metabolism. An additional control (data not shown) is an empty well control—no cells are plated in these wells, but probe and media are added as per the protocol to monitor residual probe photoactivation. The data are plotted as "magnitude of signal increase"; this represents the ratio of the time x intensity reading to the time 0 (i.e., how many times greater the new signal is from the time 0 signal), and thus accounts for plating error or differing probe uptake of the cells in the individual wells. The bars are the mean of six repeats ($n = 6$) with each repeat containing four replicates, and the error bars represent SEM. The PrP null cells show increased ROS production, indicating a lesser ability to deal with Mn-induced oxidative stress than the wild-type cells

3.5. Detection of GFP by Nondenaturing Polyacrylamide Gel Electrophoresis (PAGE)

To detect proteins in their native form, nondenaturing PAGE can be used. Unlike sodium dodecyl sulfate-PAGE, proteins separate primarily according to their isoelectric point (pI), and size has a lesser influence; therefore, it is important to know the pI of the protein before attempting nondenaturing PAGE. The pI is defined as the pH where the protein carries no net charge (18). Because electrophoresis is separation of proteins in an electrical field from negative to positive, if the pI of the protein of interest is the same as the pH of the running buffer the protein will have no net charge and so will not migrate. If the pI is lower than the pH of the running buffer, the protein will carry a net negative charge and so will migrate away from the anode toward the cathode through the gel. However, if the pI of the protein is higher than the pH of the running buffer, it will carry a net positive charge and so will migrate in the opposite direction when the electric field is applied. Therefore, proteins with a higher pI than the pH of the running buffer should be run with the electrical field reversed, running positive to negative.

The pI can be calculated by entering the amino acid sequence of the gene of interest into ExPASy proteomics server.

3.5.1. Calculating pI

1. Obtain amino acid sequence of gene of interest (EGFP).
2. Go to http://www.expasy.org/.
3. Select "primary structure analysis."
4. Select "ProtParam" or "Compute pI/MW."
5. Enter amino acid sequence into box provided.
6. Click the compute button, and the data will be displayed.

3.5.2. Native PAGE

1. This method was developed using the ATTO minigel system (supplied by GRI, Dunmow, UK) but it can be adapted to any minigel system.
2. Plate cells in six-well culture plates to be 95% confluent at the end of the assay.
3. Treat cells as per experiment requirements.
4. Before extracting cells, wash in Hanks' balanced salt solution.
5. Extract cells by incubating each well in 100 µl of native extraction buffer at 37°C for 20 min. Ensure cell monolayer is evenly covered by the extraction buffer.
6. Collect cell lysates into microcentrifuge tubes and centrifuge briefly (18,000 xg for 1 min) to separate the soluble protein fraction from the cell debris. At this stage, the lysates can be frozen at −80°C until use. Samples should not be frozen and thawed on multiple occasions, and they will need to be centrifuged again after thawing.

7. Mix the lysate (protein fraction) 3:1 with 4× native loading buffer.
8. Assemble the gel casting apparatus. Wipe the glass plates with 70% ethanol to remove residual protein, place the rubber seal around the spacer glass plate, and place the flat glass plate over the top. Clamp together using the clips provided.
9. Prepare 12% native gels by mixing 3.39 ml of dH_2O, 4 ml of 30% (w/v) acrylamide solution, 2.5 ml of 1.5 M Tris base, pH 8.8, 100 µl of 10% (w/v) ammonium persulfate, and 10 µl of TEMED per gel. Add the reagents in the order listed as the APS and TEMED cause the gel to polymerize. Invert the tube gently to mix.
10. Pipette or pour gel mix into the cast ensuring no bubbles remain around the wells of the comb and allow to set. Gels usually set within 30 min (*see* **Note 8**).
11. Remove the rubber seal from around the gel.
12. Transfer the gel into the electrophoresis tank, comb/well side inward, and fully fill the tank center and half fill the outer with native running buffer. Check no bubbles are trapped under the gel, because this will interfere with the flow of the current, gently tilting the tank will allow trapped bubbles to float from under the gel.
13. Carefully remove the comb from the wells. Ensure no gel pieces remain in the wells, gently pipetting running buffer up and down in each well should remove any fragments.
14. Carefully pipette samples into wells. The maximum capacity is 20 µl/well. A native protein ladder (e.g., Native Marker, cat. no. LC0725, Invitrogen) can be used or a suitable recombinant protein can be run alongside to identify the protein of interest.
15. Run at 30 mA/gel for approx 1 h or until the dye elutes from the bottom of the gel.
16. For proteins with intrinsic fluorescence (such as GFP), the gel can be removed from the cast and visualized immediately using a phosphorimager set to appropriate wavelengths (excitation, 488 nm; emission, 530 nm for EGFP) (*see* **Note 9**).

Acknowledgements This work was funded by the Biochemical and Biological Sciences Research Council (UK).

4. Notes

1. This is the recommended starting ratio; however, it may need to be optimized for the cell line used. Other suggested ratios include 3:2 and 6:1. See the product insert for details.
2. Note that different plasmids have different antibiotic resistance markers. G-418 is described for use here with the pEGFP-C1 plasmid.
3. A kill curve of 4 days was used here, but time this can be varied. The LD_{90} differs for each different cell line, and it should be determined for each new cell line used.

4. An enzyme-free cell dissociation buffer should be used as opposed to trypsin, because PrP^C is a cell surface protein; thus, it will be cleaved from the cell surface by trypsin. This would alter the expression levels at the cell surface and as such, the data generated.

5. If the adherence of the cells is low or if they have a tendency to lift off the coverslip under experimental conditions, precoating the wells of the chambered coverslip with poly-D-lysine or other similar coating agent may be useful.

6. Phenol red-free reduced serum media is a preferable alternative to normal growth media, because minimal interference with the light source either from absorbance or light scattering of the media is desirable.

7. The wash step is not recommended for low-adherence cell lines, because cells might be dislodged, resulting in patchy well coverage and thus variable signal. Empty well background controls should be treated as for the test and control wells to monitor any background fluorescence from residual probe.

8. Gels should not be left for long periods at room temperature, because they will start to dry and shrink away from the cast. If the gel is not required immediately after it has been poured, it can be stored wrapped in a damp paper towel and cling film at 4°C.

9. Alternatively, proteins can be transferred onto PVDF or nitrocellulose membranes for further analysis.

References

1. Nunziante, M. Gilch, S. Schatzl, HM. (2003) Essential role of the prion protein N-terminus in subcellular trafficking and half-life of PrPc. *Journal of Biological Chemistry* **278**(6), 3726–3734.

2. Winklhofer, KF., Heske, J., Heller, U., Reintjes, A., Muranyi, W., Moarefi, I., Tatzelt, J. (2003) Determinants of the in vivo folding of the prion protein: a bipartate function of helix 1 in folding and aggregation. *Journal of Biological Chemistry* **278**(17), 14961–14970.

3. Hornshaw, MP., McDermott, JR., Candy, JM., Lakey, JH. (1995) Copper binding to the N-terminal tandem repeat region of mammalian and avian prion protein: structural studies using synthetic peptides. *Biochemical and Biophysical Research Communications* **214**(3), 993–999.

4. Brown, DR., Qin, K., Herms, JW., Madlung, A., Manson, J., Strome, R., Fraser, PE., Kruck, T., Bohlen, A., Schulz-Schaeffer, W., Giese, A., Westaway, D., Kretzschmar, H. (1997) The cellular prion protein binds copper in vivo. *Nature* **390**, 684–687.

5. Hasnain, SS., Murphy, LM., Strange, RW., Grossmann, JG., Clarke, AR., Jackson, GS., Collinge, J. (2001) XAFS study of the high-affinity copper-binding site of human PrP$^{91-231}$ and its low-resolution structure in solution. *Journal of Molecular Biology* **311**, 467–473.

6. Jackson, GS., Murray, I., Hosszu, LLP., Gibbs, N., Waltho, JP., Clarke, AR., Collinge, J. (2001) Location and properties of metal-binding sites on the human prion protein. *Proceedings of the National Academy of Sciences USA.* **98**(15), 8531–8535.

7. Pauly, PC., Harris, DA. (1998) Copper stimulates endocytosis of the prion protein. *Journal of Biological Chemistry* **273**(50), 33107–33110.

8. Lee, KS., Magalhães, AC., Zanata, SM., Brentani, RR., Martins, VR. Prado, MA. (2001) Internalisation of mammalian fluorescent cellular prion protein and N-terminal deletion mutants in living cells. *Journal of Neurochemistry* **79**(1), 79–87.

9. Sunyach, C., Jen, A., Deng, J., Fitzgerald, KT., Frobert, Y., Grassi, J., McCaffrey, MW., Morris, R. (2003) The mechanism of internalisation of glycosylphosphatidylinositol-anchored prion protein. *EMBO Journal* **22**(14), 3591–3601.
10. Haigh, CL., Edwards, KE., Brown, DR. (2005) Copper binding is the governing determinant of prion protein turnover. *Molecular and Cellular Neuroscience* **30**, 186–96.
11. Brown, DR., Schmidt, B., Kretzschmar, HA. (1998) Effects of copper on survival of prion protein knockout neurons and glia. *Journal of Neurochemistry* **70**, 1686–93.
12. Wong, BS., Brown, DR., Pan, T., Whitemann, M., Liu, T., Bu, X., Li, R., Gambetti, P., Olesik, J., Rubenstein, R., Sy, MS. (2001) Oxidative impairment in scrapie-infected mice is associated with brain metal perturbations and altered anti-oxidant activities. *Journal of Neurochemistry* **79**, 689–698.
13. Thackray, AM., Knight, R., Haswell, SJ., Bujdoso, R., Brown, DR. (2002) Metal imbalance and compromised anti-oxidant function are early changes in prion disease. *Biochemical Journal* **362**, 253–258.
14. Brown, DR., Besinger, A. (1998) Prion protein expression and superoxide dismutase activity. *Biochemical Journal* **334**, 423–429.
15. Brown, DR., Clive, C., Haswell, SJ. (2001) Antioxidant activity related to copper binding of native prion protein. *Journal of Neurochemistry* **76**, 69–76.
16. Brown, DR., Hafiz, F., Glasssmith, LL., Wong, BS., Jones, IM., Clive, C., Haswell, SJ. (2000) Consequences of manganese replacement of copper for prion protein function and proteinase resistance. *EMBO Journal* **19**(6), 1180–1186.
17. Holme, A., Daniels, M., Sassoon, J., Brown, DR. A novel method of generating neuronal cell lines from gene-knockout mice to study prion protein membrane orientation. *European Journal of Neuroscience* **18**, 571–579.
18. Dawson, RM., Elliott, DC., Elliott, WH., Jones, KM. (1994) Data for Biochemical Research (third edition). Oxford Scientific Publications, Oxford, UK.
19. Martin, BD., Schoenhard, JA., Sugden, KD. (1998) Hypervalent chromium mimics reactive oxygen species as measured by the oxidant-sensitive dyes 2',7'-dichlorofluorescin and dihydrorhodamine. *Chemical Research and Toxicology* **11**(12), 1402–10.

Chapter 3
Immunodetection of PrP^{Sc} Using Western and Slot Blotting Techniques

Hanna Gyllberg and Kajsa Löfgren

Summary Prion infectivity is often linked to presence of the protease-resistant isoform of prion protein (PrP), PrP^{res}; therefore, it is of highest interest to have convenient methods for rapid detection of PrP^{res} in the research laboratory. For detection of PrP^{res} in model systems to confirm infectivity, there are several methods that can be applied. This chapter focuses on detection of PrP^{res} by proteinase K digestion followed by Western blot, which is the only method that is both quantitative and qualitative. For large-scale screening of PrP^{res} content in samples, the dot blot method offers a great advantage for detecting PrP^{res}, and this method is also thoroughly described in this chapter.

Keywords Dot blot; guanidinium thiocyanate; immunoprecipitation; nitrocellulose membrane; proteinase K digestion; PrP antibodies; PVDF membrane; reprobing; Western blot.

1. Introduction

Prions are composed largely, if not solely, of malfolded prion protein (PrP^{Sc} in scrapie). The formation of PrP^{Sc} from the cellular prion protein (PrP^{C}) is considered to involve mainly a conformational change *(1)*. PrP^{C} has high α-helix content (42%), with no prominent β-sheet inclusions (3%), whereas PrP^{Sc} has lower α-helix content (30%), and it is instead composed mainly of β-sheets (43%). Together with this conformational change, properties of PrP^{Sc} arise that differ from those of PrP^{C} *(2)*.

Although PrP^{C} is susceptible to cellular degradation processes, PrP^{Sc} is largely insensitive to proteases, including proteinase K (PK). Only an N-terminal fraction of PrP^{Sc} is removed by proteolytic cleavage, leaving a protease-resistant core protein, defined as PrP^{res} (for resistant) *(2)*. The β-sheet content of PrP^{res} is approx 50%, and due to this hydrophobic element, PrP^{res} has the tendency to polymerize into amyloid rods *(3)*.

During early attempts to purify infectious prions, brain homogenates from scrapie-infected hamsters were subjected to limited proteolysis with PK. These fractions contained enriched scrapie infectivity and an N-terminally truncated form of PrP^{Sc},

From: *Prion Protein Protocols.*
Methods in Molecular Biology, Vol. 459.
Edited by: A. F. Hill © Humana Press, Totowa, NJ

denoted PrP$^{27\text{-}30}$ (*3, 4*). The denotation refers to the size of the main glycosylation form of hamster PrP^{res}, which is revealed using sodium dodecyl sulfate-polyacrylamide gel electrophoresis (SDS-PAGE). This nomenclature, with PrP$^{27\text{-}30}$ as equivalent to the PK resistant core protein of PrP^{Sc}, is often seen, even though the migration of PrP^{res} in other model systems often includes a larger kilodalton span.

Prion infectivity is often associated with the presence of PrP^{res}, and, perhaps too often, the terms PrP^{Sc} and PrP^{res} have been used synonymously. This correlation does not always hold true. Protease resistance is a relative concept that depends on the ratio of PK to protein and also on the PrP species, prion strain, or a combination. Only minor rearrangements of the PrP^{res} structure are required to affect the protease resistance (*5*).

Cases of prion disease with no detectable PrP^{res} have proven to be infectious. Prions in those cases may be composed of protease-sensitive PrP (PrP^{sen}). Furthermore, the presence of PrP^{res} does not always guarantee transmissibility of prion disease (*4*). In this chapter, the term PrP^{Sc} states prion infectivity, whereas PrP^{res} is used as a label for PK resistance. Still, whether or not a sample contains detectable PrP^{res} upon analysis, the only really accurate way to ensure whether infectious prions are present is to use the sample to reinfect a host and to confirm clinical signs of prion disease.

Despite this discrepancy between infectivity and detectable PrP^{res}, all presently available commercial detection kits designed for routine diagnostics of scrapie and bovine spongiform encephalopathy nevertheless are based on the presence of PrP^{res} (*6*). There are several procedures to detect PrP^{res}, including, for example, enzyme-linked immunosorbent assay, dot blot, cell blot, and Western blot. Although all of them are based on the relative resistance of PrP^{res} to PK digestion, the Western blot is currently the only method that is both quantitative and qualitative. With Western blot, it is possible to determine the selective abundance of each glycosylation form of PrP^{Sc} and molecular weights of PrP^{C} and PrP^{res}. These electrophoretic profiles are used in prion strain determination (*7*).

Upon PK digestion, the N-terminal truncated PrP^{res} can be detected on Western blot by specific PrP antibodies (**Table 1**) immunoblotting, as shown in **Fig. 1**. PrP^{C} has a molecular mass of approx 20–37 kDa, depending on the degree of glycosylation (di-, mono- or unglycosylated) (**Fig. 1**), and it is best visualized after separation on a 10–15% SDS-PAGE. PrP^{res} retains the glycosylation forms of PrP^{C} from the infected cells, but due to the N-terminal truncation during cellular degradation processes, PrP^{res} bands display a size shift.

To visualize only one distinct band of PrP^{C} or PrP^{res}, peptide *N*-glycosidase F (PNGase F) treatment of cell extracts or tissue homogenate can be performed after PK digestion. PNGase F removes the attached carbohydrate molecules, by cleaving the binding between the first *N*-acetylglucoseamine on the carbohydrate molecule and the asparagine residues at positions 181 and 197 on PrP. This renders full-length PrP to migrate at approx 20 kDa on an SDS-PAGE (*8, 9*).

Dot blot is a technique for the detection and identification of proteins and analysis of protein levels. Not involving any step of protein separation, this method is faster to perform than Western blot. The dot blot method offers a great advantage over Western blot when applied in large-scale screenings of PrP^{res} content in samples (*10*). Instead of PK digesting each sample, the entire blotted membrane is treated

Table 1 Most frequently used antibodies for detection of PrP^C and PrP^{Sc}, including type of antibody, respective epitopes on PrP of each antibody, suggested dilution for Western blot (WB), and species cross-reactivity

Antibody[a]	Origin[b]	Epitope on PrP	Suggested dilution for WB	Species cross-reactivity
Anti-PrP, sc-7694	goat-Pab	Very C terminus	1:200–1:1,000	Mouse, rat
D13	rec. Fab	94–105	1:1,000	Mouse, hamster, ovine
6H4	Mab	144–150	1:5,000–1:20,000	Human, cattle, sheep, rat rabbit, mouse, hamster, mink, various primates
3F4	Mab	107–112	1:100,00–1:100,000	Human, hamster, felines
1E4	Mab	108–119	1:5,000	Human, bovine, ovine, hamster, mouse
SAF32	Mab	79–92	1:1,000	Human, mouse, sheep, bovine, hamster

[a]The antibodies listed have different specificities (e.g., 1E4 has broad species reactivity together with 6H4). All enlisted antibodies are commercially available from respective companies: sc-7694, Santa Cruz Biotechnology, Inc.; D13, InPro, South San Francisco, CA; 6H4, Prionics, Zurich, Switzerland; 3F4, Signet Laboratories, Dedham, MA; 1E4, Sanquin Reagents, Amsterdam, The Netherlands; and SAF32, Axxora Platform, Lausen, Switzerland.
[b]Pab, polyclonal antibody; Fab, fragment antigen binding monoclonal antibody; and Mab, monoclonal antibody.

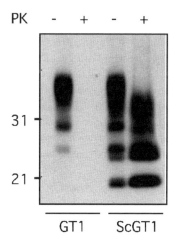

Fig. 1 PrP^C and PrP^{Sc} in GT1 and ScGT1 cells, before and after PK digestion. Samples are separated on a 12% SDS-PAGE and immunoblotted with anti-PrP (sc-7694). When stated, lanes 1 to 4 correspond to the samples from left to right on the blot. In GT1 −PK sample (lane 1), un-, mono-, and diglycosylation forms of PrP^C are visible. Nothing remains of PrP^C after PK digestion of GT1 cell extract (lane 2). The PrP^C glycosylation forms are also present in the ScGT1 −PK sample, together with the corresponding glycosylation forms of PrP^{res} (lane 3). In the ScGT1 +PK sample (lane 4), only PrP^{res} glycosylation forms are left after PK digestion, and the size shifts due to N-terminal truncation of PrP^{res} is clearly seen compared with PrP^C in lane 1. In addition, in the ScGT1 samples (lanes 3 and 4), a PrP^{res} fragment at ~18 kDa is visible, likely to derive from cellular processing of PrP^{Sc}

with PK. Alternatively, concentrations of PrP^C or PrP^{res} in crude preparations can be estimated quantitatively by using dot blot if there is purified PrP^C of known protein concentration for comparison. Dot blot is often used as a valuable technique to test for suitable solutions when designing further, more complex, experiments. However, the disadvantage of the dot blot technique is that unspecific antibody binding cannot be separated from the sample reading. High specificity of the antibodies used is thus necessary.

Dot blot is performed by filtration of the samples by using a dot blot unit. In the dot blot unit, samples of culture medium, cell extracts, or tissue homogenates are placed in separate wells, and then they are further filtered through the membrane by applying vacuum. Proteins adsorb to the membrane, and the other sample components are pulled through by the vacuum. After application of samples, membranes are air-dried, by which the proteins becomes immobilized and ready for subsequent analysis.

When preparing samples for Western blot, samples are boiled in Laemmli sample buffer that contains SDS, to denature and charge proteins before electrophoretical separation *(11)*. Boiling in SDS denatures PrP^{res}, which is a decontaminating effect because the infectivity is tightly linked to the conformation of PrP^{Sc}. However, primarily, PrP^{res} needs to be denatured to make the epitopes, which are hidden within the malfolded structure, accessible for detection by several of the PrP antibodies described in **Table 1**. When performing dot blot culture medium, plain cell extracts or tissue homogenates are applied without any pretreatment of the samples. Thus, PrP^{Sc} remains contagious when bound to the membrane, and the need for decontaminating the actual membrane arises.

Besides boiling in SDS, the most common method to denature PrP^{res} to make it recognizable for antibodies is treatment with guanidinium thiocyanate (GDN-HCN) *(10)*. Concentrations between 3 and 5 M GDN-HCN denature PrP^{res} within 5 min *(5)*. After incubation in this solution, the membrane is also decontaminated from prions.

Sometimes, a higher concentration of PrP^{Sc}, PrP^C, or both is desired, and in such cases, an immunoprecipitation (IP) can be performed on whole cell lysate or tissue homogenate *(12)*. This method is often used for purification of PrP^C, PrP^{Sc}, or both, depending on whether a PK digestion is performed on the sample before the IP. Furthermore, IP is also very applicable for studying the interaction of PrP^C with other proteins.

2. Materials

2.1. Solution Recipes and Equipment

1. Tris-HCl–buffered saline (TBS): 10 mM Tris-HCl, 150 mM NaCl, pH 7.5.
2. Phosphate-buffered saline (PBS): 150 mM NaCl, pH 7.5.
3. Wash buffer: TBS or PBS with 0.05–0.1% Tween-20.
4. Extraction buffer: 50 mM Tris-HCl, pH 7.5, 10 mM EDTA, 150 mM NaCl, 0.5% sodium deoxycholate, and 0.5% Triton X-100.
5. PK storage buffer: 50 mM Tris-HCl, pH 8.0, 5 mM CaCl$_2$, and 50% glycerol.
6. PK stock solution: dried PK powder is diluted to 20 mg/ml in PK storage buffer.

 7. Glycoprotein digestion buffer: 20 mM phosphate, pH 7.0, 25 mM EDTA, 0.6% Nonidet P-40, 1% β-mercaptoethanol, and 0.1% SDS.
 8. 4× Laemmli sample buffer: 0.25 M Tris-HCl, pH 6.8, 8% SDS, 40% glycerol, and 100 mM dithiothreitol.
 9. SDS-PAGE upper buffer: 0.5 M Tris-HCl, pH 6.8, and 0.1% SDS.
10. SDS-PAGE lower buffer: 1.5 M Tris-HCl, pH 8.8, and 0.1% SDS.
11. Protogel: 37.5:1 acrylamide:bisacrylamide solution.
12. 12% SDS-PAGE separating gel: 4 ml of Protogel; 2.5 ml of lower buffer; 3.5 ml of H$_2$O; 112 μl of 10% ammonium persulfate (APS); 30 μl of *N,N,N',N'*-tetramethylethylene-diamine; *N,N,N',N'*-di-(dimethylamino)ethane; and *N,N,N',N'*-tetramethyl-1,2-diamino-methane (TEMED). Mix immediately.
13. 4.2% SDS-PAGE stacking gel: 0.42 ml of Protogel, 0.75 ml of upper buffer, 1.8 ml of H$_2$O, 18 μl of 10% APS, and 5 μl of TEMED. Mix immediately.
14. Running buffer: 0.025 M Tris, 0.192 M glycine, and 0.1% SDS.
15. Transfer buffer: 5 ml of SDS 20%, 200 ml of methanol (or use 100 ml of methanol and 100 ml of ethanol). Fill up to 1 liter with running buffer.
16. Blocking solution: also used as incubation-buffer for antibodies: Is commonly made by dissolving 1 to 1.5% (w/v) bovine serum albumin or nonfat milk in wash buffer.
17. Dot blot buffer: TBS or PBS at pH 7.5.
18. Alternative dot blot buffer: 0.15 M NaCl, 50 mM HEPES, and 2% Triton X-100. Fill up to 250 ml with water.
19. PK solution for dot blot: PBS and 5 μg/ml PK.
20. IP buffer: 20 mM Tris-HCl, pH 7.5, 1% Nonidet-P40 (NP-40), 0.5% sodium deoxycholate (NaDOC).
21. IP buffer with protease inhibitors: 20 mM Tris-HCl, pH 7.5, 1% NP-40, 0.5% NaDOC, 5 mM phenylmethylsulfonyl fluoride (PMSF), 1 μg/ml aprotinin, and 1 μg/ml pepstatin.
22. Alternative IP buffer: PBS, 0.5% NaDOC, and 0.5% NP-40, pH 7.4.
23. Dehybridizing buffer: 0.2 M glycine, pH 2.5, and 0.05% Tween-20.
24. Alternative dehybridizing buffer: 100 mM 2-mercaptoethanol, 2% SDS, and 62.5 mM Tris-HCl, pH 6.7.
25. Proteinase K.
26. PNGase F.
27. SDS-PAGE equipment.
28. Transfer equipment for Western blot.
29. Dot blot unit.
30. Whatman 3 MM paper.

2.2. Membranes

Polyvinylidene fluoride (PVDF) membranes and nitrocellulose membranes are both commonly used in Western blot and dot blot. Protocols for blotting onto PVDF or nitrocellulose are the same, with a few exceptions that are discussed in **Note 6**. Typical binding capacity of proteins onto a nitrocellulose membrane is 80–100 µg/cm², depending on pore size. The corresponding value for PVDF membranes is 100–200 µg/cm².

2.3. Common PrP Antibodies for Immunodetection

Several of the antibodies listed in **Table 1** have broad species reactivity, as described in **Note 5**. Dilution factors for each specific antibody might vary with model system and detection method used; however, recommendations are found in **Note 5**.

3. Methods

3.1. Western Blot

3.1.1. Proteinase K Digestion of Protein Samples and Western Blot

This protocol is adjusted for 1.1 ml of cell lysate with protein concentration of 2 mg/ml.

1. Transfer 100 µl of the sample to new tubes and add 33 µl of 4× Laemmli sample buffer. Boil for 5 min and store at −20°C.
2. To remaining 1-ml (2-mg) sample, add 20 µg of PK. Thus, the ratio PK:protein should be 1:50. Alternative PK digestion conditions are described in **Note 4**.
3. Incubate at 37°C for 1 h, with rocking.
4. Put samples on ice; add PMSF to final concentration of 5 mM, vortex.
5. Centrifuge at 30,000 g for 40 min at 4°C.
6. If desired, perform PNGase F digestion, according to **step 7**. Otherwise, resuspend pellet in 4× Laemmli sample buffer and boil at 100°C for 5 min.
7. Resuspend the pellet in 40 µl of glycoprotein denaturing buffer and boil at 100°C for 10 min before addition of PNGase F (30,000 units/ml). Incubate for 24 h at 37°C. Add 13 µl of 4× Laemmli sample buffer and boil at 100°C for 5 minutes. Alternative PNGase F digestion conditions are described in **Note 1**.
8. Separate on 12% SDS-PAGE and transfer to PVDF or nitrocellulose membrane according to SDS-PAGE (*see* **Subheading 3.1.2.**).
9. Block membrane in blocking solution for 1 h.
10. Incubate membrane in primary antibody for 18 h at 4°C. For detection of PrP, we use goat polyclonal anti-PrP (sc-7694, Santa Cruz Biotechnology Inc., Santa Cruz, CA) diluted in blocking solution 1:1000.
11. Wash membrane in blocking solution four times for 15 min, in total 1 h.
12. Incubate in horseradish peroxidase-conjugated anti-goat secondary antibody, diluted in blocking solution 1:5000, for 1 h, rocking at room temperature.
13. Wash in wash buffer four times for 15 min, in total 1 h and rinse briefly in PBS.
14. Incubate in substrate solution for detection method of choice.

3.1.2. SDS-PAGE and Wet Transfer

If precast SDS-PAGE gels are used, continue to **step 9**.

1. Clean gel-plates with ethanol and mount with spacers and rubber sealing.
2. Make a mark ~1 cm below the comb.
3. Mix 12% separating gels according to recipe. Fill up to the marking.
4. Add ~1 ml of isopropanol on top to even the surface.
5. When polymerized, remove the isopropanol using Whatman 3 MM paper.
6. Mix stacking gels according to recipe. Fill up to top.
7. Insert the well comb. Leave to polymerize.
8. Remove clamps and rubber sealing.
9. Place gels in electrophoresis equipment. Pour running buffer on top and bottom of the structure.
10. Remove bubbles at the bottom of the gel.
11. Remove well comb.
12. Load molecular weight ladder in the first well and then add samples rightward.
13. Add 4× Laemmli sample buffer to any wells not containing samples.
14. Electrophorese the gel until the front has reached the bottom.
15. Remove spacers and remove one of the gel plates, leaving the gel exposed. The stacking gel can now be removed.
16. If wet transfer is applied, prepare the transfer-sandwich submerged in transfer buffer. Place pad in the bottom, and then three layers of Whatman 3 MM paper.
17. Place membrane with labeling upward.
18. Place the gel on top of the membrane in the correct orientation relative to the membrane labeling. When transferring the gel from the gel-plate to the sandwich, work with wetted gloves to avoid breakage of the gel.
19. Remove air bubbles between the membrane and gel by gently pressing on the gel.
20. Place three layers of Whatman 3 MM paper on top of the gel and on top another pad. Close the transfer cassette.
21. Place the sandwich in the transfer system. Fill up with transfer buffer.
22. Use a water cooler system and magnetic stirrer while running the transfer.
23. If desired, perform staining with Ponceau solution or Coomassie as mentioned in **Note 3** before continuing to Western blot (*see* **Subheading 3.1.1., steps 9–14**).

3.2. Dot Blot and PK Digestion on Membrane

1. Label a membrane with a grease-containing pencil, for example, a cajal pen. Duplicate settings are labeled 1'and 2' (**Fig. 2**). **Steps 2–8** are performed in a fume hood.
2. Soak the membrane in dot blot buffer before mounting it in the dot blotting unit. Apply vacuum pump to create suction across the membrane. In step 3, apply the samples as duplicates (**Fig. 2**), one set for PK digestion and one set for control.

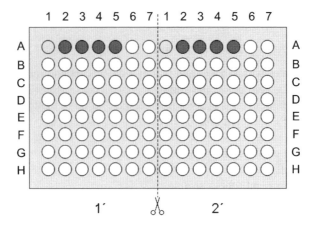

Fig. 2 Dot blot. Schematic picture of a typical dot blot setting. The membrane is divided in duplicates, 1' and 2', of which 1' serves as a control and 2' is further subjected to PK digestion according to **Subheading 3.2.** In this scheme, four samples (2–5) are applied in parallel wells on these duplicates. Extraction buffer is applied to one well on each duplicate (1':1 and 2':1) and used for background measurements. After sample application and drying, the membrane is cut, separating the duplicates 1' and 2'

3. Apply equal amounts of samples to appointed wells, we use 400 µl sample/well when analyzing cell lysates normalized to 1 mg/ml.

4. To one well on each duplicate, an equivalent volume (400 µl) of extraction buffer is applied, to serve as background control. Apply the equivalent volume (400 µl) of dot blot buffer to any empty wells.

5. Dismantle the dot blotting unit and air-dry the membrane for 30 min at room temperature.

6. Cut the membrane into the separate duplicates 1' and 2' (**Fig. 2**).

7. To membrane 2', add 2 ml of 5 µg PK/ml diluted in PBS, and incubate for 90 min, rocking, at 37°C. Digestion may be performed under alternative conditions described in **Note 2**.

8. Wash the PK-treated membrane three times for 1 min, in total 3 min, in H_2O, rocking.

9. Incubate both membranes in 3 M GDN-HCN for 10 min, rocking.

10. Wash membranes three times for 1 min in H_2O, rocking.

11. Incubate membranes in blocking solution and continue according to Western blot (*see* **Subheading 3.1.1., steps 9–14**).

3.3. Immunoprecipitation of PrP^res

If PK digestion will be performed, dilute samples using IP buffer without protease-inhibitors. When cell cultures are used, these may be extracted using the IP buffer without protease inhibitors. However, if no PK digestion will be performed, as with controls from uninfected cells or animals, normalization of sample protein concentrations may be performed using the IP buffer with protease inhibitors to avoid degradation of PrP^C. The following protocol is adjusted for a cell lysate.

1. Normalize protein concentration in samples to 2 mg/ml using IP buffer for dilutions. Prepare 1.1 ml of each sample.
2. Transfer 100 μl of the sample to new tubes and add 33 μl of 4× Laemmli sample buffer. Boil for 5 min and store at −20°C. If no PK digestion should be performed, continue to **step 5**.
3. On the remaining 1 ml (2 mg) of protein, perform PK digestion according to Western blot (*see* **Subheading protocol 3.1.1., steps 2–5**).
4. Remove supernatant and dissolve pellet in 1 ml of IP buffer.
5. To 1 ml, add approximately 20 μg of antibody depending on the efficiency of the antibody; see comments in **Note 4** and **Table 1**.
6. Rotate at 4°C for 3 h.
7. Add 20 μl of protein G-Sepharose beads.
8. Rotate at 4°C for 18 h.
9. Wash the pellet (consisting of protein G-Sepharose-antibody-protein complex) three times with PBS. Spin down beads between washes at very low speed (maximum 2,000 g); discard supernatant without removing any of the pellet.
10. If desired, perform PNGase F digestion on the pellet, according to **Subheading 3.1.1, step 7**. Otherwise, resolve pellet in 4× Laemmli sample buffer and boil at 100°C for 5 min.
11. Analyze samples on Western blot according to **Subheading 3.1.1., steps 8–14.**

3.4. Blot Storage Procedures and Reprobing of Membranes

3.4.1. Blot Storage Procedures

1. PVDF membranes: leave the blot to air dry and place membrane between two clean sheets of Whatman 3 MM paper. Nitrocellulose blots should be stored wetted in blocking solution containing 0.01% sodium azide.
2. Place the PVDF blot-filter paper sandwich or the wet nitrocellulose membrane into a clean plastic folder.
3. Close or seal the plastic envelope.
4. Store the blot according to recommendations in **Note 6**.
5. Blots stored in freezer must be brought to room temperature before being taken out of the plastic envelope to avoid breakage.

3.4.2. Reprobing of Membranes

1. Rinse the membrane briefly in PBS to remove any traces of enhanced chemiluminescence.
2. Incubate the membrane in dehybridizing buffer with glycine at room temperature for 7 min, agitating.

3. Neutralize the membrane in Tris-HCl, pH 8.0.
4. Rinse the membrane twice in PBS.
5. Continue according to Western blot according to **Subheading 3.1.1., steps 8–13**.

4. Notes

1. Proteinase K Digestion and Western Blot

PK is stable in the stock solution for up to 6 months at −20°C. The stock solution also should be used for diluting PK before use. Alternatively to what is stated in **Subheading 3.1.1.**, the PK:protein ratio can be 1:20 given the incubation time is shortened to 30 min at 37°C. Occasionally, a milder treatment is carried out, incubating the samples with PK at 4 °C for 1 h. In this case, keep the PK:protein ratio 1:50. Ten percent of brain homogenates are usually digested in PK concentrations ranging between 5 and 20 μg.

Besides being an analytical tool, PK treatment of protein samples should be performed regularly to confirm that prion-infected cell lines do not lose infectivity. Cells or tissue homogenates should be extracted, and the protein concentration of all samples normalized in extraction buffer containing no addition of any protease inhibitors. Protein concentrations recommended are 1–2 mg/ml, but PK-resistant PrP^{Sc} can be detected from lower amounts of protein depending on the PrP^{Sc} levels in the model system. In our laboratory, PrP^{res} is regularly detected starting with 100 μg of total protein in cell extracts from Rocky Mountain Laboratory (RML)-infected GT1-1 cell lines, by using Western blot of PK-digested samples. As control samples, noninfected cells or tissues should be used to confirm that the PK digestion assay is efficient.

When performed, PNGase F digestion is executed after PK digestion, because the glycoprotein digestion procedure renders the proteins denatured. Alternatively to what is described in **Subheading 3.1.1.**, the incubation time for PNGase F digestion may be shortened to 2–4 h at 37°C (**9**). Furthermore, deglycosylation of non–PK-digested material may be performed following the same conditions as described in **Subheading 3.1.1.**, although this will not allow for discrimination between PrP^{C} and PrP^{Sc} isoforms except for the shift in size due to N-terminal truncation of PrP^{res}.

After SDS-PAGE, transfer of proteins to either a PVDF or a nitrocellulose membrane is preferably performed in a wet-transfer tank system. Appropriate current and transfer time will depend on the transfer system used. We use 0.7 ampere and 45 min, respectively, in a Bio-Rad transfer system.

For some cell lines, an easier and more convenient approach to PrP^{Sc} detection not involving PK digestion exists. After separating proteins from non–PK-digested whole cell extract or tissue homogenate on a SDS-PAGE followed by immunoblotting of PrP, a PrP^{res}-specific band is clearly visible at approx 18–20 kDa. This PrP^{res}-specific protein band is probably a degradation product of

PrP^{Sc} (**Fig. 1**, lanes 3 and 4). This approach cannot be applied in all cell or tissue models, but we regularly perform PrP^{Sc} detection in RML-infected GT1-1 cell lines by using this method. However, scrapie-infected neuroblastoma cell lines, for example, ScN2a, generally do not display such detectable PrP^{res}-specific lighter fragments on Western blots; thus, this approach to PrP^{res} detection is not accessible for all model systems.

The amount of PrP^{res} in samples can be interpreted by determination of the optical density (OD) of PrP^{res}, which in turn may or may not be set in relation to the total amount of PrP isoforms measured as the OD of the whole cell extract/tissue homogenate PrP profile. We have used the software ImageGauge version 3.46 (Fujifilm, Tokyo, Japan) for measurements of this type. If PrP levels are to be compared between samples run on different gels, internal PrP standards from cell extracts or recombinant PrP^C of known PrP concentrations can be loaded on all gels.

2. Dot Blot and PK Digestion on Membrane

One of the advantages with dot blot is that large volumes can be filtered. Prion-infected cell cultures release PrP^{Sc}-containing exosomes into the culture medium. Filtering the growth medium makes it easy to detect the presence of such released PrP^{res} without having to sacrifice the actual culture.

When performing dot blot by using a dot blotting unit, there are several considerations to be made. To avoid the samples to leak out between wells, the unit often contains a sealing to place on top of the membrane during mounting. Furthermore, it is important to note that certain detergents used when preparing extracts or homogenates may inhibit adsorption of proteins to the membrane. The buffers used for dissolving samples should not contain higher levels of SDS than 0.05%, which also holds for any buffers used in the process.

In the dot blotting unit, samples are applied into separate wells followed by filtration through the mounted membrane. The sample volume should be enough to cover the membrane surface in each well. The total amount of protein applied at each well should exceed 100 µg for crude samples, but the protein contents should not exceed the binding capacity of the membrane. Protein concentrations are recommended to be in the 1 mg/ml range for nitrocellulose membranes. Each dot/sample should contain the same amount of total protein when crude samples are analyzed for variations in PrP^C or PrP^{res} contents. All samples should thus be normalized to the same total protein concentration.

Viscous solutions may reduce the flow rate across the membrane, and they, therefore, are diluted in water before application. Especially when considering homogenates, samples containing high particle loads may clog the membrane.

The dot blot membrane can be digested with PK to detect PrP^{res} levels specifically. As an alternative to what is suggested in the dot blot (**see Subheading 3.2.**), 500 PK µg/ml diluted in PBS could be used. In this case, perform digestion/incubate for 15 min at room temperature. Any PK digestion needs to be performed before GDN-HCN treatment, or PK will digest also the denatured PrP^{res}. During all procedures before GDN-HCN treatment, membrane-bound PrP^{Sc} still is contagious.

3. Membranes

Proteins were originally transferred to nitrocellulose membranes, and in our laboratory, we use this type of membrane. Nitrocellulose has a high affinity for proteins, and the membranes of higher quality consist of pure nitrocellulose that gives lower background than membranes wherein cellulose acetate is added. Nitrocellulose membranes need no other pretreatment step than wetting with dot blot buffer or water.

For a PVDF membrane to be compatible with aqueous systems, it must first be wetted in 50–100% (v/v) methanol or ethanol. The alcohol is removed by rinsing in water, and the membrane is equilibrated in dot blot buffer or transfer buffer for dot blot or Western blot, respectively. Another note is that the SDS tolerances are not equivalent for the two membrane types. Too much SDS can inhibit the protein's ability to bind to the PVDF, and it can, in fact, cause proteins already bound to the membrane to slip off. Therefore, SDS levels should never exceed 0.1%.

The presence and location of PrP^C and PrP^{res} can be detected with immunochemical methods. However, staining of the molecular weight standard marker bands on Western blots is often necessary. PVDF membranes can be stained with Coomassie blue to visualize all blotted proteins. The Coomassie blue-stained PVDF membrane is destained in a 25% acetic acid/50% methanol solution, which allows subsequent immunostaining on the same membrane. Ponceau S dye is compatible with both PVDF and nitrocellulose membranes. The Ponceau S dye binds reversible to all proteins, and they thus fade to become invisible during the blocking procedure, not interfering in any way with further probing of the membrane-bound proteins. After blotting, both these stains can give a rough estimate of relative protein levels between lanes or dots. This method is recommended to register any misfortunes in sample normalization or if air bubbles have interfered with the transfer procedure. These stains also may serve as controls for calculating relative PrP^C content, PrP^{res} content, or both, in each dot.

4. Immunoprecipitation

Depending on the model system used and what type of samples that are going to be analyzed, the adsorption of antibody-bound PrP to Sepharose beads will have to be optimized. This applies to the IP buffer in terms of incubation times and incubation temperatures, salt concentrations, pH, and the detergents that are used. Two alternative IP buffer recipe suggestions are, therefore, enlisted in **Subheading 2.1.** All PrP antibodies enlisted in **Table 1** are suitable for IP, under conditions specified by the respective supplier.

5. Common PrP Antibodies

Alternatively to the standard procedure in **Subheading 3.1.1.** of Western blot or dot blot membrane incubation with primary antibody, **step 9** could be altered to 2-h incubation at room temperature with membranes kept on rotation. This alternative incubation might result in a weaker signal compared with the standard procedure, although most antibodies still give a detectable signal. The most common commercially available antibodies directed against the prion protein together with their epitopes and suggested dilutions for Western blot are listed in **Table 1**.

6. Storage and Reprobing of Blots

Both nitrocellulose and PVDF membrane blots may be stored wet in plastic envelopes at 4°C for a week or two, if a bacteriocide such as sodium azide is added. Nitrocellulose membrane blots can be stored wet in sealed plastic envelopes at [minus]20°C for up to 6 months. In sealed plastic envelopes, PVDF membrane blots can be stored as dried at −20°C for 2 months or at −70°C for longer. A dry PVDF membrane needs to be wet before any further analysis or reprobing by immersion in 100% methanol. However, storage and dehybridization procedures render the blots fragile, especially nitrocellulose membranes. Also, some proteins may be sensitive to oxidation, hydrolysis, or other chemical changes that might occur during prolonged storages. Blots stored in freezer should be handled carefully not to cause breakage of the membrane.

Reprobing allows for several specific protein detections from the same sample and membrane. Precious samples can by this method be analyzed to a fuller extent, not to mention the economical benefits of reprobing membranes. The dehybridization of primary and secondary antibodies can be carried out up to four or five times. Some of the blotted sample proteins also detach during the dehybridization process, leaving the signal too weak after too many rounds of reprobing. The two most common methods for dehybridizing membranes of bound antibodies involve the use of low pH glycine buffer or reduction of the covalent hydrogen bonds to the membrane with, for example, mercaptoethanol. If antibodies are attached very tightly to the membrane, using glycine may not be sufficient to remove all signals. Using mercaptoethanol is then a somewhat harsher alternative of dehybridization to turn to. In this case, incubate the membrane in dehybridizing buffer with mercaptoethanol at 50°C for 30 min. Rinse thrice with PBS and continue to block the membrane as described in **Subheading 3.1.1. step 9**.

References

1. Pan, K. M., Baldwin, M., Nguyen, J., Gasset, M., Serban, A., Groth, D., Mehlhorn, I., Huang, Z., Fletterick, R. J., Cohen, F. E., et al. (1993). Conversion of alpha-helices into beta-sheets features in the formation of the scrapie prion proteins. *Proc Natl Acad Sci U S A* **90**, 10962–6.
2. Prusiner, S. B. (1998). Prions. *Proc Natl Acad Sci U S A* **95**, 13363–83.
3. McKinley, M. P., Meyer, R. K., Kenaga, L., Rahbar, F., Cotter, R., Serban, A., and Prusiner, S. B. (1991). Scrapie prion rod formation in vitro requires both detergent extraction and limited proteolysis. *J Virol* **65**, 1340–51.
4. Weissmann, C. (2004). The state of the prion. *Nat Rev Microbiol* **2**, 861–71.
5. Oesch, B., Jensen, M., Nilsson, P., and Fogh, J. (1994). Properties of the scrapie prion protein: quantitative analysis of protease resistance. *Biochemistry* **33**, 5926–31.
6. Aguzzi, A., Heikenwalder, M., and Miele, G. (2004). Progress and problems in the biology, diagnostics, and therapeutics of prion diseases. *J Clin Invest* **114**, 153–60.

 7. Collinge, J., Sidle, K. C., Meads, J., Ironside, J., and Hill, A. F. (1996). Molecular analysis of prion strain variation and the aetiology of 'new variant' CJD. *Nature* **383**, 685–90.
 8. Blochberger, T. C., Cooper, C., Peretz, D., Tatzelt, J., Griffith, O. H., Baldwin, M. A., and Prusiner, S. B. (1997). Prion protein expression in Chinese hamster ovary cells using a glutamine synthetase selection and amplification system. *Protein Eng* **10**, 1465–73.
 9. Cancellotti, E., Wiseman, F., Tuzi, N. L., Baybutt, H., Monaghan, P., Aitchison, L., Simpson, J., and Manson, J. C. (2005). Altered glycosylated PrP proteins can have different neuronal trafficking in brain but do not acquire scrapie-like properties. *J Biol Chem* **280**, 42909–18.
10. Serban, D., Taraboulos, A., DeArmond, S. J., and Prusiner, S. B. (1990). Rapid detection of Creutzfeldt-Jakob disease and scrapie prion proteins. *Neurology* **40**, 110–7.
11. Gasset, M., Baldwin, M. A., Fletterick, R. J., and Prusiner, S. B. (1993). Perturbation of the secondary structure of the scrapie prion protein under conditions that alter infectivity. *Proc Natl Acad Sci U S A* **90**, 1–5.
12. Beringue, V., Vilette, D., Mallinson, G., Archer, F., Kaisar, M., Tayebi, M., Jackson, G. S., Clarke, A. R., Laude, H., Collinge, J., et al. (2004). PrPSc binding antibodies are potent inhibitors of prion replication in cell lines. *J Biol Chem* **279**, 39671–6.

Chapter 4
Assaying Prions in Cell Culture

The Standard Scrapie Cell Assay (SSCA) and the Scrapie Cell Assay in End Point Format (SCEPA)

Sukhvir P. Mahal, Cheryl A. Demczyk, Emery W. Smith, Jr., Peter-Christian Klohn, and Charles Weissmann

Summary Prions are usually quantified by bioassays based on intracerebral inoculation of animals, which are slow, imprecise, and costly. We have developed a cell-based prion assay that is based on the isolation of cell lines highly susceptible to certain strains (Rocky Mountain Laboratory and 22L) of mouse prions and a method for identifying individual, prion-infected cells and quantifying them. In the standard scrapie cell assay (SSCA), susceptible cells are exposed to prion-containing samples for 4 days, grown to confluence, passaged two or three times, and the proportion of rPrP^{Sc}-containing cells is determined with automated counting equipment. The dose response is dynamic over 2 logs of prion concentrations. The SSCA has a standard error of ±20–30%, is as sensitive as the mouse bioassay, 10 times faster, at least 2 orders of magnitude less expensive, and it is suitable for robotization. Assays performed in a more time-consuming end point titration format extend the sensitivity and show that infectivity titers measured in tissue culture and in the mouse are similar.

Keywords 22L; CAD cells; Elispot; Me7; neuroblastoma; PK1 cells; prions; PrP^C; rPrP^{Sc}; R33 cells; RML; scrapie; scrapie cell endpoint assay (SCEPA); standard scrapie cell assay (SSCA); strains.

1. Introduction

In prion diseases, the normal host prion protein (PrP^C) is converted to rPrP^{Sc}, a largely proteinase K-resistant form of PrP *(1)*, that serves as a surrogate marker for prion disease, although its level does not necessarily correlate with the infectivity titer *(2–5)*. Infectivity is usually quantified by injecting samples intracerebrally into indicator animals, mostly mice, and determining the time to appearance of definitive clinical symptoms (incubation time method) *(6)* or by injecting serial dilutions of the sample and determining the dilution at which 50% of the animals succumb to scrapie (endpoint method) *(7)*.

From: *Prion Protein Protocols.*
Methods in Molecular Biology, Vol. 459.
Edited by: A. F. Hill © Humana Press, Totowa, NJ

Several cell lines can be infected by prions, as evidenced by the persistent accumulation of rPrP^{Sc}, infectivity, or both (8–14). Murine neuroblastoma-derived N2a cells are susceptible to certain strains of mouse prions, such as the mouse-adapted Rocky Mountain Laboratory (RML) scrapie strain (15); however, only a small proportion of cells accumulate detectable levels of rPrP^{Sc}. Selected sublines of N2a cells are more susceptible to the RML strain than the original stock, but they are nonetheless resistant to murine strains such as Me7, 301V, 22A, and others (16–20).

We have isolated two cell lines highly susceptible to RML and 22L scrapie prions: N2a-PK1 (PK1 for short) (20), derived from the murine neuroblastoma cell line N2a; and more recently, CAD2A2D5 (CAD5 for short) (S.P. Mahal, C.A. Baker, C. Demczyk, C. Julius, and C. Weissmann, published work) from the murine Cath.a-differentiated (CAD) cell line (21). We visualized individual rPrP^{Sc}-positive cells filtered onto membranes of an Elispot plate, and we established conditions under which the proportion of rPrP^{Sc}-positive cells was a function of the prion concentration. With these cell lines we could quantify RML prion concentrations as low as those that can be determined in the mouse bioassay. The assay has a standard deviation of ±20–30%; can be completed in <2 weeks, compared with 20 weeks in the most rapid mouse bioassay (22); and it is 2 orders of magnitude cheaper. The scrapie cell assay can be performed in an endpoint titration format, and although more time-consuming, it is more sensitive and robust, capable of detecting infectivity at a 10^{-8} dilution of RML-infected mouse brain, which is beyond the range of the standard mouse bioassay.

2. Materials

2.1. Standard Scrapie Cell Assay (SSCA)

Principle: Mouse prion-susceptible PK1 or CAD5 cells in wells of 96-well plates are exposed to the prion-containing sample for 4 days, split 1:8 (CAD5 cells) or 1:7 (PK1 cells), grown to confluence, and split similarly once more. When the cells have reached confluence after the second split, 20,000 cells, or when appropriate, fewer cells are filtered onto membranes of Elispot plates, and the proportion of rPrP^{Sc}-containing cells is identified by an enzyme-linked immunosorbent assay (ELISA). The dose response is dynamic between approx 50 to 1,000 "spots," but because it is nonlinear, every assay must consist of a serial dilution of titered RML as standard.

2.1.1. Safety Precautions

Experiments with mouse scrapie prions are classified as Hazard group 2, and they are carried out in laboratory containment level 2 following the procedures described in the Biosafety in Microbiological and Biomedical Laboratories Manual from the Center for Disease Control (http://www.cdc.gov/od/ohs/pdf files/4th%20BMBL.pdf).

2.1.2. Equipment

1. Zeiss KS Elispot system (Stemi 2000-C stereomicroscope equipped with a Hitachi HV-C20A color camera, a KL 1500 LCD scanner, Carl Zeiss, MicroImaging, Inc., Thornwood, NY; and Wellscan software, Imaging Associates, Bicester, Oxfordshire, UK).
2. Tissue culture/containment class 2 cabinet.
3. Standard cell culture inverted microscope (Axiovert 25, Carl Zeiss).
4. Humidified CO_2 incubator for tissue culture.
5. 37°C incubator.
6. 50°C oven.
7. Vaccu-Pette/96 multiwell pipetter (Sigma-Aldrich, St. Louis, MO).
8. Vacusafe vacuum pump or vacuum line with hand controller and trap (Integra Biosciences, Ijamsville, MD).
9. Brightline hemacytometer (VWR, Suwanee, GA).
10. Multiscreen IP 96-well 0.45-μm filter plates (Elispot plates, Millipore, Billerica, MA), or alternatively, AcroWell 96-well 0.45-μm BioTrace filter plates (Pall, East Hills, NY).
11. Twelve-channel pipette (Rainin Instruments, Oakland, CA).
12. 96-well plates (Corning, Falcon; BD Biosciences Discovery Labware, Bedford, MA).
13. Reagent reservoir, 100 ml (VWR).
    Optional for partial automation:
14. Scan Washer 300 (Molecular Devices, Sunnyvale, CA) allows automated washing of 20 96-well plates in <1 min/plate.
15. Freedom Evo (Tecan, Durham, NC): The 96-channel head allows automated resuspension of cells, dispensing of medium, dilution and filtration onto Elispot plates.

2.1.3. Reagents

Water for solutions is from the Milli-Q® Biocel Ultrapure Water Purification System (Fisher Scientific, Suwanee, GA). Reagents are filtered through a 0.2-μm Filter Unit (Millipore), to sterilize material, remove particulate material, or both.

1. OBGS: Opti-MEM, 9.1% bovine growth serum (BGS), 10,000 units/ml penicillin, and 100 μg/ml streptomycin. Mix 500 ml of Opti-MEM (Invitrogen, Carlsbad,

CA), 50 ml of BGS (cat. no. SH30541.03, HyClone Laboratories, Logan, UT; or equivalent, Fisher Scientific), and 5.5 ml of penicillin G (10,000 units/ml), streptomycin G (10 mg/ml), and 0.85% saline (Invitrogen). The BGS lot was selected by testing its capacity to sustain usual cell growth rates and spot number in the SSCA. Sterilize by filtration.

2. Lysis buffer: 50 mM Tris-HCl, pH 8.0, 150 mM NaCl, 0.5% Na deoxycholate, and 0.5% Triton X-100. Mix 10 ml of 1 M Trizma-HCl, pH 8.0, solution T3038 (Sigma-Aldrich), 6 ml of 5 M NaCl stock solution, 10 ml of 10% Na deoxycholate (>99% ; Sigma-Aldrich), and 1 ml of Triton X-100 (Sigma-Aldrich). Make up to 200 ml with H_2O. Make up fresh on day of use; filter.

3. Proteinase K (PK): 30 Hb milliunits/ml (usually approx 0.6 μg/ml) in lysis buffer. Mix 1 μl of PK (30 Hb units/mg, 20 mg/ml; Roche Diagnostics, Indianapolis, IN). Make up to 36 ml with lysis buffer.

4. TBST: 10 mM Tris-HCl, pH 8.0, 150 mM NaCl, and 0.1% Tween-20. Mix 100 ml of 1 M Tris-HCl, pH 8.0 (99.9%; Sigma-Aldrich), 87.5 g of NaCl, and 10 ml of Tween-20 (Sigma-Aldrich). Make up to 1000 ml with H_2O. Make up 1:10 dilution on day of use and filter.

5. TBST-1% nonfat dry milk: 1 g of nonfat milk powder (Bio-Rad, Hercules, CA) in 100 ml of 1× TBST.

6. Guanidinium thiocyanate (GSCN): 3 M in 10 mM Tris-HCl, pH 8. Mix 17.73 g of guanidinium thiocyanate (ISC BioExpress, Kaysville, UT) and 0.5 ml of 1 M Tris-HCl, pH 8.0 (Sigma-Aldrich). Make up to 50 ml with H_2O. Filter through a Folded Filter 5971/2 (Schleicher & Schuell Microscience, VWR).

7. Phenylmethylsulfonyl fluoride (PMSF): 2 mM in PBS. Dissolve 17.42 mg/ml PMSF (>99%; Sigma-Aldrich) in isopropanol to make a 100 mM solution. Store aliquots at −20°C. Dilute 1:50 in PBS just before use and filter.

8. Superblock: Dissolve 1 pouch Superblock dry blend blocking buffer in TBS (Pierce Chemical) in 200 ml of H_2O. Filter. Store at 4°C.

9. Anti-PrP monoclonal HuM chimeric antibody D18 (0.7 μg/ml).

   a. HuM D18 (2 mg/ml) was purified from the supernatant of Chinese hamster ovary cells (23), a gift from R. A.Williamson (The Scripps Research Institute, La Jolla, CA).

   b. One-milliliter stock aliquots are stored at −20°C.

   c. Working aliquots (100 μl) are centrifuged 10 min at 16,000 g to remove aggregates and stored at 4°C.

   d. Before use, dilute 1:3,000 into TBST-1% nonfat dry milk.

   e. D18 can be replaced by the murine monoclonal antibody ICSM 18 (D-Gen Ltd., London, UK) at 0.6 μg/ml.

10. Alkaline phosphatase (AP)-conjugated secondary anti-human IgG-AP antibody (cat. no. 9042-04, Southern Biotechnology Associates, Birmingham, AL). Centrifuge 30 s at 16,000 g to remove aggregates. Before use, dilute 1:5,000 into TBST-1% nonfat dry milk.

11. 10× PBS (cat. no. BP3994, Fisher Scientific). Dilute 1:10 with H_2O. Filter.

12. AP Conjugate Substrate kit (Bio-Rad).

a. AP color reagent A (nitroblue tetrazolium in aqueous dimethylformamide containing $MgCl_2$ (store in 130-μl aliquots at −20°C).
b. AP color reagent B (5-bromo-4-chloro-3-indolyl phosphate in dimethylformamide) (store in 130-μl aliquots at −20°C).
c. AP color development buffer (25× liquid concentrate; store at 4°C).
d. Preparation of AP reagent for one plate: AP 25× color development buffer, 240 μl; AP color reagent A, 60 μl; AP color reagent B, 60 μl; and H_2O, 5.64 ml.
e. Mix within minutes before use.

13. 0.4% Trypan blue (Sigma-Aldrich): 0.4 g of trypan blue in 100 ml of H_2O. Filter.

3. Methods

3.1. Preparation of Assay Cells

1. Four days before starting an assay, thaw a frozen aliquot (approx 3×10^6 cells in 0.5 ml of freezing medium), centrifuge off cells, and suspend in 10 ml of OBGS (*see* **Subheading 3.14.1.**).
2. Place cell suspension in a 15-cm tissue culture dish, and grow the cells to near confluence (3–5 days; change medium after 3 days). At confluence, there are 1–2×10^7 cells, enough for 20–40 assay plates.
3. If more cells are required, a 1:10 split into 15-cm dishes can be performed (*see* **Note 1**).

3.2. Preparation of Cells Chronically Infected with RML

1. A semiconfluent layer of PK1 or CAD5 cells is exposed for 4 days to a 10^{-3} dilution of RML brain homogenate.
2. The cells are split three times, 1:7 or 1:8 for PK1 and CAD5, respectively.
3. About 150 cells in 10 ml of OBGS are plated into each of three 10-cm tissue culture dishes. After 6–8 days of growth, approx 96 single clones are isolated by aspiration with a 20-μl pipette, transferred into wells of a 96-well plate, and disaggregated by pipetting.
4. After 6–8 days, the cells are suspended, counted, and 1,000 cells are filtered off into the wells of an Elispot plate. The remaining cell suspension is split into a fresh 96-well plate.
5. The Elispot plate is assayed for positive cells as described in **Subheading 3.6.4.**, and the clones with the highest proportion of positive cells (60–90%) are identified.
6. Cells from the corresponding wells are expanded and frozen down in aliquots as described in **Subheading 3.14.4.**

3.3. Sample Preparation

1. Brain and organ 10% homogenates are prepared in phosphate-buffered saline (PBS). For small volumes (one to three mouse brains), use syringes (20 strokes) in succession through 19-, 22-, 25-, and 28-gauge needles. For larger volumes, use a hand-held Ultramax T18 basic homogenizer (IKA Works Inc., Bloomington, NC) at 20–25,000 g for 10 s, on ice. Homogenates are stored in small aliquots at −80°C. After thawing, aliquots are rehomogenized by passing through a 28-gauge needle. Homogenates are not centrifuged because in the absence of detergents even low-speed centrifugation results in the sedimentation of part of the infectivity. We make our stock homogenates without addition of detergents because these are mostly toxic for the assay cells, for example, Triton-X-100 at concentrations >0.0005%.
2. Serial dilutions of infected RML brain homogenate (RML for short) are prepared in OBGS such that the final dilution is at least 10^{-3} (10% homogenate = 10^{-1}). Lower dilutions are toxic to the cells.
3. The dilutions should be such that 100 to 800 positive cells are recovered on the Elispot membrane after plating between 20,000 (maximum) and 1,000 cells.

3.4. Experimental Design

All samples are assayed in at least sextuplicate. Each experiment should contain the following controls.

1. Mouse-titered RML serially diluted 1:2 or 1:3 from 10^{-5} to about 10^{-7} serves as reference for quantitation.
2. An "inhibited control": Scrapie-infected brain homogenates contain immuno-reactive rPrP^{Sc} particles that cannot be distinguished from infected cells in the Elispot assay. To ascertain whether these inoculum-derived particles were diluted out after the final split, the most concentrated unknown, and the most concentrated standard sample, are infected in the presence of the anti-PrP monoclonal antibody D18 at 1:1,000 to prevent prion replication *(17, 24)*; D18 is omitted after first split to avoid diffuse staining in the Elispot assay. Any spots above background found after this treatment are due to residual inoculum.
3. Each Elispot plate within an experiment should contain six wells with approx 1,000 chronically RML-infected PK1 (iPK1) or CAD5 (iCAD) cells (which normally give 300–500 spots), to monitor the success of the Elispot protocol and six wells of 20,000 uninfected cells, to determine background spots.

3.5. *Exposure of Cells to Prion Samples*

1. Dilute infectious sample in OBGS in a final volume of 145 μl and add to a 96-well plate.
2. Resuspend PK1 or CAD5 cells to 5,000 cells/145 μl of OBGS and add to infectious samples in wells, giving a final volume of 290 μl (*see* **Note 2**).
3. After 4 days, suspend the confluent monolayer by gently pipetting up and down with a 12-channel pipette. Transfer a 35-μl aliquot of CAD5 or a 40-μl aliquot of PK1 cell suspension into a well of a 96-well plate containing 255 or 250 μl, respectively, of OBGS, and grow cells to confluence.
4. Split the cells one or two more times (1:7 for PK1; 1:8 for CAD) as described in **step 3** and grow to confluence.
5. After reaching confluence following the second or third split, cells are suspended and assayed by the Elispot protocol (*see* **Subheading 3.6.**).

3.6. *Elispot Reaction*

All solutions must be filtered, and plates should be protected from dust, otherwise particles may increase the background.

Decontamination procedures: Solutions are discarded into 6 M NaOH; when the final concentration reaches 2 M, the solution is kept for at least 1 h and then discarded. Note that NaOH solutions rapidly absorb CO_2 from the air, reducing the pH to below 13–14, precluding efficient decontamination.

3.6.1. Activation of Elispot Membranes

1. To "activate" the membrane, add 60 μl of ethanol (70%) to each well of the Elispot plate, and suction off the ethanol with vacuum after 2 min.
2. Rinse the wells twice with 160 μl of 1× PBS and suction with vacuum.
3. Add 50 μl of PBS to each well to prevent the membrane from drying out.

3.6.2. Plating of Cells

1. Suspend the cells in the 96-well assay plate in 290 μl of OBGS, determine the cell concentration of eight representative wells per plate by using a hemacytometer, and calculate the average. Transfer the volume containing 20,000 cells (or less, if necessary to bring the expected spot number below approx 1,000) into the wells of the Elispot plate by using a 12-channel pipette and suction.
2. Dry the plates 1 h at 50°C or until all wells are dry. Dried plates can be stored for several weeks at 4°C before assaying.

3. Optional: Because the cell number in the individual wells of a plate may vary to some extent, it may be desirable to determine the actual number of cells transferred into each well. Transfer aliquots from each well, containing approx 500 cells, into the corresponding wells of a 96-well plate and carry out the trypan blue assay (*see* **Subheading 3.7.1.**) or the ATP luminescence assay (*see* **Subheading 3.7.2.**). Correct spot numbers by multiplying with 20,000/[actual cell number].

3.6.3. PK Digestion (*see* **Note 3**)

1. Add 60 μl of PK solution (30 Hb milliunits/ml of freshly prepared 1× lysis buffer; usually approx 0.6 μg/ml; Roche Diagnostics) to each well, incubate rocking for 90 min at 37°C.
2. Wash the membrane twice with 160 μl of 1× PBS and suction by vacuum.
3. Add 150 μl 2 mM PMSF and incubate with rocking for 10 min at room temperature, suction by vacuum.
4. Wash with 160 μl of 1× PBS and suction by vacuum.
5. Add 120 μl of Tris-GSCN, and, after rocking for 10 min at room temperature, suction by vacuum.
6. Wash the wells four times with 160 μl of 1× PBS, and suction by vacuum.

3.6.4. Immunoreaction

The primary antibody is monoclonal anti-PrP HuM chimeric antibody D18, which recognizes epitope 134–158 of murine and hamster PrP^c (*25*). The secondary antibody is an anti-human IgG.

1. Add 160 μl of Superblock, rock for 60 min at room temperature, and suction off by vacuum.
2. Add 60 μl of anti-PrP D18 (0.7 μg/ml) in 1× TBST-1% milk powder, dislodge any air bubbles on the filter membrane, and rock for 60 min at room temperature.
3. Wash four times with 160 μl of 1× TBST, and suction off the last wash by vacuum.
4. Add 60 μl anti-human IgGγ-AP (1:5,000) in 1× TBST-1% milk powder, dislodge any air bubbles on the filter membrane, and rock for 60 min at room temperature.
5. Wash four times with 160 μl of 1× TBST, and suction off the last wash by vacuum.
6. Remove the plastic underdrain of the Elispot plate (applicable for Millipore plates only), blot the membrane gently on a lint-free tissue, and tap the plate upside down to remove residual droplets.
7. Dry the plate upside down at room temperature.
8. Add 60 μl of AP dye to each well, dislodge any air bubbles on the filter membrane, and incubate for 16 min at room temperature.

9. Flick the supernatant into waste container.
10. Wash twice with 160 µl of water and flick off the wash.
11. Dry the plate upside down at room temperature.
12. Store the plates at −20°C in the dark (the stain is light sensitive).

3.7. Variation: Plating of Cells and Parallel Determination of Cell Number with Trypan Blue or ATP Assay

The SSCA requires determination of the proportion of rPrP^{Sc}-positive cells. When transferring the cells to the Elispot plate, one routinely determines the cell number in a few representative wells and dispenses the volume containing, on average, 20,000 cells. This may give rise to some error if the cell number differs substantially in individual wells. It is possible to deal with this variability by determining the actual number of cells dispensed post hoc and correcting for the variation when calculating the proportion of infected cells. After suspending the cells in the donor plate, the calculated volume is dispensed into the Elispot wells, and then a further sample from each well of the donor plate is used to determine the actual cell number. Counting is done either by filtering a sample (containing approx 500 cells) onto a separate Elispot plate, staining the cells with trypan blue, and counting them; or by dispensing a sample into a white 96-well plate and determining ATP by a luminescence assay, relative to a standard curve of counted cells.

3.7.1. Determination of Cell Number by Trypan Blue Assay

1. For the determination of the cell number of the cell suspension described in 2.8.2, a 5-µl aliquot of the cell suspension is diluted (1:50) into 250 µl of PBS. Fifty microliters of the final cell suspension (approx 400 cells) is then transferred into an Elispot plate, and vacuum is applied.
2. Plates are dried for 1 h at 50°C or until all wells are dry.
3. Cells are stained by placing the plate on the vacuum device and flushing the wells with 100 µl of 0.04% trypan blue in lysis buffer (a few seconds of exposure). The wells are rinsed twice with PBS and dried. Cells are counted using the Elispot system.

3.7.2. Determination of Cell Number by Luminescence-based ATP Assay (*see* Note 4)

1. Suspend 80,000–100,000 PK1 or CAD5 cells from a confluent culture in 300 µl of OBGS in wells of a 96-well plate and determine the cell number by using a hemacytometer. Twenty microliters of the suspension is serially diluted 1:2 in OBGS, to be used as a standard. OBGS is used as negative control.
2. Ten-microliter aliquots of the cell suspension described in **Subheading 2.5, step 2** (~3,000 cells) are diluted 1:10 into PBS. Twenty microliters of the resulting cell

suspension (~600 cells) is transferred into the wells of a white 96-well plate (Costar polystyrene no. 3692, Corning Life Sciences, Acton, MA), and an equal volume of the ATP reagent mix is added.

3. After 10 min at room temperature, luminescence is measured on the Analyst GT multimode reader (Molecular Devices) or equivalent.

3.8. Guideline for Counting rPrP^{Sc}-positive Cells (Spots) and Trypan Blue-stained Cells by Using WellScan Software

The Zeiss KS Elispot system allows the automated enumeration of spots in the 96 wells of an Elispot plate in approx 2 min.

To count immunostained cells, the detection settings "object size" and "threshold" of the WellScan software (Imaging Associates, Oxfordshire, UK) are optimized for wells with rPrP^{Sc}-positive cells and wells with noninfected cells, respectively (object size is usually ~9 and threshold is usually between 10 and 30). Additional parameters (e.g., spot diameter, color, color saturation, contrast, spot shape, and edge steepness) are adjusted using the "training feature" of the detection system, by visually identifying and manually marking spots not recognized under the initial parameters, following the instructions of the manufacturer.

Briefly, a well of stained, infected cells (100-1,000 rPrP^{Sc}-positive cells) is scanned with the object size of the detection module set to 9 and the threshold (which is inversely proportional to the sensitivity of detection) adjusted to between 10 and 30, to register most visible spots of the well without however unduly increasing the background (aim to keep <20). The classifier is then reset, and positive spots are added manually, whereby the additional parameters are set automatically. After the training of the classifier has been completed, the shape factor is set to 0.4 to detect spherical objects only. Wells with noninfected cells are scanned, and, if necessary, the threshold values are adjusted to give the desired low background, normally <20 spots. The aim is to maximize signal-to-noise ratios while minimizing loss of sensitivity (**Fig. 1**).

It is usually not possible to count all the spots distinguishable by eye without incurring high backgrounds; however, the spot yield is similarly reduced in the samples and the standard curve, and it is thus accounted for when spot numbers are translated to LD$_{50}$ units (or RML brain homogenate concentrations).

A similar counting procedure is used for trypan blue-stained cells.

3.9. Dose–Response Curve for CAD5 Cells Challenged with RML Prions

Figure 1 shows the appearance of typical wells resulting from the SSCA. Intensity and size of the spots, that is, of the rPrP^{Sc}-positive cells, varies greatly, ranging from barely visible to very strong. This variability most likely reflects variable accumula-

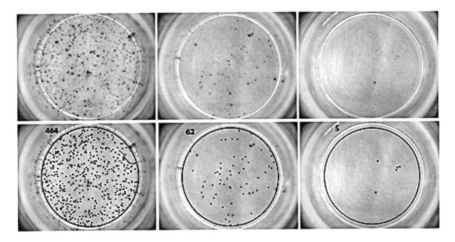

Fig. 1 Representative wells of an SSCA plate as imaged by the Zeiss KS Elispot system. *Top row*: Spots given by 20,000 CAD5 cells in an SSCA of RML-infected brain homogenate at (**A**) 10^{-6} (**B**) 10^{-7}, and (**C**) uninfected cells. *Bottom row*: Same wells as in top row; the black spots (manually highlighted from the original computer image) were recognized by the WellScan software trained to register approx 70% of the spots visible by eye (settings: threshold, 12; size, 9; and area, 0.4)

tion of rPrP^{Sc} in the individual cells. Because of this heterogeneity, the enumeration of spots depends on the settings of the software, in particular the threshold value. **Figure 2A** shows a representative plot of "spot number" as a function of serial dilutions of RML- and 22L-infected brain homogenate, as determined on CAD5 cells, and **Fig. 2B** shows the dose response to RML in 18 consecutive, unselected SSCAs carried out by two operators over 9 months. The variability in the response is attributed in part to variations in the Elispot immunoassay, which give rise to different intensities of the spots, and to different settings of the imaging system, in particular, the threshold and object size, but some variability is due to the state of the cells at the time of infection or plating. Because of this variability, it is essential that a serial dilution of a standard prion preparation is run with every assay.

3.10. Standard Deviation

1. The distribution of observed standard deviations ($n = 16$) as a function of spot number is shown in **Fig. 3**. In the range of 200–800 spots/20,000 cells, the standard deviation is ±20–30%.
2. The standard deviation may be reduced to some extent if the cell numbers of each well are determined and the spot numbers are expressed relative to measured cell numbers.
3. Unusually high standard deviations may come about if the initial infectivity is low and leads to infection of only few cells by the end of the first 4-day incuba-

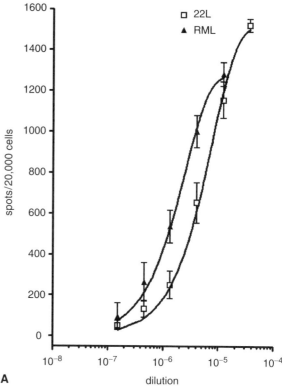

Fig. 2 Dose response of CAD5 cells. (**A**) CAD5 cells were exposed to a serial 1:3 dilution of a RML- or 22L-infected brain homogenate (RML, $10^{8.75}$ LD_{50} units/g brain; 22L, $10^{8.26}$ LD_{50} units/g brain), and then they were subjected to SSCA. (**B**) Dose response of CAD5 cells to RML, as determined in 18 consecutive, unselected experiments carried out by two operators over the past 9 months. CAD5 cells were exposed to a serial 1:2 or 1:3 dilution of the RML-infected brain homogenate described in A. The variability in the response may be due to variations in the Elispot assay, which result in different intensities and sizes of the spots, the settings of the imaging system, and possibly to variation in the response of the cells. Because of this variability it is essential that a serial dilution of a standard prion preparation is run with every assay

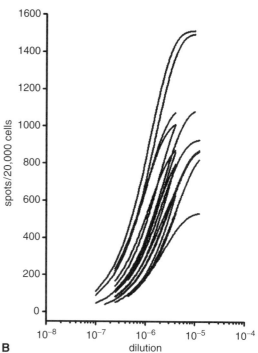

tion. Under these circumstances, a 1:10 split will transfer a very variable number of infected cells to the next well (Poisson distribution) and therefore give rise to a high standard deviation ("jackpot effect"). This can be remedied to some extent by doing two 1:3 splits followed by two 1:10 splits, or by diluting the initial sample less (to the extent that one does not apply toxic amounts of homogenate to the cells).

4. Another source of high standard deviations may be unequal processing of the individual wells, such as unequal cell splitting or uneven processing during the Elispot procedure.

3.11. Time Factor and Throughput

Infection, twice splitting, and growth to confluence take 10–13 d. The immunoassay is performed in 1 day. One operator can routinely process 10 plates manually at a time, or 20 plates or more with the help of partial automation (*see* **Subheading 2.1.2.**, optional equipment). On a routine basis, an experienced operator can initiate two assay series per week, allowing a throughput of approx 300 samples in sextuplicate per week (or 600 samples or more with partial automation).

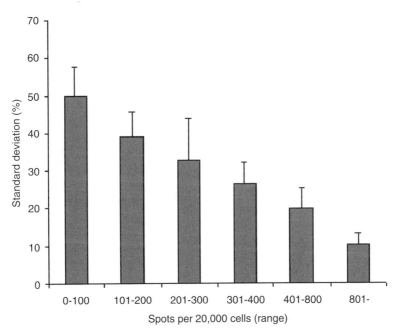

Fig. 3 Standard deviations of SSCA. The standard deviations of 16 consecutive, nonselected assays are plotted as a function of the spot number. Within the range of 200 to 800 spots/20,000 cells, the standard deviation is between 20 and 30%

3.12. SCEPA Format

Scrapie-sensitive cells are exposed to a concentration of prions so low that only a few cells per well are infected. This is insufficient to allow detection by the Elispot assay, because the value is too close to background (5–15 spots/20,000 cells). If, however, the cell population is propagated for several generations, "recruitment" of as yet uninfected cells, due to infection by secreted prions or through cell–cell contact, results in an increase of the proportion of infected cells (~20–25%/cell doubling; [20]) (Fig. 4).

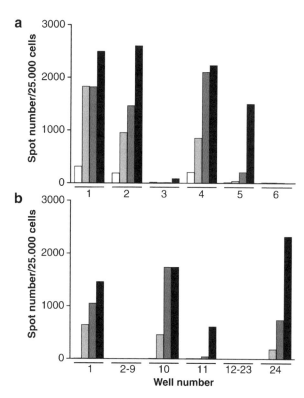

Fig. 4 SCEPA of RML-infected brain homogenate. Twenty thousand PK1 cells were exposed for 3 days to 300 μl of 10^{-7} (**A**) and 10^{-8} (**B**) dilutions of RML-infected brain homogenate in OBGS. The cells were split three times 1:10 (12 days, *white bars*), five times (18 days, *light gray bars*), eight times (27 days, *gray bars*), and 12 times (39 days, *black bars*); 25'000 cells were subjected to the Elispot assay. Mean ± SD of background values of noninfected cells were 11 ± 6, 16 ± 13, 14 ± 5, and 9 ± 2 at 12, 18, 27, and 39 days, respectively. After 39 days of culture, wells with spot numbers exceeding the mean value of the background by 5 SDs were scored as positive. The dilution leading to 50% positive wells was $10^{-7.5}$, as calculated following Reed and Munch (*7*). The brain, therefore, contained $10^{7.5}/(0.3) = 10^{8}$ tissue culture LD_{50} units/g. (Reproduced with permission from **ref.** *20)*

Susceptible cells in wells of 96-well plate are exposed to 12 or 24 replicas of a serial dilution of the prion-containing sample for 4 days, grown to confluence, and split three times 1:3 and twice or more 1:7 (PK1) or 1:8 (CAD5). After reaching confluence, 20,000 cells are transferred to membranes of Elispot plates, and rPrP^{Sc}-containing cells are identified by ELISA.

Because the background remains constant once the particles due to the inoculum have been diluted out (after two to three splits), the signal-to-noise ratio increases continuously, so that most infected wells end up with hundreds of positive cells after the multiple splits. The outcome of the assay is scored by determining the proportion of "positive" wells.

3.12.1. Preparation of Samples

1. Brain and organ homogenates are prepared and diluted in OBGS as described in **Subheading 2.4.**
2. The dilutions should be such that approx 50% of the wells will result positive; the critical range is 10^{-7} to 10^{-8} dilution for RML (specific infectivity $10^{-8.4}$ LD$_{50}$ unit/g).
3. Controls:

   a. Each SCEPA assay should be accompanied by a serial 1:10 dilution of titered RML (between 10^{-6} and 10^{-9}).
   b. Each assay should comprise 12 wells with uninfected PK1 or CAD5 cells.

3.12.2. Infection of PK1 or CAD5 Cells with Prion Samples

1. Replicate 145-μl aliquots of samples of each dilution are placed into 12 or 24 wells to which are added prion-susceptible cells, and cells are incubated for 4 days. The cells are split three times 1:3 every 3 days, and then three times 1:7 (PK1) or 1:8 (CAD5) when confluent.
2. When confluence is reached after the final split, 20,000 cells from each well are filtered onto the membranes of an Elispot plate, and rPrP^{Sc}-positive cells are determined as described for SSCA in **Subheading 3.6.**

3.12.3. Evaluation of SCEPA

1. The assay is evaluated by determining the ratio of positive to "negative" wells whereby wells are considered positive if the spot count is greater than the average background value plus 5 times the standard deviation (*see* **Note 5**).
2. The tissue culture infectivity 50% dose can be calculated by the method of Reed and Muench *(7)* or Spearman-Korber *(26)* from the values resulting from a serial dilution, of which the lowest yields all or a majority of positive wells and the highest all or a majority of negative wells.

3. General considerations: A well containing thousands of susceptible cells and one or a few infected cells would be scored as negative. However, after several splits, the infected cell will not only yield infected daughter cells but also infect surrounding cells, so that the well will contain hundreds of positive cells and be scored as positive. If cells are infected with a prion concentration that, on average, results in the order of 0.1 to a few infected cells per well, and we score 10 or 20 wells, then $P_{(0)}$, the proportion of negative wells, N_o, to total wells, N, is given by $P_{(0)} = N_o/N = e^{-m}$ (the Poisson equation), where m is the average number of infected cells per well. From this, in principle, one could determine m and thus the initial number of infected cells per well. The equivalence in LD_{50} units can be obtained from a comparison with the SCEPA of a titered RML control.

In reality, the number of infected cells is underestimated. Assume that at the end of the initial infection period, there are 10 infected cells among roughly 100,000 total cells in the well. If a 1:10 split is carried out, 1/10 of the cells will, on average, make up one infected cell; therefore, there is a probability of $P_{(0)} = e^{-m}$ $= e^{-1} = 0.37$ that there will not be an infected cell in the aliquot taken. Thus, such a well will remain negative throughout the subsequent repeated splittings. Reduction in the recovery of cells occurs at every split, however at a decreasing pace, depending on the extent of lateral spread of infection. If initially 1:3 splits rather than 1:10 splits are carried out, the probability of losing infected wells is diminished to some extent.

Also, when starting with a population containing few infected cells, the aliquots may repeatedly contain only one infected cell in consecutive splits, so that the wells apparently remain negative for several splits, but then they may become positive late in the assay (**Fig. 4**), perhaps when after the third or fourth split the aliquot happens to capture three or four positive cells. The proportion of positive wells will asymptotically approach some "final" value with increasing number of splits.

In practice, we circumvent the problem of "correcting" for these distortions by running a serial dilution of a reference sample with known titer in parallel. For example, a 10^{-8} dilution of RML with a titer of $10^{8.4}$ LD_{50} units/ml gives 20/24 negative wells. One can construct a standard curve using a series of RML dilutions.

3.13. Selection, Maintenance, Freezing, and Thawing of Cells

A frozen aliquot (~3–6 million cells) is thawed, expanded, and passaged not more than six to seven times with 1:10 splitting, to ensure that susceptibility to prion infection is preserved. However, stability as a function of passaging has not been systematically investigated.

Before the supply of frozen aliquots draws to an end, a sensitive subline is reisolated, aliquots are frozen, and an aliquot is tested relative to a sample from the preceding lot.

3.14.1. Thawing Cells

1. Thaw a vial of frozen cells by rapid agitation in a 37°C water bath (1–2 min).
2. Clean the outside of the vial in 70% ethanol and wipe dry.
3. Add dropwise a volume of prewarmed OBGS equal to that of the freezing medium (0.5 or 1 ml) and incubate for 5 min at room temperature.
4. Place the cell suspension in a centrifuge tube (15 ml or larger), add another volume of prewarmed OBGS equal to that of the freezing medium, and incubate for 5 min at room temperature.
5. Add 10 ml of prewarmed OBGS and centrifuge at 400g for 5 min.
6. Discard the supernatant and disperse the cell pellet by agitation and resuspend in 2–3 ml of prewarmed OBGS.
7. Transfer the cell suspension into a 10- or 15-cm Petri tissue culture dish or a T75 flask containing 10, 35, or 25 ml of prewarmed OBGS, respectively.

3.14.2. Cell Maintenance

1. Optimal tissue culture conditions are a prerequisite for maintaining high susceptibility of cells to prions.
2. Cells should be split 1:10 every 3–4 days or before becoming overconfluent. If cells are to be grown to confluence, a daily medium (OBGS) change at near-confluence levels is recommended.
3. After six to seven passages, we discard the cells and thaw a fresh aliquot.

3.14.3. Reisolation of Highly Susceptible Cells

1. 150 cells in 10 ml of OBGS are plated into each of six to eight 10-cm tissue culture dishes. After 6–8 days of growth, single clones are isolated by aspiration with a 20-μl pipette, transferred into wells of a 96-well plate and disaggregated by pipetting. At least 50 single clones should be screened.
2. After 6–8 days, the cells are suspended, and approx 5,000 cells are placed into corresponding wells of two 96-well plates. One of the plates (test plate) contains 145 μl of a 10^{-6} dilution of RML-infected brain homogenate in OBGS, whereas the parallel (uninfected) maintenance plate is incubated with OBGS.
3. After 4 days of incubation the cells in the test plate are split 1:8 after 3–4 days (CAD cells) or 1:7 after 4–5 days (PK1 cells). After reaching confluence after the second split, the proportion of rPrP^{Sc}-positive cells is determined.
4. Wells giving high spot numbers are identified, and the cells from the corresponding wells of the maintenance plate are expanded and aliquots frozen down (*see* **Subheading 3.14.4.**).

3.14.4. Freezing Cells

1. Change medium of confluent cell layers on a 15-cm tissue culture dish to remove any floating cells and debris.
2. Suspend cells gently in 10 ml of OBGS and determine cell number.
3. Pellet the cells at 400g for 5 min and discard supernatant.
4. Loosen the pellet by agitation, add 1 ml of freezing medium a (OBGS containing 5% dimethyl sulfoxide (Sigma-Aldrich, cell culture tested) per 6 million cells. Resuspend gently and transfer aliquots of 3 to 6 million cells into freezing vials labeled with batch designation, date, and vial number.
5. Place vials into an isopropanol-containing freezing container Mr. Frosty (Nalgene, Wessington Cryogenics, Tyne & Wear, England) and keep at −80°C overnight.
6. Transfer vials into liquid nitrogen.
7. Test one or more aliquots for sensitivity to standard RML by the SSCA, in comparison with the previous batch of sensitive cells.

4. Notes

1. We do not perform more than six to seven serial 1:10 splits, because susceptibility may decline; however, this has not been systematically explored.
2. Previously, SSCA and SCEPA were performed by plating 20,000 PK1 or 15,000 CAD5 cells in 200 µl of OBGS into wells of a 96-well plate. After 16 h at 37°C, the medium was replaced by 0.29-ml assay samples in OBGS. The cells were split after 3 days of infection. All experiments in **Figs. 1–4** were performed with this original method. The new improved method (5,000 cells in 145 µl of OBGS added to 145-µl sample) allows the sample/cell preparation to be performed in 1 day, reduces the number of cells required and has resulted in three- to fivefold higher spot number.
3. PK digestion, although not affecting rPrP^{Sc}, removes PrP^C, which may otherwise give a diffuse background, and it also enhances the immunoreaction by removing proteins. GSCN treatment is required to denature the sample and render it immunoreactive.
4. The amount of ATP per cell is very constant for cells of a particular cell line under the same growth conditions. Cell suspensions containing approx 200–5,′000 cells are placed in the wells of a white 96-well plate, and the ATP is determined by the luciferin-luciferase assay (CellTiter-Glo™ Luminescent Cell Viability Assay, Promega, Madison, WI).
5. "False positives": Under the assay conditions described in this chapter, the antibody-reactive spots are diagnostic for scrapie-infected cells. After certain treatments, PK1 (and perhaps other) cells may accumulate a form of PrP that also gives rise to spots, but they do not represent rPrPSc:

   a. N2a cells transfected with murine PrP expression plasmids giving rise to very high PrP levels (M. Messenger and S. Brandner, unpublished results).

b. N2a cells after treatment with pentosan polysulfate $10\,\mu g/ml$ (but not $1\,\mu g/ml$) or proteasome inhibitors (such as lactacystin). We think this represents PrP in the form of aggresomes.

c. The spots arising in cells overexpressing PrP are eliminated if the PK concentration used in the SSCA is increased to $10\,\mu g/ml$, whereas spots due to infected cells are only approx 40% diminished.

References

1. Prusiner SB. (1991) Molecular biology of prion diseases. Science;252(5012):1515–22.
2. Sakaguchi S, Katamine S, Yamanouchi K, et al. (1993) Kinetics of infectivity are dissociated from PrP accumulation in salivary glands of Creutzfeldt-Jakob disease agent-inoculated mice. J Gen Virol;74(10):2117–23.
3. Shaked GM, Fridlander G, Meiner Z, Taraboulos A, Gabizon R. (1999) Protease-resistant and detergent-insoluble prion protein is not necessarily associated with prion infectivity. J Biol Chem;274(25):17981–6.
4. Manson JC, Jamieson E, Baybutt H, et al. (1999) A single amino acid alteration (101L) introduced into murine PrP dramatically alters incubation time of transmissible spongiform encephalopathy. EMBO J;18(23):6855–64.
5. Lasmezas CI, Deslys JP, Robain O, et al. (1997) Transmission of the BSE agent to mice in the absence of detectable abnormal prion protein. Science;275(5298):402–5.
6. Prusiner SB, Cochran SP, Groth DF, Downey DE, Bowman KA, Martinez HM. (1982) Measurement of the scrapie agent using an incubation time interval assay. Ann Neurol;11(4):353–8.
7. Reed J, Muench H. (1938) A simple method of estimating fifty per cent endpoints. Am J Hyg;27:493–7.
8. Schätzl HM, Laszlo L, Holtzman DM, et al. (1997) A hypothalamic neuronal cell line persistently infected with scrapie prions exhibits apoptosis. J Virol;71(11):8821–31.
9. Birkett CR, Hennion RM, Bembridge DA, et al. (2001) Scrapie strains maintain biological phenotypes on propagation in a cell line in culture. EMBO J;20(13):3351–8.
10. Follet J, Lemaire-Vieille C, Blanquet-Grossard F, et al. (2002) PrP expression and replication by Schwann cells: implications in prion spreading. J Virol;76(5):2434–9.
11. Vilette D, Andreoletti O, Archer F, et al. (2001) Ex vivo propagation of infectious sheep scrapie agent in heterologous epithelial cells expressing ovine prion protein. Proc Natl Acad Sci U S A;98(7):4055–9.
12. Rubenstein R, Carp RI, Callahan SM. (1984) In vitro replication of scrapie agent in a neuronal model: infection of PC12 cells. J Gen Virol;65(12):2191–8.
13. Butler DA, Scott MR, Bockman JM, et al. (1988) Scrapie-infected murine neuroblastoma cells produce protease- resistant prion proteins. J Virol;62(5):1558–64.
14. Race RE, Fadness LH, Chesebro B. (1987) Characterization of scrapie infection in mouse neuroblastoma cells. J Gen Virol;68(5):1391–9.
15. Chandler RL. (1961) Encephalopathy in mice produced by inoculation with scrapie brain material. Lancet;1(24):1378–9.
16. Beranger F, Mange A, Solassol J, Lehmann S. (2001) Cell culture models of transmissible spongiform encephalopathies. Biochem Biophys Res Commun;289(2):311–6.
17. Enari M, Flechsig E, Weissmann C. (2001) Scrapie prion protein accumulation by scrapie-infected neuroblastoma cells abrogated by exposure to a prion protein antibody. Proc Natl Acad Sci U S A;98(16):9295–9.
18. Nishida N, Harris DA, Vilette D, et al. (2000) Successful transmission of three mouse-adapted scrapie strains to murine neuroblastoma cell lines overexpressing wild-type mouse prion protein. J Virol;74(1):320–5.

19. Bosque PJ, Prusiner SB. (2000) Cultured cell sublines highly susceptible to prion infection. J Virol;74(9):4377–86.
20. Klohn PC, Stoltze L, Flechsig E, Enari M, Weissmann C. (2003) A quantitative, highly sensitive cell-based infectivity assay for mouse scrapie prions. Proc Natl Acad Sci U S A;100(20): 11666–71.
21. Qi Y, Wang JK, McMillian M, Chikaraishi DM. (1997) Characterization of a CNS cell line, CAD, in which morphological differentiation is initiated by serum deprivation. J Neurosci; 17(4):1217–25.
22. Fischer M, Rülicke T, Raeber A, et al. (1996) Prion protein (PrP) with amino-proximal deletions restoring susceptibility of PrP knockout mice to scrapie. EMBO J;15(6):1255–64.
23. Safar JG, Scott M, Monaghan J, et al. (2002) Measuring prions causing bovine spongiform encephalopathy or chronic wasting disease by immunoassays and transgenic mice. Nat Biotechnol;20(11):1147–50.
24. Peretz D, Williamson RA, Kaneko K, et al. (2001) Antibodies inhibit prion propagation and clear cell cultures of prion infectivity. Nature ;412(6848):739–43.
25. Leclerc E, Peretz D, Ball H, et al. (2003) Conformation of PrP(C) on the cell surface as probed by antibodies. J Mol Biol;326(2):475–83.
26. Dougherty T. Animal virus titration techniques. (1964) In: Harris RJC, ed. Techniques in Experimental Virology. New York: Academic Press; 169–224.

Chapter 5
Generation of Cell Lines Propagating Infectious Prions and the Isolation and Characterization of Cell-derived Exosomes

Laura J. Vella and Andrew F. Hill

Summary Prion-propagating cell lines are an efficient and useful means for studying the cellular and molecular mechanisms implicated in prion disease. Use of cell-based models has lead to the finding that prion protein (PrP^C) and PrP^{Sc} are released from cells in association with exosomes. Furthermore, exosomes have been shown to act as vehicles for infectivity, transferring PrP^{Sc} between cell lines and providing a mechanism for prion spread between tissues. As a role for exosomes in prion disease is emerging, this chapter outlines a method for the generation of prion-infected cell lines and the isolation and characterization of PrP^C- and PrP^{Sc}-containing exosomes.

Keywords Cell culture media; exosomes; GT1-7; infection; passaging; prion.

1. Introduction

According to the protein only hypothesis, an abnormal isoform of the host-encoded prion protein (PrP^C), referred to as PrP^{Sc}, is the sole or major component of the infectious agent (the "prion") (1). The mechanism of neurodegeneration, transmission, and even the normal biological function of PrP^C are not completely clear; however, considerable progress has been made in the past 20 years due in part to the establishment of cell culture models of prion propagation. The main advantages of cell culture models have been the ability to study PrP^C and PrP^{Sc} propagation at a cellular level and to provide an alternative to lengthy bioassays in mice, enabling efficient drug screening and prion strain characterization.

The first prion-infected cell culture was reported in 1970 (2), and many groups since then have successfully propagated a variety of prion strains in cell culture (3–6), the most common being Rocky Mountain Laboratory (RML) strain in N2a cells (7, 8). Use of cell culture models recently lead to the novel finding that the culture media of RK13 cells overexpressing ovine PrP (Rov cells) contains PrP^C and PrP^{Sc} in association with exosomes (9).

From: *Prion Protein Protocols.*
Methods in Molecular Biology, Vol. 459.
Edited by: A. F. Hill © Humana Press, Totowa, NJ

Exosomes are small membrane vesicles formed by invagination of the membrane of multivesicular bodies (MVBs) *(10)*. Exosome secretion into the extracellular environment occurs upon fusion of MVBs with the cell membrane, and it was suggested to be a mechanism of releasing unnecessary proteins during the maturation of reticulocytes. However, the physiological relevance of exosomes has been established by their identification in vivo, in association with follicular dendritic cells *(11)*, urine *(12)*, and malignant tumor effusions *(13)*. The function of exosomes seem to extend beyond the simple removal of unwanted cellular proteins and a role for exosomes in mediating intercellular communication has been identified *(9, 14, 15)*. Exosomes released from dendritic cells, mast cells, tumor cells, and intestinal epithelial cells can be targeted to T-cells, bone marrow-derived cells, and splenic dendritic cells, and they can modulate immune responses *(16–22)*.

A recent study demonstrated a role for exosomes in the intercellular trafficking of human immunodeficiency virus-1 infectivity, illustrating the exploitation of a preexisting cellular pathway to increase the efficiency of viral distribution *(23)*. Endogenous PrP^C is associated with exosomes from epididymal fluid, platelets, and primary cultured cortical neurons, suggesting exosomes may contribute to spreading prions not only through the lymphoreticular system but also potentially in the brain *(24–28)*. However, the association of PrP^{Sc} with exosomes was not examined in any of these studies *(24–28)*, leaving uncertainty in our current understanding as to their role in PrP^{Sc} trafficking from endogenously expressing PrP cell lines and the speculation regarding transfer of exosomal infectivity to cells of different origin. We have demonstrated that GT1-7 cells release PrP^C and PrP^{Sc} in association with exosomes when infected with mouse-adapted strain of human prions (M1000), providing a mechanism for the interneuronal dissemination of endogenous prions within the CNS *(29)*. We also have reported that exosomes originating from either non-neuronal or neuronal cells can transmit infection between cell lines, validating the notion that exosomes may have a role in trafficking of infectious prions to different tissue types in vivo *(29)*.

Although the main consequence of prion infection is CNS degeneration, a wealth of evidence highlights the importance of prion replication in the lymphoreticular system (LRS) before the ensuing invasion of the CNS (reviewed in **ref.** *30*). Delays in neuroinvasion as a result of a functionally compromised LRS have been reported previously *(31)*. After oral challenge with prions in animal models, PrP^{Sc} is detected in the Peyer's patches of the gut and then the mesenteric lymph nodes, followed by the lymphoreticular system (such as the spleen) *(32–34)*. Within the lymphoid organs, including Peyer's patches, PrP^{Sc} accumulates and replicates in predominantly immobile cells such as mature follicular dendritic cells *(35)* and only to a lesser extent on mobile cells such as dendritic cells *(36)*. How PrP^{Sc} is transported from immobile cells to lymphoid organs and subsequently the CNS is not fully understood. It has been suggested the infection likely spreads to the CNS via the enteric nervous system, or splanchnic or vagal nerves *(37–39)*; however, it is not known how infectivity reaches peripheral nerve endings. Interestingly, dendritic and follicular dendritic cells are known to accumulate high levels of PrP and to secrete exosomes, making a relationship between PrP and exosomes from these cell types plausible.

Another enigma in prion disease distribution is PrP^{Sc} spread within the CNS. In vitro systems mimicking prion trafficking in the CNS are limited, and we are limited to neuropathological imaging performed subsequent to neurotoxicity. Based on our findings *(29)* and the findings others *(28)*, who demonstrated that cortical neurons secrete PrP^{C} containing exosomes, the relationship between prion spread within the CNS and neuronal exosomes warrants further investigation.

A role for exosomes in prion disease is gradually emerging, and research in this area is escalating to examine the relationship further. This chapter outlines a method for generating a prion-infected cell line and the isolation and characterization of exosomes from the cell culture media.

2. Materials

2.1. Reagents

1. Fetal calf serum (FCS) (cat. no. 10099-141, Invitrogen, Carlsbad, CA).
2. Opti-MEM (Invitrogen) containing 10% (w/v) FCS, 1% penicillin/streptomycin (100 μg/ml), and 1% glutamine (1 mM).
3. Phosphate-buffered saline (PBS) (cat. no. 14040-182, Invitrogen).
4. Lysis buffer: 0.5% sodium deoxycholate, 0.5% Triton X-100, 150 mM NaCl, and 50 mM Tris-HCl, pH 7.5.
5. Proteinase K; fungal (cat. no. 25530-015, Invitrogen).
6. Phenylmethylsulfonyl fluoride (PMSF) (Sigma-Aldrich, St. Louis, MO).
7. Coomassie blue stain: 10% acetic acid, 45% methanol, and 0.025% Coomassie blue R (Sigma-Aldrich).
8. Sample buffer: 125 mM Tris-HCl, pH 6.8, 4%, sodium dodecyl sulfate (SDS), 10% glycerol, 0.02% bromophenol blue, and 5% β-mercaptoethanol (added just before use).
9. Ponceau S stain: 0.1% (w/v) ponceau S (Sigma-Aldrich) and 5% acetic acid.
10. Sucrose stock solution: 2.5 M sucrose and 20 mM HEPES, pH 7.0.
11. HEPES buffer: 20 mM HEPES, pH 7.2.
12. Uranyl acetate.
13. Glutaraldehyde.
14. Bicinchoninic acid (BCA) Protein Quantitation kit (Pierce Chemical, Rockford, IL).
15. Methanol.
16. Organelle markers (organelle detector kit, cat. no. 612740 BD, BD Biosciences, Franklin Lakes, NJ).
17. Prion protein antibodies: ICSM18 (cat no. 0130-01810, D-Gen Ltd., London, UK) and SAF32 (cat. no. 189720, Cayman Chemical, Ann Arbor, MI).

2.2. *Equipment*

 1. Biological safety cabinet; class II.
 2. Tissue culture plasticware (Nalge Nunc International, Rochester, NY); six-well plates; T75-cm^2 and T300-cm^2 vented flasks.
 3. CO_2 humidified incubator.
 4. Benchtop refrigerated centrifuge.
 5. Ultracentrifuge with fixed angle and swinging bucket rotors.
 6. Polyallomer tubes or polycarbonate bottles, appropriate for the ultracentrifuge rotor.
 7. 0.22-μm filters (Millex GP, Millipore, Billerica, MA).
 8. Rotating heat block (e.g., Thermomixer, Eppendorf, Hamburg, Germany).
 9. Liquid nitrogen.
10. Water bath.
11. Syringes and needles.
12. Transmission electron microscope (e.g., Siemens Elmiskop 102).
13. Formvar electron microscopy grids (ProSciTech, QLD, Australia).
14. Water bath sonicator.

3. Methods

3.1. *Maintaining Cultured Cell Lines*

Only a few cultured cell lines of neuronal (N2a and GT1-7) and non-neuronal origin (PC12 and Rov, 3T3) are infectable with prions *(5, 6, 40–42)*. It is not known why these cells are susceptible to prion infection, whereas many others are not *(43, 44)*. It was originally thought PrP expression played a large role in susceptibility; however, many high PrP-expressing cell lines cannot support prion propagation. With this in mind, if you are inexperienced in propagating prions in cell culture, it may be best to use cell lines previously recognized for supporting prion propagation.

   For the purpose of this methods chapter, we describe maintenance and infection of the mouse hypothalamic cell line GT1-7. This cell line is described because it is often the most difficult to culture; however, the methods herein can be readily applied to other cell lines such as N2a and non-neuronal cell lines.

   In our hands, GT1-7 cells divide slowly (doubling time of ~48-72h), and they require more maintenance and care than other hardier cell lines, such as RK13 cells. A protocol for passaging GT1-7 cells is listed below. Ensure fresh media and PBS are prewarmed to 37°C.

 1. Wash adhered cells twice with warm (37°C) PBS. Add 10 ml of PBS to the cells; use this solution to detach cells from the flask surface by repeated washing of the flask surface. Use the pressure generated by the pipettor to force cells from the flask surface with PBS. When one quarter of the cells are detached,

place the PBS-containing cells into a 50-ml sterile tube. Repeat detaching with fresh PBS. Repeat this process until all cells have deattached and have been collected. (*see* **Note 1**).
2. Spin the cells at 600*g* for 5 min.
3. Remove PBS and gently resuspend cells in media and replate into a fresh flask. (*see* **Note 2**).
4. Passage the cells every 3–4 days. Cells should not be left longer than 5 days (*see* **Note 3**).

3.2. Infection of Cell Lines with Prions

The GT1-7 cell line has been shown to support propagation of a variety of prion stains, including RML *(5)*, 22L, 139A *(45)*, Fukuoka, and mouse-adapted sporadic CJD (SY) *(46)*, making them advantageous over other cell lines. The most common approach to propagating prions in cell culture is to put infectious brain homogenate in contact with the cells for a certain period (1–3 days), and then wash and passage the cells extensively before testing for *de novo* prion propagation. Other methods have been recently used such as infection with prion propagating cell lysate or cell media. Importantly, the presence of infectious PrP^{Sc} in cell lines must be confirmed by mouse bioassay before the commencement of biological studies. To successfully maintain prion propagation, see **Note 8** and **Note 9**.

3.2.1. Infection of Cell Lines with Brain Homogenate (*see* Note 4)

1. Plates cells into six-well plates (*see* **Note 5**) the day before infection. Cells should be ~50% confluent on the day of infection.
2. On the day of infection, wash the cells with PBS and incubate with 600 μl of 1% brain homogenate, either prion infected or uninfected. Ensure the surface of the well is completely covered and place the plate in an incubator (37°C, 5% CO_2) for 5 h (*see* **Note 6**).
3. After incubation, add 2.0 ml of fresh Opti-MEM to the cells and incubate for a further 72 h.
4. Remove brain homogenate, wash cells twice with PBS, and split cells into a new six-well plate. Passage an additional 8–10 times before testing for the expression of protease resistant PrP (*see* **Note 7**).

3.2.2. Infection of Cell Lines with Cell Lysate or Media

A simple way to generate an infected cell line, without always resorting to brain homogenate, is to use lysate or cell culture media from prion-propagating cell lines. In our hands, this method is very efficient, and it results in less toxicity than brain homogenate infections.

1. Grow non-infected and prion-infected cells to 80% confluence in a T75-cm^2 flask.
2. Wash cells twice with PBS and then remove cells from flask surface with ice-cold PBS.
3. Pellet cells and remove PBS. To prepare the cell lysate, expose the cell pellet to repeated freeze-thawing with liquid nitrogen.
4. Place the cell pellet into liquid nitrogen for 30 s and then into a 37°C water bath until thawed. Repeat four to five times.
5. The lysate is then passed through a 28-gauge needle to ensure complete homogenization of whole cells.
6. Spin lysate 2,000g for 3 minutes, to remove any large, insoluble particles.
7. Dilute the lysate in 5 ml of fresh Opti-MEM and then incubate 2.5 ml with GT1-7 cells (prepared the day before) in one well of a six-well plate and incubate for 72 h in a 37°C 5% CO_2 incubator.
8. Remove inoculum, wash cells twice with PBS, and split cells into a new six-well plate. Passage cells an additional 8–10 times before testing for the expression of protease resistant PrP.

3.2.3. Infection of Cells with Media Obtained from Previously Infected Cells

1. Plate out cultures of noninfected and prion-infected cells and incubate for 4 days.
2. Remove cell media and spin at 3,000g for 10 min, to remove detached cells and cellular debris.
3. Take the supernatant and incubate it for 72 h in a 37°C 5% CO_2 incubator with GT1-7 cells prepared the day prior.

3.3. Bioassay of Prion-infected Cell Lysates (see Note 10)

1. Resuspend cells (3×10^6) in 1 ml of PBS and subject to five consecutive cycles of freeze-thawing and sonication. (If the bioassay is being conducted using exosomes as the inoculum, ensure sterility has been maintained by filtering through a 0.22-μm filter.).
2. Inoculate 30 μl into the left parietal region of anaesthetized mice.
3. Monitor mice for typical signs of murine prion disease, such as hunched posture, hindlimb paresis, ataxia, and reduced motor activity, and euthanize mice in the terminal stage of disease.
4. Examine the brain of every animal for the presence of PrP^{Sc} by immunoblotting or immunohistochemistry.

3.4. Preparation of Exosomes from Prion-infected Cell Cultures

Protocols for isolating exosomes vary based on whether they are isolated from cultured cell lines or biological fluid. The common method involves ultrafiltration and a series of ultracentrifugation spins *(47, 48)*; however, novel methods such as coupling exosomes on beads *(49)* and use of nanomembrane ultrafiltration concentrators *(50)* have been reported recently. In in vitro cell culture, exosomes are secreted into the media, where they can be isolated and studied.

1. Plate out cells (2×10^7) in T300-cm^2 flasks in serum-depleted media (*see* **Note 11**).
2. Incubate the cells for 4 days; ideally, cells should be at 100% confluence (80% for GT1-7 cells) by day 4 (*see* **Notes 12** and **13**).
3. On day 4, collect medium from the cells and centrifuge at $2,000\,g$ for 20 min to remove cellular debris (*see* **Note 14**).
4. Transfer the supernatant to fresh tubes and filter using 0.22-μm filters into polycarbonate bottles. Fill tubes to the top, adding PBS if necessary (*see* **Note 15**).
5. Ultracentrifugate the supernatant at $10,000\,g$ for 30 min at 4°C with medium deacceleration.
6. Immediately after spin completion, transfer the supernatant to fresh centrifuge tubes.
7. To isolate exosomes, ultracentrifugate the supernatant at $100,000\,g$ for 1 h at 4°C with medium deacceleration.
8. Immediately after spin completion, take note of where the exosomal pellet should be (because they are sometimes difficult to see) and pipette out the supernatant (discard it), taking care to avoid the pellet. When removing the supernatant, hold the tube at the same angle as when it was positioned in the rotor.
9. Resuspend the pellet in PBS (large volume) and ultracentrifuge at $100,000\,g$ for 1 h at 4°C.
10. Immediately after centrifugation, pipette out the supernatant, mark the position of the exosomal pellet and then turn the tube upside down to drain excess liquid.
11. Resuspended the exosome pellet in 100–200 μl of PBS (or sucrose if performing a gradient) and either use immediately or stored at −80°C (*see* **Note 16**).

3.5. Exosome Characterization

3.5.1. Sucrose Density Gradient

To confirm the identity of microvesicles as exosomes, continuous sucrose density gradient centrifugation is performed. Exosomes float in sucrose gradients with a density ranging from 1.13 to 1.19 g/ml depending on the cell type from which they were prepared *(47, 48, 51)*.

1. Resuspend the exosomes in 5 ml of 2.5 M sucrose/20 mM HEPES buffer and place into the bottom of a swing out rotor tube. For a tube balance, repeat this step excluding the exosomes.

2. Layer a 6-ml linear sucrose gradient (2.0–0.25 M sucrose and 20 mM HEPES, pH 7.2) on top of the exosome suspension, taking care not to mix the layers.

3. Centrifuge the samples at 70,000 g for 16 h at 4°C (e.g., SW41 rotor; Beckman Coulter, Fullerton, CA).

4. After spin completion, remove 1-ml gradient fractions from the top of the tube (11 fractions) and measure the refractive index of each fraction.

5. Dilute each fraction in 10 ml of PBS and ultracentrifuge each fraction for 1 h at 200,000 g (e.g., SW41 rotor) (see **Note 17**).

6. Remove the supernatant and solubilize the proteins in the bottom of the tube with 1× sample buffer (without β-mercaptoethanol). The vesicles may not be visible at this stage.

7. Boil sample at 100°C for 5 min, centrifuge for 2 min at 14,000 g, and electrophorese on SDS-polyacrylamide gel electrophoresis (PAGE) gels.

8. Immunoblot with known exosome markers such as flotillin or tsg101 and then probe for PrP by using a C-terminal antibody, such as ICSM18 (see **Notes 18** and **19**).

3.5.2. Electron Microscopy

Purfied extracellular particles can be examined by negative staining electron microscopy (EM). Under EM, exosomes look membrane-bound and "cup-shaped," and they have a similar size (50–80 nm in diameter) to previously described exosomes *(47, 48)*.

1. Purify exosomes as described in **Subheading 3.5.** and resuspend in filtered PBS.

2. Fix the exosome suspension in 2.5% glutaraldehyde, and as a negative control, replace exosomes with PBS.

3. Apply 5 µl of the fixed sample to a 200 mesh copper grid supported with Formvar/carbon (ProSciTech) and leave to dry at room temperature for 30 min.

4. Wash the grid by floating it on a drop of PBS 3 × 1 min.

5. Wash the grid by floating it on a drop of distilled H_2O 6 × 2 min.

6. Negatively stain the grid with 3% saturated aqueous uranyl acetate for 20 min in the dark.

7. Removing excess liquid, air dry the grid for 30 min at room temperature, and view with a transmission electron microscope (e.g., Siemens Elmiskop 102).

3.5.3. Western Blot Analysis

In addition to or instead of mass spectrometry analysis (see **Note 20**), immunoblotting of exosomal proteins can be performed to confirm the presence or absence of exosomes. It is important to compare equal concentrations of cell lysate with exosomes, to illustrate protein enrichment in exosome samples. Conversely, samples can be probed for proteins that are clearly absent in exosomes, such as most endoplasmic reticulum and nuclear markers.

1. Quantify the protein content (e.g., BCA assay) of the exosome pellet and cell lysate from which the exosomes are derived (*see* **Note 21**).
2. Add 2× sample buffer, boil for 10 min, and spin 13,000 g for 2 min. Samples can be stored at −20°C before performing the Western blot; however, avoid repeated freeze-thawing.
3. Electrophorese samples on an SDS-PAGE gel and transfer to either nitrocellulose or polyvinylidene difluoride (PVDF) membrane.
4. Detect PrP by immunoblotting with anti-PrP antibodies (or detect other proteins of interest; *see* **Note 22**).

3.6. Exosomal PrP^{Sc} Detection

To examine whether released exosomes from prion-infected cells contain PrP^{Sc} and associated infectivity, animal bioassays (as described above for cell lysate) can be performed. In addition, Western blot of proteinase K (PK)-treated exosomal proteins can be carried out to assay for PrP^{Sc} in recipient cells.

3.6.1. PK Digestion

1. Lyse confluent cells and released exosomes for 15 min in ice-cold lysis buffer.
2. Centrifuge 5,000 g for 2 min.
3. Samples that are PK treated: Add 50 µg/ml PK for 30 min at 37°C, with agitation. Stop protease digestion by the addition of PMSF (Sigma-Aldrich) to a final concentration of 5 mM.
4. Samples that are not PK treated: Methanol precipitate by adding 9 volumes of −20°C 100% methanol to each sample and incubate at −20°C for at least 1 h.
5. Spin all samples at 20,000 g for 45 min at 4°C.
6. Resuspend the pellets in sample buffer, boil for 10 min, centrifuge at 13,000 g for 2 min, and electrophorese on SDS-PAGE gels.

3.6.2. Transmission of Exosome PrP^{Sc} to Further Cells

Exosome-associated PrP^{Sc} is transmissible between homologous cell types (9) in addition to heterologous cells (29). A method for testing exosomal prion infectivity is listed below.

1. Isolate exosomes (from 1×10^7 cells) under sterile conditions. If this is not possible, then resuspend your exosome pellet in Opti-MEM and filter through a 0.22-µm filter.

2. Incubate the Opti-MEM containing exosomes with cells (50% confluent) plated the day prior. Exosomes isolated from 1×10^7 cells should be ample to establish highly efficient conversion of PrP^C on cells plated in two wells of a six-well tissue culture plate (Nalge Nunc International).
3. Incubate for 72 h in a 37°C 5% CO_2 incubator.
4. Wash and passage the cells as normal.
5. After 15 passages, test recipient cells for protease resistant PrP expression.

4. Notes

1. The reason for collecting cells as they detach is that repeated pipetting of cells results in cell death. Scraping cells to remove them from the flask surface is optional; however, it will probably result in a large number of dead cells. Trypsinizing is also optional; however, it is not recommended when PrP experiments are to be performed due to shedding of PrP from the cell surface after tryspin treatment.
2. GT1-7 cells are generally split at a 1:3 ratio. If the cells are plated at too low confluence, they will not grow and should be replated into a smaller flask. The cells can discharge cellular debris into the culture media, so it may be necessary to replace the media 1–2 days after passaging.
3. Ideally, GT1-7 cells should be no >80% confluent at the time of passaging.
4. Brain homogenates are prepared from Balb/C mice in the terminal phase of prion infection with the M1000 (Fukuoka) strain or matched uninfected mice. Note that it is important to maintain sterility after removal of the brain from the skull, to avoid contamination of cell cultures. Brain 10% homogenates are prepared in sterile PBS by passing through 18-, 22-, and 28-gauge needles in succession (~20 times through each needle). Homogenates are snap-frozen in liquid nitrogen, stored at −80°C, and repeated freeze-thawing is avoided. After thawing, the appropriate dilution of brain homogenate is made in Opti-MEM.
5. To avoid using large volumes of brain homogenate, infections are usually done in a six-well plate. Plates with smaller wells also can be used.
6. Varying dilutions of brain homogenate can be used. However, we would recommend no >1% for GT1-7 cells, due to associated toxicity.
7. This ensures complete removal of the original inoculum. Methods for detecting PrP^{Sc} are covered in detail in other chapters of this book. After detection of PrP^{Sc}, some cell lines may need to be subcloned to isolate high PrP^{Sc}-expressing cells; however, this is unnecessary for GT1-7 cells.
8. Cell lines can vary in susceptibility to infection, and they can lose their PrP^{Sc} expression very easily. We and others (4) have noted the importance of media and FCS on the establishment of infection in cell lines and also the longevity of their PrP^{Sc} expression. Bosque et al. (4) grew scrapie-susceptible N2a cells in high-glucose Dulbecco's modified Eagle's medium (DMEM), low-glucose DMEM, or minimal essential medium (Invitrogen). Although the cells were morphologically identical and PrP^{Sc} was detected in all exposed cultures, N2a cells grown in low-glucose media grew more slowly and had lower amounts of

PrP^{Sc} than cells grown in high-glucose DMEM. The study concluded that the concentration of glucose was critical for sensitivity to inoculated prions. In addition, we have tested GT1-7 and mouse PrP expressing RK13 (moRK13) cells grown in DMEM versus Opti-MEM, and we found that cells grown in DMEM and subsequently inoculated with M1000 prions exhibit less susceptibility to the initial infection and are unable to propagate PrP^{Sc} for as many passages as cells grown and infected in Opti-MEM media.

9. In our experience, the quality of the FCS used in the media is very important. Several brands of FCS should be tested and a few batches of each of the brands should also be tested. It is recommended growing infected cells in a variety of calf sera, testing for PrP^{Sc} expression over time, and choosing the FCS that results in the least lost of signal over several passages. It is best to use the same batch number once a suitable FCS brand is chosen to avoid having to test the suitability of every bottle.

10. It is essential to perform mouse bioassays to confirm that your cell line is in fact producing infectious prions. Prion-infected or control cell lysate should be derived from cells passaged at least 15 times to ensure the original inoculum is no longer present.

11. The FCS used in cell culture media contains bovine exosomes, which must be removed before incubation with cells. There is the option of growing cells in serum-free media; however, in GT1-7 cells, this is not an option. Instead, deplete the exosomes from the FCS by overnight (16-h) ultracentrifugation at $100,000\,g$ (45Ti rotor, Beckman Coulter) and then add 10% to Opti-MEM containing penicillin/streptomycin and glutamate. This media is referred to as serum-depleted medium.

12. It is important that cells are not over confluent by day 4, because they will undergo cell death, resulting in apoptotic bleb release into the cell media.

13. Depending on the application of the final exosome pellet, it may be necessary to use sterile ultracentrifugation tubes and perform the following steps in an aseptic environment. For example, if the exosomes are going to be used for in vitro or in vivo bioassays.

14. There are reports of storing exosome containing in vivo fluids at −80°C before exosome isolation; however, in our experience storage of in vitro medium results in protein degradation and loss of exosomes and is not recommended.

15. Be careful what type of tube is used for the isolation. Exosomes tend to bind less to the walls of polyallomer tubes, resulting in decreased yield.

16. If the exosomes are to be used for EM analysis, or any other application that requires vesicle integrity, do not resuspend the exosome pellet by scratching it from the tube wall.

17. This sediments the protein/vesicles out of each fraction. It may be necessary to perform two ultracentrifugations, because there are 11 fractions, so keep the other tube/fractions at 4°C until they can be centrifuged.

18. After transfer of the proteins onto a PVDF or nitrocellulose membrane, it is useful to stain the membrane with ponceau S, to visualize which fractions contain protein and the protein profile of the fractions.

19. If PrP is associated with exosomes, it will be detected in the same fractions as tsg101 and/or flotillin, which will be in fractions with a density ranging from 1.13

to 1.19 g/ml. Any contaminating proteins or larger vesicles will be in the densest/bottom fractions. Also, take note that exosomes isolated from prion-infected cell lines tend to have heavier densities, presumably due to the presence of aggregated prion protein.

20. The protein composition of lysed cells, exosomes or apoptotic blebs has previously been analyzed by SDS-PAGE and Coomassie blue staining *(52)*. To confirm the molecular identity of exosomal proteins, separated proteins are commonly excised from gels, and analyzed by mass spectrometry *(48, 53, 54)*. The protein composition of exosomes will reflect the cell type from which they have originated and there endosomal origin. The common proteins identified to date include membrane-associated proteins *(48)*, cytosolic, adhesion proteins, and lipids *(54–57)*.

21. Approximately 2×10^7 GT1-7 cells plated for 72 h will yield ~80 μg of protein, compared with other reported yields of ~1.5 μg/10^6 cells *(58)*.

22. We have reported previously that exosome-associated PrP (PK untreated) is N-terminally distinct and shows altered abundance of glycoforms in comparison with cell lysate *(29)*. Immunoblotting with antibodies raised against different PrP epitopes was performed to characterize the PrP species associated with exosomes isolated from GT1-7 and moRk13 cells. Using two different antibodies, we found that exosomal PrP is nonimmunoreactive at the N-terminal region, residues 23–30. Furthermore, residues 37–44 were undetectable in the moRK13 exosomes. No difference was detected in the molecular weight of unglycosylated full-length PrP in exosome samples compared with cell lysate, suggesting exosomal PrP could either not be truncated through the entire epitopes that are detected by these antibodies, or may contain modified N-terminal epitopes, which abrogate detection with the two different antibodies. This phenomenon was observed in both control and prion-infected cell-derived exosomes from both cell lines. We would, therefore, suggest probing exosomal PrP with C-terminal antibodies to ensure detection of PrP.

References

1. Prusiner, S.B. (1982) Novel proteinaceous infectious particles cause scrapie. Science **216**(4542): 136–44.
2. Clarke, M.C. and D.A. Haig. (1970) Evidence for the multiplication of scrapie agent in cell culture. Nature **225**(5227): 100–1.
3. Race, R.E., L.H. Fadness, and B. Chesebro. (1987) Characterization of scrapie infection in mouse neuroblastoma cells. J Gen Virol **68**(5): 1391–9.
4. Bosque, P.J. and S.B. Prusiner. (2000) Cultured cell sublines highly susceptible to prion infection. J Virol **74**(9): 4377–86.
5. Schatzl, H.M., et al. (1997) A hypothalamic neuronal cell line persistently infected with scrapie prions exhibits apoptosis. J Virol **71**(11): 8821–31.
6. Vorberg, I., et al. (2004) Susceptibility of common fibroblast cell lines to transmissible spongiform encephalopathy agents. J Infect Dis **189**: 431–9.

7. Priola, S.A., et al. (1994) Prion protein and the scrapie agent: in vitro studies in infected neuroblastoma cells. Infect Agents Dis **3**(2–3): 54–8.

8. Supattapone, S., et al. (2001) Branched polyamines cure prion-infected neuroblastoma cells. J Virol **75**(7): 3453–61.

9. Fevrier, B., et al. (2004) Cells release prions in association with exosomes. Proc Natl Acad Sci U S A **101**(26): 9683–8.

10. Stoorvogel, W., et al. (2002) The biogenesis and functions of exosomes. Traffic **3**(5): 321–30.

11. Denzer, K., et al. (2000) Follicular dendritic cells carry MHC class II-expressing microvesicles at their surface. J Immunol **165**(3): 1259–65.

12. Pisitkun, T., R.F. Shen, and M.A. Knepper. (2004) Identification and proteomic profiling of exosomes in human urine. Proc Natl Acad Sci U S A **101**(36): 13368–73.

13. Andre, F., et al. (2002) Tumor-derived exosomes: a new source of tumor rejection antigens. Vaccine **20**(Suppl 4): A28–31.

14. Gastpar, R., et al. (2005) Heat shock protein 70 surface-positive tumor exosomes stimulate migratory and cytolytic activity of natural killer cells. Cancer Res **65**(12): 5238–47.

15. Abusamra, A.J., et al. (2005) Tumor exosomes expressing Fas ligand mediate CD8+ T-cell apoptosis. Blood Cells Mol Dis **35**(2): 169–73.

16. Cho, J.A., et al. (2004) Exosomes: a new delivery system for tumor antigens in cancer immunotherapy. Int J Cancer **114**(4): 613–22.

17. Hee Kim, S., et al. (2005) Exosomes derived from genetically modified DC expressing FasL are anti-inflammatory and immunosuppressive. Mol Ther **13**(2): 289–300.

18. Skokos, D., et al. (2001) Mast cell–dependent B and T lymphocyte activation is mediated by the secretion of immunologically active exosomes. J Immunol **166**(2): 868–76.

19. Wolfers, J., et al. (2001) Tumor-derived exosomes are a source of shared tumor rejection antigens for CTL cross-priming. Nat Med **7**(3): 297–303.

20. Skokos, D., et al. (2003) Mast cell-derived exosomes induce phenotypic and functional maturation of dendritic cells and elicit specific immune responses in vivo. J Immunol **170**(6): 3037–45.

21. Van Niel, G., et al. (2003) Intestinal epithelial exosomes carry MHC class II/peptides able to inform the immune system in mice. Gut **52**(12): 1690–7.

22. Kim, S.H., et al. (2005) Exosomes derived from IL-10-treated dendritic cells can suppress inflammation and collagen-induced arthritis. J Immunol **174**(10): 6440–8.

23. Wiley, R.D. and S. Gummuluru. (2006) Immature dendritic cell-derived exosomes can mediate HIV-1 trans infection. Proc Natl Acad Sci U S A **103**(3): 738–43.

24. Ecroyd, H., et al. (2006) An epididymal form of cauxin, a carboxylesterase-like enzyme, is present and active in mammalian male reproductive fluids. Biol Reprod **74**(2): 439–447.

25. Ecroyd, H., et al. (2004) Compartmentalization of prion isoforms within the reproductive tract of the ram. Biol Reprod **71**(3): 993–1001.

26. Gatti, J.L., et al. (2002) Prion protein is secreted in soluble forms in the epididymal fluid and proteolytically processed and transported in seminal plasma. Biol Reprod **67**(2): 393–400.

27. Robertson, C., et al. (2006) Cellular prion protein is released on exosomes from activated platelets. Blood **107**(10): 3907–11.

28. Faure, J., et al. (2006) Exosomes are released by cultured cortical neurones. Mol Cell Neurosci **31**(4): 642–8.

29. Vella, L.J., et al. (2007) Packaging of prions into exosomes is associated with a novel pathway of PrP processing. J Pathol **211**(5): 582–90.

30. Aguzzi, A. (2003) Prions and the immune system: a journey through gut, spleen, and nerves. Adv Immunol **81**: 123–71.

31. Brown, K.L., et al. (1999) Scrapie replication in lymphoid tissues depends on prion protein-expressing follicular dendritic cells. Nat Med **5**(11): 1308–12.

32. Kitamoto, T., et al. (1991) Abnormal isoform of prion protein accumulates in follicular dendritic cells in mice with Creutzfeldt-Jakob disease. J Virol **65**(11): 6292–5.

33. Andreoletti, O., et al. (2000) Early accumulation of PrP(Sc) in gut-associated lymphoid and nervous tissues of susceptible sheep from a Romanov flock with natural scrapie. J Gen Virol **81**(12): 3115–26.

34. van Keulen, L.J., et al. (2000) Pathogenesis of natural scrapie in sheep. Arch Virol Suppl (16): 57–71.
35. McBride, P.A., et al. (1992) PrP protein is associated with follicular dendritic cells of spleens and lymph nodes in uninfected and scrapie-infected mice. J Pathol 168(4): 413–8.
36. Clarke, M.C. and R.H. Kimberlin. (1984) Pathogenesis of mouse scrapie: distribution of agent in the pulp and stroma of infected spleens. Vet Microbiol 9(3): 215–25.
37. Kimberlin, R.H. and C.A. Walker. (1989) Pathogenesis of scrapie in mice after intragastric infection. Virus Res 12(3): 213–20.
38. Beekes, M. and P.A. McBride. (2000) Early accumulation of pathological PrP in the enteric nervous system and gut-associated lymphoid tissue of hamsters orally infected with scrapie. Neurosci Lett 278(3): 181–4.
39. Baldauf, E., M. Beekes, and H. Diringer. (1997) Evidence for an alternative direct route of access for the scrapie agent to the brain bypassing the spinal cord. J Gen Virol 78(5): 1187–97.
40. Butler, D.A., et al. (1988) Scrapie-infected murine neuroblastoma cells produce protease-resistant prion proteins. J Virol 62(5): 1558–64.
41. Rubenstein, R., R.I. Carp, and S.M. Callahan. (1984) In vitro replication of scrapie agent in a neuronal model: infection of PC12 cells. J Gen Virol 65(12): 2191–8.
42. Vilette, D., et al. (2001) Ex vivo propagation of infectious sheep scrapie agent in heterologous epithelial cells expressing ovine prion protein. Proc Natl Acad Sci U S A 98(7): 4055–9.
43. Clarke, M.C. and G.C. Millson. (1976) Infection of a cell line of mouse L fibroblasts with scrapie agent. Nature 261(5556): 144–5.
44. Gibson, P.E., T.M. Bell, and E.J. Field. (1972) Failure of the scrapie agent to replicate in L5178Y mouse leukaemic cells. Res Vet Sci 13(1): 95–6.
45. Nishida, N., et al. (2000) Successful transmission of three mouse-adapted scrapie strains to murine neuroblastoma cell lines overexpressing wild-type mouse prion protein. J Virol 74(1): 320–5.
46. Arjona, A., et al. (2004) Two Creutzfeldt-Jakob disease agents reproduce prion protein-independent identities in cell cultures. Proc Natl Acad Sci U S A 101(23): 8768–73.
47. Raposo, G., et al. (1996) B lymphocytes secrete antigen-presenting vesicles. J Exp Med 183(3): 1161–72.
48. Thery, C., et al. (1999) Molecular characterization of dendritic cell-derived exosomes. Selective accumulation of the heat shock protein hsc73. J Cell Biol 147(3): 599–610.
49. Clayton, A., et al. (2001) Analysis of antigen presenting cell derived exosomes, based on immuno-magnetic isolation and flow cytometry. J Immunol Methods 247(1-2): 163–74.
50. Cheruvanky, A., et al. (2007) Rapid isolation of urinary exosomal biomarkers using a nanomembrane ultrafiltration concentrator. Am J Physiol Renal Physiol 292(5): F1657–61.
51. van Niel, G. and M. Heyman. (2002) The epithelial cell cytoskeleton and intracellular trafficking. II. Intestinal epithelial cell exosomes: perspectives on their structure and function. Am J Physiol Gastrointest Liver Physiol 283(2): G251–5.
52. Thery, C., et al. (2001) Proteomic analysis of dendritic cell-derived exosomes: a secreted subcellular compartment distinct from apoptotic vesicles. J Immunol 166(12): 7309–18.
53. Hegmans, J.P., et al. (2004) Proteomic analysis of exosomes secreted by human mesothelioma cells. Am J Pathol 164(5): 1807–15.
54. Mears, R., et al. (2004) Proteomic analysis of melanoma-derived exosomes by two-dimensional polyacrylamide gel electrophoresis and mass spectrometry. Proteomics 4(12): 4019–31.
55. Bard, M.P., et al. (2004) Proteomic analysis of exosomes isolated from human malignant pleural effusions. Am J Respir Cell Mol Biol 31(1): 114–21.
56. Wubbolts, R., et al. (2003) Proteomic and biochemical analyses of human B cell-derived exosomes. Potential implications for their function and multivesicular body formation. J Biol Chem 278(13): 10963–72.
57. Segura, E., S. Amigorena, and C. Thery. (2005) Mature dendritic cells secrete exosomes with strong ability to induce antigen-specific effector immune responses. Blood Cells Mol Dis 35(2): 89–93.
58. Amzallag, N., et al. (2004) TSAP6 facilitates the secretion of translationally controlled tumor protein/histamine-releasing factor via a nonclassical pathway. J Biol Chem 279(44): 46104–12.

Chapter 6
Neurotoxicity of Prion Peptides on Cultured Cerebellar Neurons

Giuseppe D. Ciccotosto, Roberto Cappai, and Anthony R. White

Summary Prion peptide (PrP) neurotoxicity has been modelled in vitro by using synthetic peptides derived from the PrP^C sequence. The major region of neurotoxicity has been localized to the hydrophobic domain located in the middle of the PrP sequence. The neurotoxicity assays are typically performed on cultured mouse cerebellar neurons derived from neonatal pups, and viability can be monitored by a variety of assays, including MTT (3-(4,5-dimethylthiazol-2-yl)-2,5-diphenyltetrazolium); MTS (3-(4,5-dimethylthiazol-2-yl)-5-(3-carboxymethoxyphenyl)-2-(4-sulfophenyl)-2H-tetrazolium, inner salt) lactate dehydrogenase release; and apoptotic assays. These neurotoxicity studies have been useful in identifying cofactors, such as PrP^C and metals as modulators of PrP peptide-mediated neurotoxicity. Given the biosafety issues associated with handling and purifying infectious prions, the use of synthetic peptides that display a dependence upon PrP^C expression for toxicity, as per the PrP^{Sc} agent for infectivity, supports the relevance of using these synthetic peptides for understanding PrP-mediated neurotoxicity.

Keywords Cell viability; cultured cerebellar neurons; immunofluorescence; lipid peroxidation; neurotoxicity; prion peptide (PrP); PrP106-126.

1. Introduction

A range of peptides derived from the prion protein (PrP^C) display toxicity against cultured neurons in vitro *(1–3)* and in vivo *(4)*. The majority of these neurotoxic peptides are derived from the hydrophobic region encompassed by residues 113–135 *(5)*. The most widely studied synthetic neurotoxic PrP peptide is PrP(106-126) *(2)*. The major data supporting its relevance for PrP toxicity are that PrP106-126 requires endogenous PrP^C expression to exert its neurotoxic activity *(6)*. Cerebellar granule neurons from Prnp gene knockout mice are resistant to PrP106-126 neurotoxic activity *(6, 7)*. The mechanism of PrP106-126 toxicity indicates it can activate proapoptotic pathways with upregulation of Annexin V binding and caspase-3, -6, and -8–like activity *(8, 9)*. Moreover, PrP106-126 activates arachidonic acid metabolism, and it requires the 5-lipoxygenase pathway to mediate toxicity *(10)*.

From: *Prion Protein Protocols.*
Methods in Molecular Biology, Vol. 459.
Edited by: A. F. Hill © Humana Press, Totowa, NJ

The nature of the PrP106-126 toxic species indicates neurotoxicity seems to correlate with PrP106-126 being in a nonamyloid/nonfibrillar state *(11)*. This indicates amyloid formation is not necessary for PrP106-126–mediated neurotoxicity. Our own sequence–activity studies have shown that the hydrophobic residues in the C-terminal hydrophobic region of PrP106-126 are necessary for PrP106-126 neurotoxicity *(12)*. Moreover, PrP106-126 neurotoxic activity is dependant upon the metal ions copper and zinc to be neurotoxic *(13, 14)*, and metal binding and toxicity can be attenuated by mutating either His 111, Met 109, or Met 112 *(13)*.

2. Materials

All chemical reagents were purchased from Sigma-Aldrich (St. Louis, MO) unless otherwise specified.

2.1. Cell Culturing

1. Rats at 6–8 days postnatal or mice at 5–7 days postnatal.
2. MgSO$_4$ stock (3.85%): Dissolve 1.54 g of MgSO$_4$•7H$_2$O in 50 ml of water. Filter sterilize the solution. Store at 4°C.
3. CaCl$_2$ (1.2%): 120 mg of CaCl$_2$•2H$_2$O in 10 ml of distilled water (dH$_2$O).
4. Poly-D-lysine (100×): Dissolve 5 mg in 10 ml of dH$_2$O by using polystyrene tubes. Store in 1-ml aliquots at −20°C. Thaw out 1 ml of stock and dilute in 100 ml sterile water. Store in the dark at 4°C.
5. Trypsin stock (20×): Dissolve 25 mg of trypsin (~7,500 units/mg) in 10 ml of Krebs/ buffer. Filter sterilize. Store in 0.75-ml aliquots at −20°C.
6. DNase/soybean trypsin inhibitor (SBTI) (10×): Dissolve 8 mg of DNase and 26 mg SBTI in 10 ml of Krebs' buffer. Filter sterilize. Store in 0.5-ml aliquots at −20°C.
7. Cytosine-β-D-arabinofuranoside (AraC) (1,000×): Dissolve 4.8 mg of AraC in 2 ml of dH$_2$O. Filter sterilize. Store 50-μl aliquots at −20°C.
8. Krebs' stock buffer (10×): Dissolve 36.25 g of NaCl, 2.0 g of KCl, 0.7 g of NaH$_2$PO$_4$•H$_2$O, 13.0 g of D-glucose, 0.05 g of phenol red, and 29.7 g of HEPES acid in 450 ml of dH$_2$O. Adjust pH to 7.4. Bring volume to 500 ml. Filter sterilize. Store at 4°C.
9. Krebs' buffer (1×): Dissolve 1.5 g of bovine serum albumin (fraction V) in 446 ml of dH$_2$O. Add 50 ml of 10× Krebs' stock and 4 ml of 3.85% MgSO$_4$. Filter sterilize. Store at 4°C.
10. Dulbecco's phosphate-buffered saline (D-PBS) (10×): Liquid contains no calcium or magnesium. Dissolve 2 g of KCl, 2 g of KH$_2$PO$_4$, 80 g of NaCl, and 21.6 g of Na$_2$HPO$_4$•7H$_2$O in 1 liter of dH$_2$O. The pH of 10× is 6.7 to 7.0. The pH of 1× should be checked after 1:10 dilution to ensure that it is 7.1 ± 0.1.

11. Basal medium Eagle's with Earle's salts (BME): To 500 ml of BME media (Invitrogen, Carlsbad, CA), add 10% fetal calf serum (heat inactivated), 5 ml of 200 mM GlutaMAX (Invitrogen), and 0.5 ml of 10 mg/ml gentamicin (Invitrogen). Store media at 4°C.
12. Primary neuronal culture medium: Combine 50 ml of neurobasal (NB) media (Invitrogen) with 1 ml of B27 supplements (Invitrogen), 50 µl of 10 mg/ml gentamicin (Invitrogen), and 125 µl of 200 mM GlutaMAX (Invitrogen). Store unused media at 4°C. Discard unused media after 3 to 5 days of storage.
13. Primary neuronal experimental culture medium: Combine 50 ml of neurobasal media (Invitrogen) with 1 ml of B27 supplements minus antioxidants (Invitrogen), 50 µl of 10 mg/ml gentamicin (Invitrogen), and 125 µl of 200 mM GlutaMAX (Invitrogen). Store unused media at 4°C. Discard unused media after 3 to 5 days of storage (*see* **Notes 1** and **2**).

2.2. Harvesting and Growing Cerebellar Neurons

1. Dissecting instruments: pair of small sharp scissors, large sharp scissors, tweezers with rat teeth grip, two fine tip watchmaker forceps (e.g., Inox #5), and a razor blade.

    a. Dissecting instruments should be sterilized by autoclaving cleaned utensils or by soaking instruments in 70% ethanol before using them.

2. Sterile disposable dishes (60 and 90 mm).
3. Sterile 50- and 15-ml polystyrene tubes.
4. Hemocytometer (Neubauer, 0.1-mm depth).
5. 0.4% trypan blue.
6. Sterile well dishes.

    a. 48-well dishes are used for toxicity assays.
    b. Four-well plates are used to place 12-mm coverslips on for immunohistochemistry.
    c. 24-well plates are used for caspase assays.
    d. 12- or six-well plates are used for experiments where protein extraction is required to run samples on gels.

2.3. Immunofluorescence Staining of Cells (see **Note 5***)*

1. Preparation of 4% paraformaldehyde in D-PBS.

    a. To make 50 ml of fixative, weigh out 2 g of paraformaldehyde into a 50-ml disposable tube and add 45 ml of water.
    b. Place tube into a 60°C water bath for 30 min.
    c. Add 8 drops of 1 N NaOH (solution should become clear).

    d. Allow the mixture to cool to room temperature (RT); check pH and adjust to 7.5.
    e. Bring volume final to add 50 ml.
    f. Mix and dispense into 10-ml aliquots.
    g. Store at −20°C and thaw fresh lot on day of staining experiments. If stored at 4°C, use within 10 days.

2. Permeabilization buffer: 0.075% (w/v) Triton X-100 in blocking buffer: To 100 ml of blocking buffer, add 7.5 µl of Triton X-100. Mix and make 10-ml aliquots. Store at −20°C and thaw fresh lot on day of staining experiments.
3. Blocking buffer: Prepare either 2 mg/ml bovine serum albumin in D-PBS or 1–10% normal goat serum in D-PBS. Use higher percentage serum when high background levels are encountered.
4. Mounting media (*see* **Note 3**): To a 30-ml bottle of Permafluor (Thermo Electron Corporation, Waltham, MA), add 1.3 g of 1,4-diazabicylo(2.2.2)octane and tumble at RT for 2 h. Add 0.5 ml of 2 M glycine, pH 10.0, and adjust to pH 10.0. Store in dark at 4°C.

3. Methods

3.1. Setup Procedures on Day before Culturing

1. Plates, dishes, and coverslips are coated with poly-D-lysine, preferably overnight for best results, or for a minimum of 2 h.
2. Add enough poly-D-lysine solution to totally cover the surface area.
3. Before plating neurons, aspirate off solution and let plates air dry, with lids partially removed. For the glass coverslips, wash the wells with PBS and then aspirate off PBS; do not air dry.

3.2. Setup Procedures on Day of Culturing

1. Thaw out frozen aliquots of trypsin and DNase/SBTI.
2. Place the BME media in a 37°C water bath.
3. Ensure that the dissecting instruments (small sharp scissors, tweezers with rat teeth grip, 2 fine tip forceps, and razor blade) are clean and sterile and placed in the laminar flow hood.
4. Add 5 ml of Krebs' buffer to the 60-mm dish.
5. Label the following tubes and add the following solutions (*see* **Note 4**):

    a. Tube 1 (50-ml tube): Add 15 ml of Krebs' buffer and 0.75 ml of trypsin.
    b. Tube 2 (15-ml tube): Add 7.5 ml of Krebs' buffer and 0.75 ml of DNase/SBTI.
    c. Tube 3 (15-ml tube): Take 1.5 ml from tube 3.
    d. Tube 4 (15-ml tube): Add 6 ml of Krebs' buffer and 0.0075 ml of $CaCl_2$.
    e. Tube "CELLS" (15-ml tube): Leave blank for now.

3.3. Cerebellar Granule Neuron Isolation Procedures

1. The neonatal mouse pups are isolated from the mother. They are brought to the laboratory and kept warm (e.g., shine light directly into box but shielded from mice) while waiting to be killed.
2. The animals are picked up with the head held over the sterile 90-mm dish. Using sterile scissors dipped in 70% alcohol, the animal is decapitated with the head falling into the sterile dish. The remaining body is placed in a biohazard bag and discarded appropriately.
3. Cut skull off from base of the head by inserting the sharp-pointed scissors into the spinal cord opening and cut away at the skull in the direction of the ears and around to the front of the head. Repeat on both sides of the head and then gently lift skull off. Remove whole brain (**Fig. 1A**), including the cerebellum from the cavity of the skull. Using the sharp scissors, cut away the cerebellum (**Fig. 1B**, *right*) and then place cerebella into the 60-mm dish containing Krebs' buffer. Repeat this procedure for all pups before going on to the next step. Remove meninges from the cerebellum with fine tip forceps. Repeat for all cerebella.
4. Transfer cleaned cerebellum to new 60-mm dish. Using a sterile blade, chop tissue into <1-mm pieces. Aspirate approx 1 ml of media from tube 1 and add to chopped tissue. Collect all of the chopped tissue and media and transfer into tube 1 (trypsin). Place tube 1 into a gently shaking water bath set at 37°C and incubate for 20 min.
5. Remove tube 1 from water bath and spray with 70% ethanol. Wipe tube down. Return tube to the laminar flow hood and then pour the contents of tube 2 (DNase/SBTI) into tube 1. Mix gently (invert tube several times) until DNA is digested and tissue pieces do not stick together. Place tube 1 into a centrifuge and spin down cells at 200 g for 3 min. Discard supernatant (manually) without disturbing pellet.

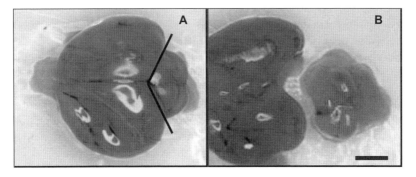

Fig. 1 (**A**) Whole mouse brain removed from skull of mouse with lines indicating dissection required to isolate cerebellum. (**B**) Isolated cerebellum cut away from whole brain. Bar = 2 mm

6. Making single cell suspension by trituration: transfer the contents of tube 3 to tube 1. Using a 1-ml plugged pipette tip that has been flamed to remove the sharp edges, or plugged pipette tips with rounded edges, triturate (i.e., pass up and down a pipette tip slowly and avoid bubbling suspension) until tissue pieces are hard to see. This step may require up to 30 triturations (do not over-triturate, because this will harm the cells). Add the contents of tube 4 to tube 1 and let the tube stand for 10 min at RT. Any undigested clumps should settle to the bottom of tube during this time. Carefully transfer the upper single cell suspension (avoiding the clumps) to a new tube labelled CELLS.

7. Pellet cells by centrifugation of tube CELLS for 5 min at 300 g at RT. Spray tube with 70% ethanol. Wipe down the tube and return it to the hood. Aspirate off and discard the supernate. Resuspend the cell pellet initially in 1 ml of BME and then dilute out to 10 ml with BME plating medium.

8. The total number of isolated cells is counted using a hemocytometer. First, add 0.4 ml of BME to a microfuge tube and then add 0.1 ml of cell suspension to tube. Mix cells and transfer 50 µl to another microfuge tube and mix in 50 µl of 0.4% trypan blue. Mix cells and add 13 µl to the cleaned hemocytometer. Let the cells settle for a minute, and then place the hemocytometer on an inverted microscope. Using the 10× objective, count the outer four corner squares, with the live cells looking round and opaque colored, whereas the dead cells will have absorbed the trypan blue and be blue colored. Calculate the average number of cells per square quadrant. (Note: The ideal number of cells to count is between 50 and 100 cells per quadrant. If there are more cells, dilute out accordingly to minimize counting errors.).

9. Total cell number = average number of cells × 10,000 × dilution × cell suspension volume.

10. Resuspend cells at a density of 1.5×10^6 cells/ml in BME.

11. For toxicity experiments, plate cells at a density of 375,000 cells/cm². Add 0.25 ml/well in 48-well plate. (Note: Plate cells only the inner 24 wells and add 0.2 ml of sterile PBS to the outer wells of the plate to help minimize evaporation in the wells that contain cells.).

12. For immunofluorescence histochemistry, plate cells at a density of 37,500 cells/well. Add 25 µl/well in a four-well dish and top with 0.5 ml of BME media.

13. For lipid peroxidation or glutathione assays, plate cells at a density of 600,000 cells/cm². Add 0.5 ml/well in 24-well plate.

14. For the growth and maintenance of cell cultures, place neuronal cell cultures in a humidified 5% CO_2 culture incubator set at 37°C to allow cells to attach to the surface in a high-serum environment. After 2 h, aspirate off 80% (e.g., 0.2 ml from 0.25 ml) of the BME plating media carefully (trying to avoid dislodging attached cells) and replace with 60% freshly prepared neurobasal culture media (NB/B27) and return to incubator (e.g., 0.15 ml). To help minimize astrocyte and microglial cell growth, AraC is added after day 1 in culture. Dilute AraC in NB/B27 media and add 20% of final cell volume to well (e.g., 0.05 ml).

3.4. Preparation of Toxic Peptides and Treating Neuronal Cultures

1. The synthetic PrP106-126 peptide was purchased from Keck Laboratories (New Haven, CT).
2. Prepare a 2 mM stock solution by dissolving the peptide directly into neurobasal medium supplemented with B27 lacking antioxidant supplements (NB/AO) and aged overnight at 37°C. The synthetic peptide aggregates rapidly in the conditioned media. The overnight treatment allows the peptide to grow into fully formed fibrils.
3. After 6 days in culture, the cells are ready to be treated with the synthetic PrP peptide.

3.5. MTS and MTT Assay: A measure of cell viability

1. The 3-(4,5-dimethylthiazol-2-yl)-2,5-diphenyltetrazolium (MTT) and 3-(4,5-dimethylthiazol-2-yl)-5-(3-carboxymethoxyphenyl)-2-(4-sulfophenyl)-2H-tetrazolium, inner salt (MTS) reagents are used to measure the redox potential (electron transfer activity) of the cell. Because redox potential is energy-dependent, it is a good indicator of how healthy the cell is, rather than measuring dead cells. The MTT and MTS tetrazolium compounds are bioreduced by live healthy cells into colored formazan by-products. MTT and MTS differ substantially in that the MTS-formed by-product is soluble, whereas the MTT formazan product deposits on the cells as a crystalline precipitate that requires an additional step to dissolve the crystals before recording the absorbance readings.
2. MTS assay, otherwise called the CellTiter96 Aqueous one solution cell proliferation assay by Promega (Madison, WI), is supplied as a 10× stock solution. Store unused 5- to 10-ml aliquots at −20°C and keep one aliquot at 4°C with the tube wrapped in foil.
3. MTT comes in a yellow powder. Make a 10× stock by dissolving 5 mg/ml in D-PBS and store at 4°C, with the tube wrapped in foil. If you plan to use MTT, you also need dimethyl sulfoxide (DMSO) to dissolve the formazan crystal deposits on the cells (after aspirating off the conditioned media).
4. To induce neurotoxicity, 6-day-old cerebellar neuronal cells are treated with 20 µM PrP106-126 aggregated peptide for at least 3 days. When treating cells with 20 µM PrP, you need to add 6.4 µl of 2 mM PrP and then make the volume up to 640 µl with freshly prepared NB/AO.
5. For the 48-well plates, a working volume of 0.2 ml/well is used. It is recommended that a minimum of triplicate repeats are done for each experiment. Therefore, there is an ability to have up to eight treatments when cells are plated in the inner 24 wells of a 48-well plate.
6. After the 3 days of PrP106-126 treatment, cell viability is determined by MTS assay.

7. For each 480-well plate, you need to prepare a working volume of 4.8 ml of fresh NB/AO media plus MTS (4.32 ml of NB/AO media plus 0.48 ml of 10× MTS).
8. Aspirate off the experimental media from the cells.
9. Add 0.2 ml of freshly prepared NB/AO/MTS reagent to the cells in culture.
10. Return the cells to the humidified incubator for 1 to 4 h, depending cell density. For the amount of cerebellar cells plated, 1 to 2 h is generally sufficient to see enough color change to get a reasonable data set.
11. When media are ready to be read, transfer 0.15 ml to a clean flat-bottomed 96-well plate. Ensure that background media (NB/AO/MTS) not exposed to cells is included on the plate to represent the background control value (the background value is subtracted from vehicle and treated absorbance readings).
12. Ensure that no bubbles are present in the wells while reading the samples. The plates can be read on any plate reader which has a filter setting at 490-nm absorbance.
13. Control wells are the vehicle-treated cells, and they should have maximum absorbance readings. The data set is now normalized to the vehicle-treated cells, which is set to 100%, and the absorbance values for treated cells is adjusted accordingly.

3.6. Lactate Dehydrogenase (LDH) Assays: Cytotoxicity Detection

1. The LDH assay is used to measure and quantitate cell death. LDH is a stable cytoplasmic enzyme present in all cells. It is rapidly released into the cell culture supernatant upon damage of the plasma membrane. The increase in the amount of enzyme activity indirectly correlates to the amount of formazan formed during a limited time. This formazan dye is water-soluble, and it can be detected using a plate reader. Therefore, the assay kit (1 644 793, Roche Diagnostics, Indianapolis, IN) allows for the detection of LDH activity, which can be measured in supernatants, at a single time point, using an enzyme-linked immunosorbent assay (ELISA) plate reader.
2. The following three controls have to be performed in each experimental setup:

   a. Background: Provides information on LDH activity in assay medium alone.
   b. Low control: Provides information on LDH activity released from the untreated normal cells.
   c. High control: Provides information on max releasable LDH activity from tetrodotoxin (TTX)-treated cells.

3. All controls and experimental wells should be done in triplicate.

   a. Background control wells: These wells have only assay medium in them.
   b. Low control wells: These wells will have untreated cells growing in assay medium.

    c. High control wells: These wells have untreated cells growing in assay medium with added 2% Triton X-100 solution (i.e., add 20 µl of TTX-100/ml in NB/AO). Incubate at 37°C for at least 1 h.

4. To make the LDH color reaction mixture, take bottle 1 and dissolve contents in 1 ml of water. Incubate for 10 min at 4°C and mix thoroughly; keep at −20°C.
5. Table 1 shows a modified protocol that will help minimize the amount of reagent waste and help your assay kit go further. Calculate the number of wells you have to assay.
6. Add 100 µl of the prepared color reaction mixture to each well (containing 100 µl of sample).
7. Wrap plate in foil (keep in dark) and leave at RT for 30 min.
8. Read plate using ELISA plate reader at 490 nm absorbance for 0.1 s.
9. To calculate the percentage of cytotoxicity of a sample, calculate the average absorbance values of the triplicates and subtract from each of these absorbance values obtained in the background controls: cytotoxicity (%) = (sample – low control) × 100/(high control – low control) (**Table 1**).

3.7. *Immunofluorescence Histochemistry*

1. Immunofluorescence studies can be done on neuronal cells grown on either glass coverslips or chambered glass slides. For glass coverslips, use 12-mm circular glass coverslips, which fit in 24-well tissue culture dishes or four-well dishes. Larger rectangular or square coverslips up to ~24 mm will fit into six-well dishes.
2. To treated neuronal cells (on 12-mm coverslips in four-well plates), aspirate off media and wash 1× with D-PBS.
3. Aspirate off D-PBS. Add 0.5 ml of 4% paraformaldehyde and incubate for 20 min at RT.
4. Aspirate off 4% paraformaldehyde. Wash with 0.5 ml of D-PBS.
5. Aspirate off D-PBS. Add 0.5 ml of permeabilization buffer and incubate for 20 min at RT.

Table 1 Quantitating Cytotoxicity

No. of samples	1	To conserve reagents:
Bottle 1 (red cap)	2.5 µl	Calculate the number of samples you need to assay (e.g., $n = 100$)
Bottle 2 (blue cap)	112.5 µl	
Total volume	115 µl	Make a stock mix of $100 \times 100\,µl = 10{,}000\,µl$ for the assay
		Color reaction mix = 10,000/115 = 86.9
		Make a mix for 88 samples
		Bottle 1 (red cap) 88 × 2.5 = 220
		Bottle 2 (blue cap) 88 × 112.5 = 9,900
		Total volume = 10,120 µl

6. Aspirate off permeabilization buffer. Wash cells with 0.5 ml of D-PBS.

7. Aspirate off D-PBS. Add 0.5 ml of block buffer and incubate for 60 min at RT.

8. Aspirate off block buffer. Add primary antibody (in block) and incubate for 120 min at RT (or overnight at 4°C; 30 min at 37°C). To conserve precious antibody, place a sheet of parafilm in a moisture box (e.g., a plastic container containing wet piece of tissue paper). Label the parafilm to indicate where coverslips are placed. Place a 50-μl drop of diluted antibody on the parafilm (for a 12-mm coverslip, more is needed for larger coverslips). Gently pick up a coverslip using fine forceps and place the coverslip on top of the droplet with the CELLS coming into contact with the antibody. Seal the container and incubate for the desired time.

9. Aspirate off antibody (transfer coverslips back into plates with cells facing up again). Wash cells with 0.5 ml of D-PBS three times at 5-min intervals.

10. To cells, add appropriate secondary antibody in block (fluorescently tagged) and incubate for 60 min at RT in the dark. (To conserve antibody, you can repeat **step 8**.).

11. Aspirate off antibody. Wash cells with 0.5 ml of D-PBS three times at 5-min intervals.

12. Place coverslip facing up on a piece of paper and add a drop of prewarmed DABCO/permount onto the coverslip. Place two to three coverslips in a row and then place a glass slide on top of the coverslips. Invert glass slide and allow mountant to dry overnight before looking at images (*see* **Note 5**).

3.8. Measurement of Lipid Peroxidation Levels in Cultures

1. Lipid peroxidation: Lipid peroxidation refers to the oxidative degradation of lipids, and it is a well-established mechanism of cellular injury. It is used as an indicator of oxidative stress in cells and tissues. Polyunsaturated fatty acids are the most vulnerable source of decomposition by the free radicals, resulting in the generation of secondary products of malondialdehyde (MDA) and 4-hydroxyalkenals (HNE) by-products. Therefore, MDA and HNE can be measured using the LPO 586 lipid peroxidation kit (OXIS Research, Inc., Portland, OR). Another common method for measuring MDA, referred to as the thiobarbituric acid reactive substances (TBARS) assay, is to react it with thiobarbituric acid (TBA) to form fluorescent red adducts and record the absorbance at 532 nm (*see* **Note 6**).

2. Harvesting cells for MDA and HNE assay: This assay requires a homogenate concentration of 50 million cells/ml, with a reaction volume of 0.1 ml. Therefore, plate out 6 million cells in six-well plates for experimental treatments. At the end of the experiments, wash treated cells three times with D-PBS. Scrape cells using a rubber policeman into 1 ml of D-PBS, collect dissociated cells, and place cells into a microcentrifuge tube. Centrifuge the

tube at 5,000 g for 5 min to pellet cells. Remove all supernatant and add 120 μl of 20 mM Tris buffer, pH 7.4, to homogenate. Add 1.2 μl of 500 mM butylated hydroxytoluene (to prevent oxidation; make 500 mM stock in acetonitrile). Gently mix the homogenate, and then centrifuge the homogenate at 3,000 g at 4°C for 10 min to remove large particles. Take a 10-μl sample aliquot for bicinchoninic acid (BCA) protein assay measurement and transfer the remaining supernatant to a clean tube. Freeze the supernate sample and store at −70°C or use immediately (keep on ice until use).

3. Assay procedure for MDA and HNE assay: Add 100 μl of sample, 4-HNE (S1), TMOP (S2) standard, or water blank to a microcentrifuge tube. Add 5 μl of 500 mM butylated hydroxytoluene. Add 325 μl 10.3 mM N-methyl-2-phenylindole, in acetonitrile (R1). Gently vortex samples for a few seconds. Add 75 μl of 15.4 M methanesulfonic acid (R2). Mix well by vortexing. Incubate sample mix at 45°C for 45 min. centrifuge samples at 15,000 g for 10 min to clarify sample. Transfer 200 μL of clear supernate to a single well in a 96 well plate (ensure that plate is resistant to acetonitrile and strong acid). Read samples on a plate reader set at 586 nm (or any wavelength between 580 and 590 nm).

4. Calculations for MDA and HNE assay: Determine the absorbance values at 586 nm for the unknown samples (A), sample blank (Asb), and reagent blank (Ao) for every assay. The calculated molar extinction coefficient (ε) at 586 nm is approximately 110,000 for MDA and HNE. To calculate the concentration of your unknown sample analyte, use the following formula: [A − Asb − Ao] × 5/ε.

5. Harvesting cells and assay procedure for TBARS assay: This assay requires a plating density at 2 million cells/treatment group. This is best done using a 24-well plate. After treating cells with peptides or drugs, add 400 μl of TBA solution to each culture well containing 600 μl of medium (the media may go cloudy due to protein precipitation). Incubate the plate for approx 10 min at RT, and then give the plate a bit of a shake. The supernatant from each well is transferred to a clean 15-ml tube and heated at 95°C for 20 min, cooled to RT, and then centrifuged at 3,000 g for 5 min to pellet precipitated protein. Transfer the clarified supernatant to a spectrophotometer cuvette and read absorbance at 532 nm.

6. Calculations for TBARS assay: Cell-free medium alone is incubated with the TBA solution as described in point 5 (Subheading 3.8) above and subtracted from test readings. The TBARS values are given as optical density units (× 10^{-3})/well. Cell numbers are determined by cell viability and total protein (BCA) assays.

3.9. Measurement of Glutathione Peroxidase (GPx) and Glutathione Reductase (GR) Levels in Cultures

1. GPx is important for inhibiting lipid peroxidation of the cell membrane. The function of GPx is to reduce lipid hydroperoxides to their corresponding alcohols and to reduce free hydrogen peroxide to water. For example, GPx catalyzes 2GSH +

$H_2O_2 \rightarrow GSSG + 2H_2O$, where GSH represents reduced monomeric glutathione and GSSG represents glutathione disulfide. GR then reduces the oxidized glutathione to complete the cycle: $GSSG + NADPH + H^+ \rightarrow 2\ GSH + NADP^+$. Therefore, the production of GSH is essential, because it acts as an antioxidant protecting the neurons from cellular oxidative stress events. Determination of GPx and GR levels in CGN cultures is performed using the colorimetric assay kit as per the manufacturer's instructions (GPx-340 and GR-340, respectively, OXIS Research, Inc.). **Steps 2–8** provide a brief summary of the methods and calculations used to determine GPx and GR levels (*see* **Note 6**).

2. Harvesting cells for GPx assay: CGN experimental cultures are washed with PBS and then harvested in ice-cold 50 mM Tris-HCl containing 1 mM 2β-mercaptoethanol, pH 7.5. Next, cells are homogenized for 10 s in a polytron homogenizer at 8500 g for 10 min at 4°C. The supernatants are stored at −80°C until ready for assay.

3. Assay procedure for GPx: Immediately before assay, warm all reagents and sample to RT and dilute sample into assay buffer. Into a 1-ml cuvette, add 350 μl of assay buffer, 350 μl of NADH reagent, and 70 μl of sample. Add cuvette to spectrophotometer, and then add 350 μl of 0.007% *tert*-butyl hydroperoxide, mix well, and record A_{340} for 3 min at 30-s intervals.

4. Calculations for GPx assay: Calculate the rate of decrease in the A_{340} per minute, which is a measure of GPx activity. This can be done by performing a linear regression of the A_{340} as a function of time. The molar extinction coefficient (ε) for NADPH is 6,220 M/cm. Then, mU/ml = $(A_{340}/\text{min})/\varepsilon$.

5. Harvesting cells for measuring GR activity: CGN experimental cultures are washed with PBS and then harvested in ice-cold 60 mM KPO_4 buffer, pH 7.5. Next, they are homogenized for 10 s in a polytron homogenizer and centrifuged at 8,500 g for 10 min. The supernatants are stored at −80°C until ready for GR activity assay.

6. Assay procedure for measuring GR activity: Combine 200 μl of sample with 400 μl of GSSG into a cuvette and place into a spectrophotometer. Add 400 μl of NADH and mix well. Record the A_{340} for a minimum of 5 min in 1-min increments.

7. Calculations for measuring GR activity: Calculate the rate of decrease in the A_{340}/min, which is a measure of GR activity. This can be done by performing a linear regression of the A_{340} as a function of time. The molar extinction coefficient (ε) for NADPH is 6,220 M/cm. Then, mU/ml = $(A_{340}/\text{min})/\varepsilon$.

8. Results for both GR and GPx assays were then adjusted to give data as mU/mg protein. To do this, the protein concentrations of each sample were determined using the BCA protein assay kit (Pierce Chemical) on aliquots of homogenized cultures before centrifugation. One GR unit reduces 1 μmol of GSSG per minute at 25°C and pH 7.6. One GPx unit consumes 1 μmol of NADPH per minute at 25°C and pH 7.6.

4. Notes

1. Unless stated otherwise, all solutions should be prepared in water that has a resistivity of 18.2 mOhm/cm and total organic content of <5 ppb. This standard is referred to as "dH₂O" in this text.
2. All solutions that are filter sterilized are passed through a 0.22-µM membrane filter (Millipore Corporation, Billerica, MA).
3. Mounting media: Traditional reagent mounting media for immunofluorescence media is composed of 90% glycerol and 10% PBS (v/v) and contains 2.6% DABCO (w/v), pH 8.6. Although this reagent mounting media works well, the coverslips need to be sealed to the glass slide by using a solidifying agent, such as nail polish. Alternatively, there are several companies that make specific mounting media for immunofluorescence work. Permafluor, which is supplied by Thermo Electron Corporation is great, because it has the advantage of drying hard compared with the glycerol-based mounting media solution, and it is suitable for water-based fixation conditions. It is relatively inexpensive and it lasts a long time. Note: If the mounting media goes hard while it is stored at 4°C for a long time, just warm to 37°C, and this should help dissolve the media before applying it to cells.
4. If you are planning to harvest cells from more than 10 pups, then you should double the volumes used for harvesting cells.
5. Nuclear counterstaining: For control nuclear staining of cells, 4,6-diamidino-2-phenylindole (DAPI), Hoechst 33258, or ToPro3 (Invitrogen) are great markers. Of the three dyes, DAPI fades the least, it is relatively stable, and stained neurons will fluorescence for many months to years. ToPro3 is ideal if your imaging setup does not have the capacity to take images by using UV, but it does have far-red filters. For DAPI and Hoechst, it is also just as easy to dilute these dyes to final working concentrations in the mounting media rather than having to add dye to the cells en block (**Table 2**).
6. The methods described for measuring lipid peroxidation and glutathione peroxidase and glutathione reductase levels are not comprehensive, and it is advisable to read the manufacturers' instructions for details.

Table 2 Reagents for Nuclear Counterstaining

	Excitation Max. (nm)	Emission Max. (nm)	Stock Concn.	Working Concn.
DAPI–DNA complex	364	454	10 mg/ml in D-PBS	2 µg/ml
Hoechst 33258	356	465	1 mg/ml in D-PBS	10 µg/ml
ToPro3	642	661	1 mM in DMSO	1 µM

References

1. Thompson, A., White, A. R., et al. (2000) Amyloidogenicity and neurotoxicity of peptides corresponding to the helical regions of PrP(C) *J Neurosci Res* **62,** 293–301.
2. Forloni, G., Angeretti, N., et al. (1993) Neurotoxicity of a prion protein fragment *Nature* **362,** 543–6.
3. Brown, D. R. (2000) Prion protein peptides: optimal toxicity and peptide blockade of toxicity *Mol Cell Neurosci* **15,** 66–78.
4. Bergstrom, A. L., Cordes, H., et al. (2005) Amidation and structure relaxation abolish the neurotoxicity of the prion peptide PrP106-126 in vivo and in vitro *J Biol Chem* **280,** 23114–21.
5. Hegde, R. S., Mastrianni, J. A., et al. (1998) A transmembrane form of the prion protein in neurodegenerative disease *Science* **279,** 827–34.
6. Brown, D. R., Herms, J., and Kretzschmar, H. A. (1994) Mouse cortical cells lacking cellular PrP survive in culture with a neurotoxic PrP fragment *Neuroreport* **5,** 2057–60.
7. White, A. R., Collins, S. J., et al. (1999) Prion protein-deficient neurons reveal lower glutathione reductase activity and increased susceptibility to hydrogen peroxide toxicity *Am J Pathol* **155,** 1723–30.
8. Forloni, G., Bugiani, O., et al. (1996) Apoptosis-mediated neurotoxicity induced by beta-amyloid and PrP fragments *Mol Chem Neuropathol* **28,** 163–71.
9. White, A. R., Guirguis, R., et al. (2001) Sublethal concentrations of prion peptide PrP106-126 or the amyloid beta peptide of Alzheimer's disease activates expression of proapoptotic markers in primary cortical neurons *Neurobiol Dis* **8,** 299–316.
10. Stewart, L. R., White, A. R., et al. (2001) Involvement of the 5-lipoxygenase pathway in the neurotoxicity of the prion peptide PrP106-126 *J Neurosci Res* **65,** 565–72.
11. Salmona, M., Malesani, P., et al. (1999) Molecular determinants of the physicochemical properties of a critical prion protein region comprising residues 106-126 *Biochem J* **342(1),** 207–14.
12. Jobling, M. F., Stewart, L. R., et al. (1999) The hydrophobic core sequence modulates the neurotoxic and secondary structure properties of the prion peptide 106-126 *J Neurochem* **73,** 1557–65.
13. Jobling, M. F., Huang, X., et al. (2001) Copper and zinc binding modulates the aggregation and neurotoxic properties of the prion peptide PrP106-126. *Biochemistry* **40,** 8073–84.
14. Brown, D. R., Schmidt, B., and Kretzschmar, H. A. (1996) Role of microglia and host prion protein in neurotoxicity of a prion protein fragment *Nature* **380,** 345–7.

Chapter 7
Understanding the Nature of Prion Diseases Using Cell-free Assays

Victoria A. Lawson

Summary A central event in the transmission and pathogenesis of transmissible spongiform encephalopathy diseases is the misfolding of the prion protein. Considerable progress has been made in our understanding of this misfolding event through the development of cell-free assays that mimic the molecular features of prion propagation. This chapter reviews the contribution of cell-free assays to our understanding of prion propagation.

Keywords Cell-free conversion (CFC); protein misfolding cyclic amplification assay (PMCA); prion; recombinant.

1. Introduction

Attestation of the protein only hypothesis in the aetiology of transmissible spongiform encephalopathy diseases and acceptance of the prion as the infectious agent of disease requires the *de novo* generation of infectivity from a recombinant protein source. Although the direct induction of disease in wild-type animals infected with synthetic prions has yet to be shown, considerable progress has been made in our understanding of prion diseases through the development of cell-free assays (**Fig. 1**).

2. Autocatalytic Propagation of a Protease-resistant Prion Protein

The protein only hypothesis posits that an abnormal isomer of the prion protein (PrP^{Sc}) is autocatalytically propagated from the normal host encoded prion protein (PrP^{C}) and that PrP^{Sc} is the sole or principal component of the infectious agent or prion. The seminal work of Caughey and colleagues confirmed the autocatalytic ability of PrP^{Sc} in the cell-free conversion (CFC) assay (*1*). The CFC assay uses a purified source of PrP^{Sc} derived from infected animal tissue to generate protease-resistant PrP (PrP^{res})

From: *Prion Protein Protocols.*
Methods in Molecular Biology, Vol. 459.
Edited by: A. F. Hill © Humana Press, Totowa, NJ

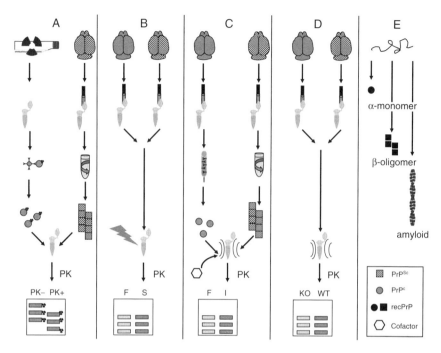

Fig. 1 CFAs for the study of prion propagation. (**A**) In the CFC assay, radiolabeled PrP^C is immunoprecipitated from lysates of mammalian cells and mixed with PrP^{Sc} purified from the tissue of a prion-infected animal. Samples are incubated for 48 h in a reaction mix (+/− guanidine) and protease-sensitive PrP digested with proteinase K (PK) before sodium dodecyl sulfate-polyacrylamide gel electrophoresis (SDS-PAGE) analysis and development by autoradiography. Conversion activity is presented as the percentage of radiolabeled PrP^C (PK−) converted to a protease-resistant product (PK+). (**B**) In PMCA, the PrP^C is prepared by homogenization of uninfected tissue in a conversion buffer containing protease inhibitors. The substrate is mixed with PrP^{Sc} derived from a homogenate prepared from prion-infected tissue. Samples are subjected to alternating rounds of sonication and incubation before PK digestion, SDS-PAGE analysis, and Western blot. Conversion activity is presented as an increase in signal in samples subjected to sonication (*S*) relative samples that are frozen at −80°C (*F*). (**C**) Amplification of purified prions in vitro is performed using immunopurified PrP^C derived from a homogenate of uninfected brain tissue and partially purified PrP^{Sc} derived from prion-infected tissue. PrP^C and PrP^{Sc} are mixed in a reaction buffer containing synthetic polyanions. The reaction is incubated for 16 h with agitation before PK digestion, SDS-PAGE analysis, and Western blot analysis. Conversion activity is presented as an increase in signal in incubated samples (*I*) relative to samples frozen at −80°C (*F*). (**D**) In CAA, a homogenate derived from uninfected tissue (PrP^C) is mixed with a homogenate derived from prion-infected tissue. The reaction is incubated for 16 h with agitation before PK digestion, SDS-PAGE analysis, and Western blot. Conversion activity is presented as an increase in signal in samples prepared from homogenates of WT versus Prnp$^{0/0}$ (*KO*) mice. (**E**) Recombinant PrP isolated from solubilized inclusion bodies can be folded into α-monomers, β-oligomers, and amyloid fibrils

from PrP^C purified from mammalian tissue culture *(2)*. The radioactive labelling of PrP^C enables the detection of the *de novo*-generated PrP^{res} without detection of the input PrP^{Sc} seed. The proportionally large amount of PrP^{Sc} seed required to drive the CFC assay (PrP^{Sc}:PrP^C = 50:1) has thus far prevented bioassay of *de novo*-generated infectivity, despite the use of the species barrier to circumvent the problem *(3)*.

Even with this limitation the contribution of the CFC assay to our understanding of prion diseases cannot be understated. It recapitulates the strain *(4)* and species *(5)* specificities of the diseases, and it has been used to model the risk of cross-species transmission in vivo *(6, 7)*. Mechanistically, it has identified the importance of amino acid sequence *(8)* and glycosylation state *(9)* as modulators of the species barrier, and it has shown that prion strain formation is influenced by both PrP^{Sc} glycoform ratios and the available pool of PrP^C *(10)*. Optimal PrP^{res} formation occurs at pH 6.0 *(11)*, and it is consistent with cell-based data indicating that PrP^{res} formation occurs at the plasma membrane or in early endocytic compartments *(12)* and most efficiently takes place when both PrP^C and PrP^{Sc} are present in the same contiguous membrane *(13)*. Therapeutically, the CFC assay has been used as a screening device to identify compounds that directly inhibit the interaction of PrP^C with PrP^{Sc} or that prevent the subsequent conversion of PrP^C to PrP^{Sc} *(14)*. From a practical perspective, it has the potential to be converted to a high-throughput screening device *(15)*, and PrP^{res} has been successfully generated from a recombinant source of PrP^C *(16)*. It was recently shown that conversion activity in the CFC assay and infectivity in vivo coincided with small nonfibrillar PrP^{Sc} particles rather than the amyloid fibrils typically associated with the disease *(17)*.

3. Ultraefficient Propagation of Prions

The relatively low efficiency of the CFC assay, requiring a 50-fold excess of PrP^{Sc} to PrP^C to convert at most 20% of the available PrP^C population was recently overcome by Soto and coworkers by using the protein misfolding cyclic amplification assay (PMCA) *(18)*. In this assay, somewhat akin to a DNA polymerase chain reaction reaction, rounds of amplification and sonication are used to generate a greater than tenfold increase in PrP^{res} by using a modest PrP^{Sc}-to-PrP^C ratio of 1:100 *(19)*. The efficiency of the PMCA assay indicated that the fragmentation of PrP^{Sc} achieved through the sonication process increases the availability of replication-competent PrP^{Sc}, and it supports the conversion activity of small PrP^{Sc} particles *(17)*. Using successive rounds of PMCA to dilute the initiating seed of PrP^{Sc} to undetectable levels, Soto and coworkers were able to generate *de novo* infectivity that was detected by bioassay in hamsters *(20)*. However, as reported by others, the levels of infectivity remained less than that associated with a similar quantity of brain-derived PrP^{Sc} *(21)*. Using a similar approach, PrP^{Sc} was detected in the blood of presymptomatic prion-infected hamsters *(22)*. With automation, this assay holds the most promise for a diagnostic test for presymptomatic blood screening.

4. Role for Accessory Factors in Development of Prion Diseases

The relative efficiency of the PMCA versus CFC assay has led to the suggestion that factors in the crude brain homogenates used in the PMCA are required for efficient generation of PrP^{res}, which is supported by the enhanced conversion activity in the CFC by the addition of heparin sulfate proteoglycans *(23)*. The role of accessory factors in the conversion of PrP^C to PrP^{res} has been further investigated by Suppatapone and colleagues who have used a simplified version of the PMCA assay to identify factors present in the brain homogenates that contribute to conversion efficiency. This approach has identified a role for polyanionic compounds (RNA or proteoglycans) in supporting PrP^{res} formation *(24)* (*see* Chap. 9).

The importance of cellular accessory factors in the conversion process is further supported by the observation that despite the efficiency of this assay, only 5–10% of the available PrP^C population is converted to PrP^{res} *(25)*. This efficiency is similar to that of the CFC, with a 10–20% conversion of PrP^C reported. Again, this points to a limiting cellular cofactor or a specific subpopulation of PrP^C that is able to support conversion. This is supported by our own work using an assay based on that of Suppatapone and colleagues who used crude brain homogenates. In this "conversion activity assay" (CAA), the increase in conversion activity observed when PrP^C is derived from PrP transgenic (Tg) versus wild-type (WT) mouse tissue is not proportional to the increased PrP^C pool available in the Tg mouse tissue (**Fig. 2**), and thus suggests that PrP^C is not the cellular factor that limits PrP^{res} formation.

5. Synthetic Prions

It is difficult to dispute the role of PrP^{Sc} in prion diseases. However, irrefutable proof requires the generation of infectivity in the absence of an infectious tissue-derived seed. Legname and coworkers were able induce a prion-like disease in Tg mice expressing a truncated (89–231) form of PrP^C (Tg9949) by inoculation with amyloid fibrils formed from recombinant PrP89-231 *(26)*. The subsequently in vivo-generated prions were able to transmit infectivity to WT mice. However, to date it has not been possible to generate synthetic prions with sufficient infectivity to infect WT mice at first passage. One explanation may be that rather than generate infectivity *de novo*, the recombinant amyloid simply accelerates a disease in Tg mice, which is associated with prion protein overexpression *(27)*. Alternatively, the in vivo environment may exert a selection pressure for disease-specific PrP conformations, which we are yet to mimic in vitro. Regardless, the ability to generate infectivity from an apparent subpopulation of synthetic prions has as of itself contributed to our understanding of the nature of prion diseases, and it will facilitate the identification and biophysical characterisation of infectious prions.

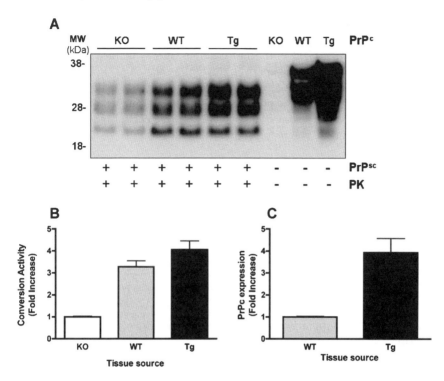

Fig. 2 Role of accessory factors in cell-free propagation of prions. (**A**) Western blot analysis of PrPres generated in the CAA in the presence of brain homogenate derived from Prnp0/0 (*KO*), Balb/c (*WT*), and Tga20 PrPC-overexpressing mice (*Tg*). Signal in KO lane represents input PrPSc seed derived from brain homogenate of M1000-infected mice. The conversion activity (**B**), normalized to the KO lane, is not significantly increased ($P > 0.05$) when PrPSc seed is incubated in the presence of a WT or Tg-derived brain homogenate, despite a fourfold increase in PrPC expression in Tg mice (**C**)

6. Toward Understanding the Molecular Nature of Prion Diseases

Importantly, CFAs have provided a molecular understanding of observations made in vivo. Novel protease resistance C-terminal fragments of PrP identified in sporadic Creutzfeldt–Jakob disease patients (*28*) have also been generated in the CFC assay from truncated forms of PrPC and in assays that mimic cross species transmission (*29*) and from the amyloid form of recombinant PrP-encompassing residues 89–230 (*30*). Expansions within the octapeptide repeat region of PrPC associated with familial prion diseases were shown to increase the rate of amyloid fibril formation by using recombinant PrP, and by using the CFC assay the rate of binding was found to be the limiting step in the process (*31*). Using recombinant PrP, the kinetics of amyloid formation was not shown to differ for PrP encoding methionine or valine at codon 129, despite the impact of this polymorphism on susceptibility to human prion diseases. However, recombinant PrP encoding methionine at position 129 favored a misfolding pathway that led to the formation of soluble β-sheet–rich oligomers over amyloid formation *(32)*. This observation may explain some of the disease characteristics associated with the codon 129 polymorphism. CFAs are not only key to our understanding of the molecular nature of prion diseases but also undoubtedly will be central in the development of successful diagnostic and therapeutic methodologies.

Acknowledgements This work was supported by the National Health and Medical Research Council (NHMRC) of Australia fellowship 209163 and project grants 400229 and 454546.

References

1. Kocisko, D. A., Come, J. H., Priola, S. A., Chesebro, B., Raymond, G. J., Lansbury, P. T., and Caughey, B. (1994) Cell-free formation of protease-resistant prion protein. *Nature* **370,** 471–4.
2. Caughey, B., Raymond, G. J., Priola, S. A., Kocisko, D. A., Race, R. E., Bessen, R. A., Lansbury, P. T., Jr., and Chesebro, B. (1999) Methods for studying prion protein (PrP) metabolism and the formation of protease-resistant PrP in cell culture and cell-free systems. An update. *Mol Biotechnol* **13,** 45–55.
3. Hill, A. F., Antoniou, M., and Collinge, J. (1999) Protease-resistant prion protein produced in vitro lacks detectable infectivity. *J Gen Virol* **80 (1),** 11–4.
4. Kocisko, D. A., Priola, S. A., Raymond, G. J., Chesebro, B., Lansbury, P. T., Jr., and Caughey, B. (1995) Species specificity in the cell-free conversion of prion protein to protease-resistant forms: a model for the scrapie species barrier. *Proc Natl Acad Sci U S A* **92,** 3923–7.
5. Bessen, R. A., Kocisko, D. A., Raymond, G. J., Nandan, S., Lansbury, P. T., and Caughey, B. (1995) Non-genetic propagation of strain-specific properties of scrapie prion protein. *Nature* **375,** 698–700.
6. Raymond, G. J., Bossers, A., Raymond, L. D., O'Rourke, K. I., McHolland, L. E., Bryant, P. K., 3rd, Miller, M. W., Williams, E. S., Smits, M., and Caughey, B. (2000) Evidence of a molecular barrier limiting susceptibility of humans, cattle and sheep to chronic wasting disease. *EMBO J* **19,** 4425–30.
7. Raymond, G. J., Hope, J., Kocisko, D. A., Priola, S. A., Raymond, L. D., Bossers, A., Ironside, J., Will, R. G., Chen, S. G., Petersen, R. B., Gambetti, P., Rubenstein, R., Smits, M. A., Lansbury, P. T., Jr., and Caughey, B. (1997) Molecular assessment of the potential transmissibilities of BSE and scrapie to humans. *Nature* **388,** 285–8.
8. Priola, S. A., Chabry, J., and Chan, K. (2001) Efficient conversion of normal prion protein (PrP) by abnormal hamster PrP is determined by homology at amino acid residue 155. *J Virol* **75,** 4673–80.
9. Priola, S. A., and Lawson, V. A. (2001) Glycosylation influences cross-species formation of protease-resistant prion protein. *EMBO J* **20,** 6692–9.
10. Vorberg, I., and Priola, S. A. (2002) Molecular basis of scrapie strain glycoform variation. *J Biol Chem* **277,** 36775–81.
11. Horiuchi, M., and Caughey, B. (1999) Specific binding of normal prion protein to the scrapie form via a localized domain initiates its conversion to the protease-resistant state. *EMBO J* **18,** 3193–203.
12. Caughey, B., Raymond, G. J., Ernst, D., and Race, R. E. (1991) N-terminal truncation of the scrapie-associated form of PrP by lysosomal protease(s): implications regarding the site of conversion of PrP to the protease-resistant state. *J Virol* **65,** 6597–603.

13. Baron, G. S., Wehrly, K., Dorward, D. W., Chesebro, B., and Caughey, B. (2002) Conversion of raft associated prion protein to the protease-resistant state requires insertion of PrP-res (PrP(Sc)) into contiguous membranes. *EMBO J* **21,** 1031–40.

14. Caughey, W. S., Raymond, L. D., Horiuchi, M., and Caughey, B. (1998) Inhibition of protease-resistant prion protein formation by porphyrins and phthalocyanines. *Proc Natl Acad Sci U S A* **95,** 12117–22.

15. Maxson, L., Wong, C., Herrmann, L. M., Caughey, B., and Baron, G. S. (2003) A solid-phase assay for identification of modulators of prion protein interactions. *Anal Biochem* **323,** 54–64.

16. Kirby, L., Birkett, C. R., Rudyk, H., Gilbert, I. H., and Hope, J. (2003) In vitro cell-free conversion of bacterial recombinant PrP to PrPres as a model for conversion. *J Gen Virol* **84,** 1013–20.

17. Silveira, J. R., Raymond, G. J., Hughson, A. G., Race, R. E., Sim, V. L., Hayes, S. F., and Caughey, B. (2005) The most infectious prion protein particles. *Nature* **437,** 257–61.

18. Saborio, G. P., Permanne, B., and Soto, C. (2001) Sensitive detection of pathological prion protein by cyclic amplification of protein misfolding. *Nature* **411,** 810–3.

19. Castilla, J., Saa, P., Morales, R., Abid, K., Maundrell, K., and Soto, C. (2006) Protein misfolding cyclic amplification for diagnosis and prion propagation studies. *Methods Enzymol* **412,** 3–21.

20. Castilla, J., Saa, P., Hetz, C., and Soto, C. (2005) In vitro generation of infectious scrapie prions. *Cell* **121,** 195–206.

21. Bieschke, J., Weber, P., Sarafoff, N., Beekes, M., Giese, A., and Kretzschmar, H. (2004) Autocatalytic self-propagation of misfolded prion protein. *Proc Natl Acad Sci U S A* **101,** 12207–11.

22. Castilla, J., Saa, P., and Soto, C. (2005) Detection of prions in blood. *Nat Med* **11,** 982–5.

23. Wong, C., Xiong, L. W., Horiuchi, M., Raymond, L., Wehrly, K., Chesebro, B., and Caughey, B. (2001) Sulfated glycans and elevated temperature stimulate PrP(Sc)-dependent cell-free formation of protease-resistant prion protein. *EMBO J* **20,** 377–86.

24. Deleault, N. R., Geoghegan, J. C., Nishina, K., Kascsak, R., Williamson, R. A., and Supattapone, S. (2005) Protease-resistant prion protein amplification reconstituted with partially purified substrates and synthetic polyanions. *J Biol Chem* **280,** 26873–9.

25. Supattapone, S. (2004) Prion protein conversion in vitro. *J Mol Med* **82,** 348–56.

26. Legname, G., Baskakov, I. V., Nguyen, H. O., Riesner, D., Cohen, F. E., DeArmond, S. J., and Prusiner, S. B. (2004) Synthetic mammalian prions. *Science* **305,** 673–6.

27. Nazor, K. E., Kuhn, F., Seward, T., Green, M., Zwald, D., Purro, M., Schmid, J., Biffiger, K., Power, A. M., Oesch, B., Raeber, A. J., and Telling, G. C. (2005) Immunodetection of disease-associated mutant PrP, which accelerates disease in GSS transgenic mice. *EMBO J* **24,** 2472–80.

28. Zou, W. Q., Capellari, S., Parchi, P., Sy, M. S., Gambetti, P., and Chen, S. G. (2003) Identification of novel proteinase K-resistant C-terminal fragments of PrP in Creutzfeldt-Jakob disease. *J Biol Chem* **278,** 40429–36.

29. Lawson, V. A., Priola, S. A., Meade-White, K., Lawson, M., and Chesebro, B. (2004) Flexible N-terminal region of prion protein influences conformation of protease-resistant prion protein isoforms associated with cross-species scrapie infection in vivo and in vitro. *J Biol Chem* **279,** 13689–95.

30. Bocharova, O. V., Breydo, L., Salnikov, V. V., Gill, A. C., and Baskakov, I. V. (2005) Synthetic prions generated in vitro are similar to a newly identified subpopulation of PrPSc from sporadic Creutzfeldt-Jakob disease. *Protein Sci* **14,** 1222–32.

31. Moore, R. A., Herzog, C., Errett, J., Kocisko, D. A., Arnold, K. M., Hayes, S. F., and Priola, S. A. (2006) Octapeptide repeat insertions increase the rate of protease-resistant prion protein formation. *Protein Sci* **15,** 609–19.

32. Baskakov, I., Disterer, P., Breydo, L., Shaw, M., Gill, A., James, W., and Tahiri-Alaoui, A. (2005) The presence of valine at residue 129 in human prion protein accelerates amyloid formation. *FEBS Lett* **579,** 2589–96.

Chapter 8
Methods for Conversion of Prion Protein into Amyloid Fibrils

Leonid Breydo, Natallia Makarava, and Ilia V. Baskakov

Summary Misfolding and aggregation of prion protein (PrP) is related to several neurodegenerative diseases in humans such as Creutzfeldt–Jacob disease, fatal familial insomnia, and Gerstmann–Straussler–Sheinker disease. Amyloid fibrils prepared from recombinant PrP in vitro share many features of the infectious prions. These fibrils can be used as a synthetic surrogate of PrP^{Sc} for development of prion diagnostics, including generation of PrP^{Sc}-specific antibody, for screening of antiprion drugs, or for development of antiprion decontamination procedures. Here, we describe the methods of preparation of prion protein fibrils in vitro and biochemical assays for assessing physical properties and the quality of fibrils.

Keywords Amyloid fibrils; conformational transition; prion diseases; recombinant prion protein.

1. Introduction

For the past few years, several protocols for producing amyloid fibrils from recombinant prion protein (PrP)-encompassing residues 90–231 have been developed by various groups (*1–6*). Here, we describe the first protocol that was recently developed by our group for converting of full-length recombinant PrP (PrP 23-230) into amyloid form (*7*). As judged from electron microscopy, atomic force microscopy, Fourier-transform infrared (FTIR) spectroscopy, and proteinase K (PK) digestion assays, physical properties of fibrils produced from the full-length PrP closely resembled that of infectious prions (*7–9*). Moreover, the immunoconformational assay that used conformational PrP^{Sc}-specific antibodies revealed that the fibrils produced from PrP 23-230 acquired a surface structure similar to that of PrP^{Sc} (*10*). In this respect, amyloid fibrils generated in vitro seem to be a suitable synthetic surrogate of PrP^{Sc}. PrP fibrils can be used for development of prion diagnostics, high-throughput screening of antiprion drugs, for the development of antiprion decontamination procedures, and for other important applications in the prion field.

From: *Prion Protein Protocols.*
Methods in Molecular Biology, Vol. 459.
Edited by: A. F. Hill © Humana Press, Totowa, NJ

2. Materials

Unless otherwise noted, all reagents are from Sigma-Aldrich (St. Louis, MO). Water for all solutions is deionized with Synergy 185UV Ultrapure Water System (Millipore Corporation, Billerica, MA).

2.1. Amyloid Fibril Formation (Manual Setup)

1. 0.5 M, 2-(N-Morpholino)ethanesulfonic acid (MES) buffer, pH 6.0.
2. Sodium acetate buffer, 10 mM, pH 5.0.
3. 0.5 M, Thiourea, pH adjusted to 6.0.
4. 6 M, Guanidine hydrochloride, pH adjusted to 6.0.
5. Dialysis tubing (Spectrapor, molecular weight cut-off [MWCO] 2,000) and dialysis clips.
6. Thioflavin T (ThT) (Invitrogen, Carlsbad, CA), 1 mM stock in water (store in the dark at 4°C).
7. Delfia plate shaker (PerkinElmer Life and Analytical Sciences, Boston, MA, or Wellmix, Thermo Labsystems, Frankfurt, Germany, or similar) with a micro-centrifuge tube rack attached (**Fig. 1**).
8. Bath sonicator (Branson 2510, Branson Ultrasonics Corporation, Danbury, CT).
9. Plastic tubes, 1.5 ml (Fisherbrand, Fisher Scientific. Pittsburgh, PA).
10. Recombinant full-length PrP produced as described by Makarava and Baskakov in Chapter 10.

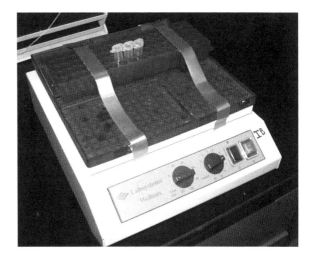

Fig. 1 Plate shaker with a microcentrifuge tube rack used for the conversion reaction. Place plastic tubes next to each other to prevent their rotation inside the rack during shaking

2.2. Amyloid Fibril Formation (Semiautomated Setup)

In addition to the reagents described in **Subheading 2.1.**, the following are needed:

1. Teflon spheres (3/32-in. diameter, McMaster-Carr, Los Angeles, CA).
2. 96-well flat bottom nontreated polystyrene assay plates (Corning Life Sciences, Acton, MA).
3. Mylar plate sealers (Thermo Labsystems, Franklin, MA).
4. Microplate fluorescence reader (Fluoroscan Ascent CF, Thermo Electron Corporation, Waltham, MA) equipped with 444-nm excitation and 485-nm emission filters.

2.3. PK Digestion of PrP Fibrils

1. Prestained molecular weight markers.
2. PK (Novagen, Madison, WI).
3. Phenylmethylsulfonyl fluoride (PMSF), 50 mM, in acetonitrile (store at −20°C).
4. Loading buffer (2×): 125 mM TRIZMA BASE, 4.5 M urea, 20% glycerol, 1.25 M β-mercaptoethanol, 4% (w/v) sodium dodecyl sulfate (SDS), 0.02% (w/v) bromophenol blue (*see* **Note 1**) (store frozen).

2.4. Epifluorescence Microscopy

1. Inverted fluorescent microscope (Nikon Eclipse TE2000-U) equipped with the illumination system; sets of objectives, filters (excitation filter 485DF22, beam splitter 505DRLPO2, and emission filter 510LP (Omega Optical Inc., Brattleboro, VT); and a charge-coupled device (CCD) camera.
2. Immersion oil type FF (Cargille Laboratories, Cedar Grove, NJ).
3. Glass Coplin staining jar.
4. Microscope cover glass No. 1 (Fisherbrand, Fisher Scientific).

3. Methods

3.1. Conversion of Full-Length PrP into Amyloid Fibrils in Manual Setup

Full-length PrP is capable of forming a variety of aggregated forms in vitro *(7, 11, 12)*. Formation of amyloid fibrils is highly dependent on the reaction conditions. Ideal reaction conditions for fibril formation combine neutral or slightly acidic pH (between 5.0 and 7.5) and moderate concentrations of denaturants such as guanidine hydrochloride (GdnHCl) (up to 2 M) or urea (up to 4 M). Possible complications during fibril formation include inhibition of fibrillization by Cu^{2+} *(11)*, copper-mediated self-cleavage of the protein *(13, 14)*, and side chain oxidation *(12)*. To minimize these problems, copper ions are removed from the protein during purification, and thiourea (10 mM) is added during fibril formation. Additional problem may arise if recombinant PrP used for conversion is of low purity. The conversion conditions described here and referred to as standard are 2 M GdnHCl, pH 6.0.

1. Prepare stock solution of recombinant PrP (130 µM, 3 mg/ml) in 6 M GdnHCl, pH 6.0. This solution can be stored at −20°C for up to 1 week (*see* **Note 2**). Alternatively, the protein can be dissolved directly in the MES buffer, pH 6.0 (50 mM) at lower concentration (0.5–1 mg/ml). This solution, however, cannot be stored and must be used for fibril formation immediately (within several hours).

2. To prepare 500-µl reaction (*see* **Note 3**), mix the following reagents in the conical plastic tube: water (273.3 µl), GdnHCl (6 M, 83.4 µl), MES buffer, pH 6.0 (0.5 M, 50 µl), and thiourea (0.5 M, 10 µl). Then, add stock solution of PrP in 6 M GdnHCl (3 mg/ml, 83.3 µl). Mix reagents gently, avoid introducing air bubbles.

3. If you are using previously formed fibrils as seeds, sonicate them for at least 10 s in the bath sonicator and add to the reaction mixture before PrP stock. Seeding capacity of fibrils decreases upon prolonged storage. Small amounts of seeds (as little as 0.1% of the amount of protein) are sufficient to significantly decrease the lag phase of conversion.

4. Incubate the tube with continuous shaking at 600 rpm by using a plate shaker at 37°C (*see* **Note 4** and **Fig. 1**).

5. Monitor the kinetics of fibril formation by using a thioflavin T binding assay. For this assay, withdraw the aliquots during the incubation. Before taking aliquots, pipette the reaction mixture gently each time (strong pipetting might perturb the kinetics of fibril formation). Dilute each aliquot into 10 mM Na-acetate buffer, pH 5.0, to a final concentration of PrP of 0.3 µM, and then add thioflavin T to a final concentration of 10 µM. Record three emission spectra (from 460 to 520 nm) for each sample in 0.4-cm rectangular cuvettes, with excitation at 445 nm on a FluoroMax-3 fluorimeter (Jobin Yvon, Edison, NJ), keeping excitation and emission slits at 4 nm. Average the spectra and determine the fluorescence intensity at emission maximum (usually ~482 nm). Emission will remain low for a few hours or days, and then it will start rising and eventually it will rise 10- to 40-fold (*see* **Note 5**).

6. After the fibril formation has reached plateau, the fibrils should be dialyzed for prolonged storage. Place the suspension of fibrils in the bag prepared from the dialysis tubing (MWCO 2,000) and dialyze against a large volume of 10 mM Na-acetate buffer, pH 5.0, with several buffer changes. Fibrils should be stored at 4°C. Prolonged storage of fibrils at room temperature or at higher pH may lead to their aggregation and copper-mediated protein self-cleavage. Freezing and thawing may cause fragmentation of fibrils into short pieces.

3.2. Conversion of Full-Length PrP into Amyloid Fibrils in Semiautomated Setup

1. Perform the conversion of PrP into amyloid fibrils in semiautomated setup at least in triplicate to ensure reproducibility. Add three Teflon spheres (3/32-in. diameter) per well of 96-well assay plate. Mix the following reagents in the conical plastic tube: water (268.3 µl), GdnHCl (6 M, 163.7 µl), MES buffer, pH 6.0 (0.5 M, 50 µl), thioflavin T (1 mM, 5 µl), and thiourea (0.5 M, 10 µl). Add stock solution of PrP in 6 M GdnHCl (3 mg/ml, 5 µl). After thorough mixing divide the reaction mixture between three wells of the 96-well plate (160 µl/ well) and cover the plate with the plate sealer. If previously formed fibrils are used as seeds, they should be sonicated for at least 10 s in the bath sonicator and added to the reaction mixture before PrP stock.
2. Insert the 96-well plate into the microplate reader. Set up incubation at 37°C, with shaking at 900 rpm (shaking diameter 1 mm), and fluorescence measurements every 5 or 10 min, with excitation at 444 nm and emission at 485 nm (**Fig. 2**).
3. After the completion of the experiment, transfer the data to Origin (OriginLab Corp., Northampton, MA) and fit to the following equation:

$$F = A + (B + c \times t)/(1 + \exp(k \times (t_m - t)))$$

where A is the initial level of thioflavin T fluorescence, B is the increase in thioflavin T fluorescence during conversion, k is rate constant of amyloid accumulation, t_m is the midpoint of conversion of PrP to amyloid, and c is an empirical parameter describing changes in fluorescence after fibril formation. The lag time (t_l) can be calculated as $t_l = t_m - 2/k$.

3.3. Proteinase K Digestion of PrP Fibrils

Upon conversion of PrP into amyloid fibrils, PK resistance of the protein significantly increases. Amyloid fibrils produced according to the standard conversion

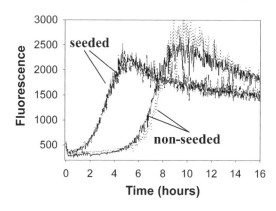

Fig. 2 Kinetics of fibril formation from PrP (1 µM) carried out in semiautomated format under standard solvent conditions. Two repeats for nonseeded and seeded conversion reactions are shown. After reaching a plateau, the ThT fluorescence often shows a decline that seems to be due to sorption of fibrils to plate walls and partial aggregation of fibrils into clumps

protocol display three major PK-resistant bands of 8, 10, and 12 kDa *(15)*. The appearance of protease-resistant 23-kDa band (that corresponds to the full-length PrP), however, indicates the presence of nonfibrillar, aggregated isoforms of PrP. These nonfibrillar, PK-resistant PrP isoforms are typically formed when PrP is exposed to Cu^{2+} *(11)*. To distinguish amyloid fibrils from nonfibrillar aggregates, we have used a procedure referred to as maturation. Maturation involves brief heating of a sample at 80°C in the presence of detergent, as described in our previous studies *(16)*. Upon maturation, the PK-resistant core of fibrils expands from 10–12 kDa to 16 kDa (**Fig. 3**) Nonfibrillar aggregates do not show 16 kDa band.

1. Prepare three samples for each batch of fibrils (sample 1, control, no PK digestion; sample 2, fibrils after PK digestion; and sample 3, fibrils subjected to maturation and PK digestion). To prepare samples, place 2 µl of suspension of PrP fibrils (0.5 mg/ml) into the plastic tube. Add water (4.8 µl), Tris-HCl buffer, pH 7.5 (1 M, 0.8 µl), and Triton X-100 (1% in water, 0.4 µl). Incubate sample 3 in the water bath at 80°C for 15 min and spin it at tabletop centrifuge for 2 s. Add PK (0.5 µl, 20 µg/ml, 1:100 PK:PrP ratio) to samples 2 and 3. Incubate samples 2 and 3 for 1 h at 37°C.
2. Add 0.5 µl of PMSF (50 mM in acetonitrile) to samples 2 and 3.
3. Add 2× loading buffer (8 µl) to all samples and incubate them in the water bath at 95°C for 10 min.
4. Prepare SDS-polyacrylamide gel electrophoresis and load the samples. Run the gel and stain it with silver nitrate or Coomassie blue (*see* **Note 6** and **Fig. 3**).

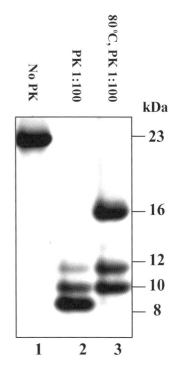

Fig. 3 Analysis of PrP fibrils in PK digestion assay: sample 1, control, no PK digestion; sample 2, fibrils after PK digestion; and sample 3, fibrils subjected to maturation and PK digestion. PK-resistant fragments of 8, 10, 12, and 16 kDa consist of residues 162–230, 152/153–230, 138/141–230, and 97–230, respectively *(15, 16)*

3.4. Epifluorescence Microscopy

Although ThT fluorescence assay is convenient for measuring the kinetics of fibril formation, this assay is not sufficient for providing proof as to whether amyloid fibrils were formed in the reaction mixture. Relatively small (two- to threefold) increases in ThT fluorescence may indicate formation of nonfibrillar PrP isoforms, such as soluble β-oligomers, which also bind ThT. Several techniques, including electron microscopy and atomic force microscopy, have been broadly used for confirming the formation of fibrils; however, their application is laborious and it requires costly equipment. In our experience, epifluorescence microscopy in the presence of ThT serves as a rapid and reliable test for the presence of the amyloid fibrils in the sample and for assessing their quality *(7, 11)*. Here, we describe the experimental procedure for imaging amyloid fibrils by using inverted fluorescence microscopy in the presence of ThT.

1. Deposit several cover glasses into staining jar clean them by sonicating it in isopropanol (1 min), then in acetone (1 min), and again in isopropanol (1 min). Wash the coverglass with water and incubate it in the mixture of sulfuric acid (70%) and hydrogen peroxide (30%) for at least 1 h. Rinse with water several times. Store in water, dry with compressed air before use.
2. Add 0.5 µl of suspension of PrP fibrils (0.5 mg/ml) in 10 mM acetate buffer, pH 5.0, to the same buffer (99.5 µl) containing 10 µM ThT. Incubate the solution in the plastic tube at 25°C for 5 min.
3. Place 10–20 µl of the solution on a coverglass. Allow fibrils to sediment on the glass surface for 1–2 min. Examine the sample with an inverted fluorescence microscope (Nikon Eclipse TE2000-U) by using 60× or 100× objective. The emission can be isolated from Rayleigh and Raman-shifted light by a combination of filters: an excitation filter 485DF22, a beam splitter 505DRLPO2, and an emission filter 510LP (Omega Optical Inc.). Digital images can be acquired using a CCD camera.
4. Shape and size of PrP fibrils vary depending on the conditions of their formation and purity of the protein (*see* **Note 5**). In the standard conditions, PrP of high purity forms long (>1-µm) and straight fibrils (**Fig. 4A**). Fibrils formed at higher pH (>7.0) tend to aggregate into large clusters. Exposure of fibrils formed at pH <6.0 to pH >6.5 or addition of salt induces aggregation into clumps of various sizes (**Fig. 4B**). Presence of impurities in the protein or substantial methionine oxidation can result in formation of shorter fibrils (<1 µm) that often seem bent or very short, dot-like fibrils (**Fig. 4C**).

3.5. FTIR Spectroscopy

FTIR spectra can serve as alternative assay of the yield and quality of PrP fibrils. The description assumes the use of ATR FTIR instrument (Bruker Tensor 27 or similar).

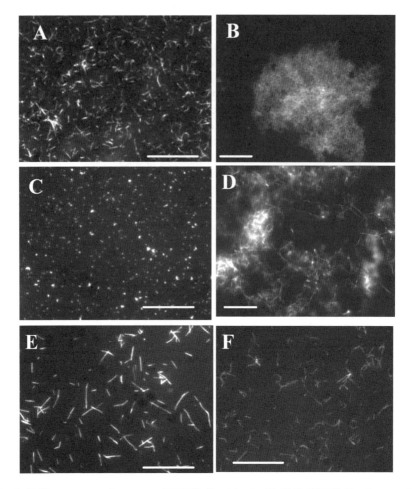

Fig. 4 Epifluorescence microscopy of PrP fibrils stained with ThT. (**A**) Fibrils of mouse PrP prepared under standard conditions (shaking, pH 6.0). (**B**) Fibrils of mouse PrP prepared under standard conditions and incubated for 24 h in PBS, pH 7.2. (**C**) PrP (mouse) of insufficient purity or with substantial methionine oxidation converts into very short or dote-like fibrils under standard conditions. (**D**) Fibrils of mouse PrP prepared under standard conditions and digested with PK for 1 h at 37°C. (**E**) Fibrils of mouse PrP prepared under standard solvent conditions upon rotation. (**F**) Fibrils of Syrian hamster PrP prepared under standard solvent conditions upon rotation. When stained with ThT, mouse fibrils show substantially brighter fluorescence than hamster fibrils. Bars = 10 μm

1. Purge the FTIR instrument with dry air or nitrogen for several hours. Monitor the level of water vapors by measuring the spectra of the blank cell.
2. Wash the surface of FTIR cell with 6 M GdnHCl to remove possible contamination or protein deposits and then several times with water. Load 20 μl of sample (in buffer, protein concentration >0.2 mg/ml) into the cell. Measure the spectrum (2/cm resolution, 1,024 scans).

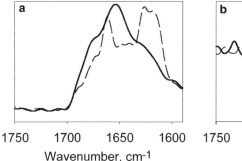 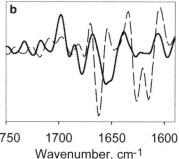

Fig. 5 (**A**) FTIR spectra of PrP fibrils (dashed) and monomer (solid). (**B**) Second derivatives of FTIR spectra from A

3 Wash the cell with water several times, load buffer (20 μl), measure the spectrum.
4. Subtract buffer spectrum from the sample spectrum. Subtract the spectrum of the water vapors to achieve flat baseline in 1,700–1,800 cm^{-1} region. Subject the spectrum to Fourier deconvolution using a Lorentzian line shape and parameters equivalent to 20 cm^{-1} bandwidth at half height and a noise suppression factor of 0.3 (**Fig. 5A**).
5. Calculate the second derivative of the spectrum (**Fig. 5B**). The shape of the second derivative of the spectrum in the region between 1,600 and 1,700 cm^{-1} can be used as an indicator of the yield and quality of PrP fibrils. Peaks in the 1,610–1,630 cm^{-1} region (β-sheets) are characteristic of amyloid fibrils, whereas the peak at 1,651 cm^{-1} (α-helices) is most prevalent in monomeric prion protein. Exact shape of the spectrum of PrP fibrils may depend on purity of the protein used for conversion, conditions of fibril formation and other variables. Fibrils formed from mouse PrP 23-230 under conditions described here give major peaks at 1,615, 1,625, and 1,661 cm^{-1}. The intensity of the first two peaks are approximately equal, however, may slightly vary from preparation to preparation.

4. Notes

1. Urea in the loading buffer helps to dissolve PrP fibrils or products of PK digestion. In the absence of urea, protein aggregates may show up as bands on the top of the gel.
2. After PrP is purified, it is preferable to store the protein as a lyophilized powder at −20°C. Storage in 6 M GdnHCl at −20°C is possible for up to 1 week. However, prolonged storage at 6 M GdnHCl leads to formation of the β-oligo-

meric nonfibrillar form. Prolonged storage (>24 h) in solution at neutral pH and in the absence of denaturants is not advisable due to self-cleavage and nonspecific aggregation.

3. Fibril formation can proceed sluggishly if the volume of the solution in the tube is too low (<300 μl).

4. Make sure that the plastic tube fits tightly within the attached to the platform rack and that it does not rotate inside the rack during orbital shaking (**Fig. 1**). End-over-end rotation on Nutator (VWR, West Chester, PA) or similar device also can be used as an alternative way of mixing solution for growing fibrils.

5. Exact value of ThT fluorescence depends on several factors, including purity of the protein and conditions of fibril formation. Differences in intensity of ThT fluorescence could be due to several factors, including differences in ThT binding affinity for fibrils prepared at different solvent conditions; different yield of fibrils formation; or different propensity of fibrils to aggregate into large clumps. The yield of ThT fluorescence of fibrils prepared under nonstandard conditions is often lower than that for standard fibrils. ThT fluorescence of fibrils prepared from different PrP variants (e.g., mouse vs hamster PrP) also can vary (**Fig. 4E,F**). Values of ThT fluorescence threefold or more above background usually indicate the presence of amyloid fibrils. In addition to ThT-fluorescence assay, fluorescence microscopy helps to analyze the yield and quality of fibrils (**Fig. 4**).

6. Intense bands in 8- to 12-kDa range after PK digestion are consistent with the presence of amyloid fibrils in the sample, although some nonfibrillar aggregates also may give rise to these bands. Appearance of the 16-kDa band after maturation (sample 3) is a specific characteristic for amyloid fibrils (**Fig. 3**).

References

1. Baskakov IV, Legname G, Baldwin MA, Prusiner SB, Cohen FE. (2002) Pathway complexity of prion protein assembly into amyloid. J Biol Chem; 277:21140–21148.
2. Jackson GS, Hosszu LLP, Power A, Hill AF, Kenney J, Saibil H, et al. (1999) Reversible conversion of monomeric human prion protein between native and fibrilogenic conformations. Science; 283:1935–1937.
3. Lee S, Eisenberg D. (2003) Seeded conversion of recombinant prion protein to a disulfide-bonded oligomer by a reduction-oxidation process. Nat Struct Biol; 10(9):725–730.
4. Kazlauskaite J, Young A, Gardner CE, Macpherson JV, Venien-Bryan C, Pinheiro TJT. (2005) An unusual soluble b-turn-rich conformation of prion is involved in fibril formation and toxic to neuronal cells. Biochem Biophys Res Commun; 328:292–305.
5. Torrent J, Alvarez-Martinez MT, Harricane MC, Heitz F, Liautard JP, Balny C et al. (2004) High pressure induces scrapie-like prion protein misfolding and amyloid fibril formation. Biochemistry; 43:7162–7170.
6. Leffer K-W, Wille H, Stohr J, Junger E, Prusiner SB, Riesner D. (2005) Assembly of natural and recombinant prion protein into fibrils. Biol Chem; 386:569–580.
7. Bocharova OV, Breydo L, Parfenov AS, Salnikov VV, Baskakov IV. (2005) In vitro conversion of full length mammalian prion protein produces amyloid form with physical property of PrP^{Sc}. J Mol Biol; 346:645–659.

8. Anderson M, Bocharova OV, Makarava N, Breydo L, Salnikov VV, Baskakov IV. (2006) Polymorphism and ultrastructural organization of prion protein amyloid fibrils: an insight from high resolution atomic force microscopy. J Mol Biol; 358:580–596.

9. Makarava N, Bocharova OV, Salnikov VV, Breydo L, Anderson M, Baskakov IV. (2006) Dichotomous versus palm-type mechanisms of lateral assembly of amyloid fibrils. Protein Sci; 15:1334–1341.

10. Novitskaya V, Makarava N, Bellon A, Bocharova OV, Bronstein IB, Williamson RA, et al. (2006) Probing the conformation of the prion protein within a single amyloid fibril using a novel immunoconformational assay. J Biol Chem; 281:15536–15545.

11. Bocharova OV, Breydo L, Salnikov VV, Baskakov IV. (2005) Cu(II) inhibits in vitro conversion of prion protein into amyloid fibrils. Biochemistry; 44:6776–6787.

12. Breydo L, Bocharova OV, Makarava N, Salnikov VV, Anderson M, Baskakov IV.(2005) Methionine oxidation interferes with conversion of the prion protein into the fibrillar proteinase K-resistant conformation. Biochemistry; 44:15534–15543.

13. McMahon HEM, Mange A, Nishida N, Creminon C, Casanova D, Lehman S. (2001) Cleavage of the amino terminus of the prion protein by reactive oxygen species. J Biol Chem; 276:2286–2291.

14. Mange A, Beranger F, Peoc'h K, Onodera T, Frobert Y, Lehmann S. (2004) alpha- and beta-Cleavages of the amino-terminus of the cellular prion protein. Biol Cell; 96:125–132.

15. Bocharova OV, Breydo L, Salnikov VV, Gill AC, Baskakov IV. (2005) Synthetic prions generated in vitro are similar to a newly identified subpopulation of PrPSc from sporadic Creutzfeldt-Jakob Disease PrP^{Sc}. Protein Sci; 14:1222–1232.

16. Bocharova OV, Makarava N, Breydo L, Anderson M, Salnikov VV, Baskakov IV. (2006) Annealing PrP amyloid fibrils at high temperature results in extension of a proteinase K resistant core. J Biol Chem; 281:2373–2379.

Chapter 9
Amplification of Purified Prions In Vitro

Surachai Supattapone, Nathan R. Deleault, and Judy R. Rees

Summary The infectious agents of prion diseases are unorthodox, and they seem to be composed primarily of a misfolded glycoprotein called the prion protein (PrP). Replication of prion infectivity is associated with the conversion of PrP from its normal, cellular form (PrP^C) into a pathogenic form (PrP^{Sc}), which is characterized biochemically by relative detergent insolubility and protease resistance. Several techniques have been developed in which PrP^C molecules can be converted into the PrP^{Sc} conformation in vitro *(1–8)*. These biochemical techniques recapitulate several specific aspects of in vivo prion propagation *(1–3)*, and one method, the protein misfolding cyclic amplification technique, also has been shown to amplify infectivity *(5)*. In this chapter, we describe a method for amplifying PrP^{Sc} molecules from hamster prions in vitro using purified substrates. Specific protocols for substrate preparation, reaction mixture, and product detection are explained. Purified PrP^{Sc} amplification assays are currently being used to study the biochemical mechanism of prion formation.

Keywords Amplification; prion, PrP^{Sc}; purification; scrapie.

1. Introduction

Mammalian prions are the infectious agents of transmissible neurodegenerative diseases affecting humans and other animals such as Creutzfeldt–Jakob disease, bovine spongiform encephalopathy, chronic wasting disease, and scrapie *(9)*. Unlike conventional infectious agents, prions lack informational nucleic acids, and therefore their mechanism of propagation has aroused great interest *(10)*. The observation that transformation of a normal glycoprotein (PrP^C) to a misfolded isoform (PrP^{Sc}) accompanies disease progression led Prusiner to propose that self-induced protein conformational change is the mechanism of prion propagation *(11)*.

Several laboratories have used a biochemical approach to study the mechanism of PrP^{Sc} formation. Caughey and colleagues first showed that purified PrP^C molecules could be converted into PrP^{Sc} molecules in a cell-free system *(1)*. Furthermore,

From: *Prion Protein Protocols.*
Methods in Molecular Biology, Vol. 459.
Edited by: A. F. Hill © Humana Press, Totowa, NJ

the specificity of the in vitro conversion process induced by template PrP^{Sc} molecules recapitulates the species and strain specificity of prion transmission in vivo *(1–3)*. Soto and colleagues showed that PrP^{Sc} molecules and prion infectivity could be amplified more efficiently by subjecting scrapie-infected and normal brain homogenates to the protein misfolding cyclic amplification (PMCA) technique *(5)*, which involves cycles of sonication and incubation and generates amplification of PrP^{res} molecules in a manner analogous to the polymerase chain reaction for DNA molecules *(4)*. Our laboratory has used biochemical purification and reconstitution techniques to show that (1) accessory polyanionic molecules facilitate efficient amplification of PrP^{Sc} in vitro *(8)*, and (2) copper ions potently inhibit PrP^{Sc} formation *(12)*. The ability to generate PrP^{Sc} molecules from purified and synthetic substrates provides a unique opportunity to study the composition and structure of prions.

2. Materials

2.1. Making the Immunoaffinity Column

1. 1 M RNase-free Tris-HCl, pH 8.0 (Ambion, Austin, TX).
2. RNase-free 5 M NaCl (Ambion).
3. RNase-free 0.5 M EDTA, pH 8.0 (Ambion).
4. Phosphate-buffered saline without calcium or magnesium (PBS) (Mediatech, Herndon, VA).
5. 200 mM triethanolamine, pH 8.0 (Acros, Geel, Belgium).
6. 1 M 3-(*N*-morpholino)-propanesulfonic acid (MOPS), pH 8.0 (Sigma-Aldrich, St. Louis, MO).
7. 0.2 M glycine, pH 2.5 (Fisher Scientific, Pittsburgh, PA).
8. PBS, 1% Triton X-100.
9. PBS, 1% Triton X-100, 20 mM Tris-HCl, pH 8.0.
10. 20 mM MOPS, pH 8.0, 0.25 M NaCl, 5 mM EDTA, 1% Triton X-100.
11. 2% sodium azide (Sigma-Aldrich).
12. Protein A agarose 50% slurry (6–7 mg/ml resin binding capacity for mouse IgG) (Pierce Chemical, Rockford, IL, distributed by Fisher Scientific, cat. no. PI20333).
13. Dimethyl pimelimidate × 2HCl (DMP) (Pierce Chemical, distributed by Fisher Scientific, cat. no. PI21666).
14. 3F4 monoclonal antibody (1 mg/ml, Signet, Dedham, MA, cat. no. 9620).
15. 1.5-cm-i.d. Econo-Pac column (Bio-Rad, Hercules, CA, cat. no. 7321010), with upper bed support.
16. 1.5-cm-i.d. Econo-Pac column adapter (Bio-Rad cat. no. 7380019).
17. End-over-end rotator.
18. Low-speed centrifuge (Centra CL3R, with rotor 243, Thermo Electron Corporation, Waltham, MA).

2.2. *Immunopurification of PrP^C*

1. 3F4 immunoaffinity column (1-ml packed volume) fitted with column adapter (prepared as described in **Subheading 3.1**).
2. Protein A agarose (Fisher Scientific cat. no. PI20333) packed in a 1.5-cm i.d. Econo-Pac column (Bio-Rad cat. no. 7321010) (1-ml packed volume), with upper bed support (included).

   a. Use forceps and the top end of a 5-ml serological pipette to insert the upper bed support flush against the top of the resin.
   b. Be careful not to crush the resin, which would impede flow.

3. SP-Sepharose (Sigma-Aldrich cat. no. S1799) ion exchange resin (1.5-ml ml packed volume) in a 1.5-cm-i.d. Econo-Pac column (Bio-Rad catalog no. 7321010), with upper bed support.
4. Zeba buffer-exchange columns (4 × 10 ml) (Pierce Chemical, distributed by Fisher Scientific, cat. no. NC9286535).
5. 500-ml Stericup-GP filtration system (0.22-μm pore diameter), polyethersulfone, radiosterilized (Millipore Corporation, Billerica, MA, cat. no. SCGPU05RE).
6. Low-speed centrifuge (Centra CL3R, with rotor 243, Thermo Electron Corporation).
7. Ultracentrifuge (Avanti J30I with rotor JA-30.50Ti, Beckman Coulter, Fullerton, CA), with appropriate tubes.
8. Biohomogenizer Mixer (Biospec Products, Bartlesville, OK, distributed by Fisher Scientific, cat. no. 15-338-51).
9. 50-ml Dounce homogenizer (Kontes Glass, Vineland, NJ).
10. Peristaltic pump and 1.6-mm-i.d. tubing.
11. Column stopcocks.
12. Molecular biology grade water (Ambion) to make solutions.
13. PBS.
14. 0.5 M MOPS stock solutions, pH 7.0, 7.5, and 8.0.
15. 0.5 M 2-morpholinoethanesulfonic acid (MES) (Sigma-Aldrich cat. no. M5287) stock solution, pH 6.4.
16. 10% Triton X-100, 10% sodium deoxycholate (DOC) (Sigma-Aldrich cat. no. D6750).
17. Immunoaffinity equilibration buffer: PBS 1% Triton X-100, 1% DOC.
18. Immunoaffinity wash buffer 1: 20 mM MOPS, pH 8.0, 0.5 M NaCl, 5 mM EDTA, 0.5% Triton X-100.
19. Immunoaffinity wash buffer 2: PBS, 0.5% Triton X-100.
20. 0.2 M glycine, pH 2.5.
21. 1 M Tris-HCl, pH 9.0.
22. SP equilibration buffer: 20 mM MES, pH 6.4, 0.15 M NaCl, 0.5% Triton X-100.
23. SP wash buffer: 20 mM MOPS, pH 7.0, 0.25 M NaCl, 0.5% Triton X-100.
24. SP elution buffer: 20 mM MOPS, pH 7.0, 0.50 M NaCl, 0.5% Triton X-100.
25. Exchange buffer: 20 mM MOPS, pH 7.5, 0.15 M NaCl, 0.5% Triton X-100.

26. PBS, 0.02% sodium azide.
27. Complete™ protease inhibitor cocktail tablets (containing EDTA) (Roche Diagnostics, Indianapolis, IN, cat. no. 1169498001).
28. Complete protease inhibitor cocktail tablets (without EDTA) (Roche Diagnostics catalog no. 11873580001).
29. Ice in buckets.
30. Six Syrian hamsters ~8–12 weeks of age, either sex.
31. Dissecting tools to remove brains from skulls (Fisher Scientific: spatulas, cat. no. 14-375-20; forceps, cat. no. 13812-36; scissors, cat. no. 13-808-4).

2.3. Preparation of PrP27-30 Template

1. 10% (w/v) Sc237-infected hamster brain homogenate in PBS.
   a. The hamster should be in the terminal stages of disease, showing clinical signs of scrapie.
2. PBS, 1% Triton X-100.
3. 2 mg/ml proteinase K diluted from stock solution into water (~48 U/mg) (Roche Diagnostics catalog no. 03115828001).
4. 0.3 M (0.0522 g/ml) phenylmethylsulfonyl fluoride (PMSF) (Roche Diagnostics catalog no. 11359061) in methanol (made immediately before use to avoid degradation).
5. 3000MPD sonicator with deep microplate horn (Misonix, Farmingdale, NY) with custom acrylic tube holder designed to keep a 1.5 ml microfuge tube ~ 1 cm from the center of the horn and ~3 mm off the surface of the horn.
6. Micro-ultracentrifuge (Discovery M120SE with rotor S45A, Sorvall, Hamburg, Germany) and 1.5 ml Safe-Lock® tubes (Eppendorf, Westbury, NY, cat. no. 2236320-4).
7. Eppendorf Thermomixer (Brinkmann, Westbury, NY, cat. no. 05-400-200).
8. Cold water.
9. Ice in buckets.

2.4. In Vitro *PrP^{Sc} Amplification*

1. RNase-free TE buffer, pH 8.0 (Ambion).
2. Exchange buffer: 20 mM MOPS, pH 7.5, 0.15 M NaCl, 0.5% Triton X-100.
3. PBS, 1% Triton X-100.
4. Sc237 PrP27-30 diluted 1:10 in PBS, 1% Triton X-100.
5. Purified hamster PrP^C preparation (*see* **Subheading 3.2**).
6. Reaction buffer: 20 mM MOPS, pH 7.0, 0.5 M imidazole (Sigma-Aldrich cat. no. I-5513), 0.15 M NaCl, 50 mM EDTA. Use RNase-free stock solutions and water purchased from Ambion. Imidazole is a base, and it should be neutralized

before being added to the reaction buffer; we suggest preparing a stock solution of 1 M imidazole and titrating the pH to 7.0 by using concentrated HCl.

7. Poly(A) RNA (polyadenylic acid potassium salt) (Sigma-Aldrich cat. no. P9403) resuspended in RNase-free water at a concentration of 1–5 mg/ml (confirmed by A_{260} measurements) and stored as aliquots in liquid nitrogen.

8. Eppendorf Thermomixer.

9. Microcentrifuge (Eppendorf model 5417-C).

10. Proteinase K.

11. SDS sample buffer, 2× stock solution: 100 mM Tris-HCl, pH 6.8, 4% sodium dodecyl sulfate (SDS), 10% glycerol, 0.01% bromophenol blue, 10% β-mercaptoethanol.

12. Tube heating block set to 95°C.

13. Vortexer.

2.5. SDS-Polyacrylamide Gel Electrophoresis (PAGE) and Western Transfer

1. Stacking gel mix: 0.6 M Tris-HCl, pH 8.8, 0.16% SDS.

2. Resolving gel mix: 0.15 M Tris-HCl pH 6.8, 0.12% SDS.

3. Acrylamide solution: 40% Acrylamide/Bis solution 29:1 (Bio-Rad cat. no. 161-0147).

4. N,N,N[prime],N[prime]-tetramethylethylenediamine (TEMED) (Bio-Rad cat. no. 161-0801).

5. 10% ammonium persulfate in water.

6. Water-saturated butanol (butanol is the top phase).

7. 10× Laemmli buffer (13): 60 g of Tris base, 288 g of glycine, 200 ml of 10% SDS.

   a. Make up to 2 liter total volume with deionized water.

   b. Final pH should be 8.8

8. Anode buffer I: 300 mM Tris-HCl, pH 10.4, 10% (v/v) methanol.

9. Anode buffer II: 25 mM Tris-HCl, pH 10, 10% (v/v) methanol.

10. Cathode buffer: 25 mM Tris-HCl, 40 mM glycine, pH 9.4, 10% (v/v) methanol.

11. Methanol.

12. Polyvinylidine fluoride (PVDF) membrane (Immobilon-P, Millipore).

13. Filter paper: 3 mm, 15 × 17 cm (Whatman, Florham Park, NJ).

14. Electrophoresis apparatus: Hoefer SE 600 Ruby Vertical Standard (GE Healthcare, Piscataway, NJ), with 18- × 16-cm glass plates and 1.5-mm spacers and 15-well comb.

15. Semi-dry transfer apparatus: Trans-Blot SD Semi-Dry Transfer Cell (Bio-Rad).

16. Power supply for electrophoresis and transfer apparatuses.

17. Prestained molecular weight markers: Precision Plus™ All Blue Protein Standard (Bio-Rad cat. no. 161-0373) and PageRuler™ Prestained Protein Ladder (MBI Fermentas, Hanover, MD, cat. no. SM0671).

18. Pipette loading tips (USA Scientific, Ocala, FL, cat. no. 1022-0600).
19. Ruler.
20. Scissors.

2.6. Protein Slot Blot

1. Ethanol.
2. Methanol.
3. Immobilon-P PVDF membrane.
4. Filter paper: 3 mm, 15 × 17 cm (Whatman).
5. Minifold-2 slot blot apparatus (Whatman Schleicher and Schuell, Keene, NH, cat. no. 10447800).
6. House vacuum attached to trap and regulated by adjustable valve.
7. Scissors.
8. Membrane-marking pen (Whatman Schleicher and Schuell cat. no. 10499001).

2.7. Blot Development and Analysis

1. 3 M guanidine thiocyanate, 20 mM Tris-HCl, pH 8.0.
2. 1× Tris-buffered saline/Tween-20 (TBST): 10 mM Tris-HCl, pH 7.2, 0.15 M NaCl, 0.1% Tween-20 (made from 10× TBST stock solution).
3. Nonfat milk (Hood, Chelsea, MA) in 1× TBST (made by mixing 9 volumes milk with 1 volume 10× TBST stock solution).
4. 3F4 monoclonal primary antibody (Signet).
5. Horseradish peroxidase sheep anti-mouse IgG secondary antibody conjugate (GE Healthcare cat. no. NA931).
6. Chemiluminescence reagent: Supersignal® West Pico Chemiluminescence Substrate (Pierce Chemical).
7. Saran wrap cling film.
8. Fujifilm LAS-3000 photodocumentation system attached to computer with Image Reader version 2.0 and Image Gauge version 4.22 software (Fujifilm USA, Valhalla, NY) installed.

3. Methods

3.1. Making the Immunoaffinity Column

1. Mix 2 ml of protein A slurry with 8 ml (8 mg) of 3F4 antibody in a capped 15-ml polypropylene tube.

2. Rotate the tube in an end-over-end rotator at ~6 rpm for 30–60 min at room temperature.
3. Centrifuge tube at 1,800 g for 2 min.
4. Remove supernatant and save. Optional: the supernatant should be depleted of 3F4 antibody, and this can be confirmed using an anti-mouse IgG antibody as a probe.
5. Wash the resin 2 × 12 ml 20 mM MOPS, pH 8.0, 0.25 M NaCl, 5 mM EDTA, 1% Triton X-100. For this and all other resin washes described in this protocol, centrifuge at 1,800 g for 2 min to sediment the resin after each wash.
6. Wash the resin 2 × 12 ml 200 mM triethanolamine, pH 8.0.
7. Add 5 ml of 10 mM DMP in 200 mM triethanolamine, pH 8.0 (this solution should be made fresh, and unused DMP should be stored in a desiccated container in the dark at 4°C).
8. Rotate the tube in an end-over-end rotator at ~6 rpm for 2 h at room temperature.
9. Quench crosslinking reaction by adding 50 µl of 1 M Tris, pH 8.0.
10. Centrifuge tube at 1,800 g for 2 min and discard supernatant.
11. Wash the resin with 10 ml of PBS, 1% Triton X-100, 20 mM Tris, pH 8.0.
12. Wash the resin 2 × 10 ml with PBS.
13. Wash the resin with 10 ml of 0.2 M glycine, pH 2.5.
14. Wash the resin with 10 ml of PBS, 1% Triton X-100.
15. Wash the resin 2 × 10 ml PBS.
16. Resuspend the resin in 25 ml of deionized water and pack into the empty Econo-Pac column.
17. Attach the column adapter to the packed column, following the manufacturer's instructions.
18. Store the column in PBS, 0.02% sodium azide at 4°C in between uses.

3.2. Immunopurification of PrP^C

1. Prechill all buffers, tubes, equipment, and rotors to 4°C. All steps of this protocol, except for brain dissection, should be performed at 4°C.
2. Attach stopcocks to all three columns and a peristaltic pump to the immunoaffinity column.
3. Clean and equilibrate the protein A agarose preclearing column by sequentially passing through the following solutions: (1) 2 ml of 0.2 M glycine, pH 2.5; (2) 10 ml of PBS; (3) 10 ml of immunoaffinity equilibration buffer. Close stopcock when the fluid level is ~0.5 ml above the upper bed support (**Notes 1–2**).
4. Prime the peristaltic pump line attached to the 3F4 immunoaffinity column. Clean and equilibrate the immunoaffinity column by sequentially pumping through the following solutions at a flow rate of 0.75 ml/min: (1) 2 ml of 0.2 M glycine, pH 2.5; (2) 10 ml of PBS; (3) 10 ml of immunoaffinity equilibration buffer. Close stopcock when the fluid level is ~0.5 ml above the upper bed support.
5. Clean and equilibrate the SP-Sepharose ion-exchange column by sequentially passing through the following solutions: (1) 10 ml of 2 M NaCl; (2) 25 ml of

water; (3) 10 ml of SP equilibration buffer. Close stopcock when the fluid level is ~0.5 ml above the upper bed support.

6. Dissolve two Complete protease inhibitor (+EDTA) cocktail tablets separately into 2 × 50 ml volumes of ice-cold PBS.

7. Euthanize six hamsters by using carbon dioxide gas.

8. Rapidly collect brains by using dissecting tools. Remove and discard dura mater and quickly rinse blood off brains. The total time spent to collect tissue should not exceed 5 min.

9. Homogenize brains for 30–60 s in 40 ml total volume of PBS plus protease inhibitors (+EDTA) by using a Biohomogenizer Mixer set at 7,000 rpm. Keep the sample tube on ice during homogenization.

10. Centrifuge the sample at 3,200 rpm for 20 min at 4°C and discard the supernatant.

11. To the pellet, add 4 ml of 10% Triton X-100, 10% DOC, plus PBS with plus protease inhibitors (+EDTA) to make a total volume of 40 ml. Resuspend the pellet using an ice-chilled Dounce homogenizer with pestle B (10 strokes).

12. Transfer the sample to an ultracentrifuge tube and incubate on ice for 30 min to solubilize membrane proteins.

13. Centrifuge at 100,000 g for 30 min at 4°C.

14. Filter the solubilized supernatant through 0.2-μm Stericup to remove particulate matter. This step helps to prevent clogging of the chromatographic columns.

15. Pour the filtered sample over the pre-equilibrated, free gravity flow protein A agarose column, and collect the flow-through fraction.

16. Apply the protein A flow-through fraction to the immunoaffinity column at a flow rate of 0.75 ml/min.

17. Meanwhile, pre-equilibrate two 10-ml Zeba buffer exchange columns according to manufacturer's instructions with 3 × 5 ml SP equilibration buffer.

18. Wash the immunoaffinty column sequentially with 15 ml of immunoaffinity wash buffer 1 and 10 ml of immunoaffinity wash buffer 2 at the same flow rate.

19. Disconnect the immunoaffinity column from the peristaltic pump and manually elute column using a 10-cc syringe filled with 0.2 M glycine, pH 2.5. Elute at a flow rate of approximately 2–3 ml/min into the collection tube containing 0.8 ml of 1 M Tris, pH 9.0, until the total volume reaches 8 ml (**Note 3**).

20. Mix contents of collection tube and apply 4 ml of sample to each pre-equilibrated Zeba buffer exchange column. Centrifuge according to manufacturer's instructions, collect the buffer-exchanged samples, and recombine.

21. Apply sample to pre-equilibrated, gravity flow SP-Sepharose column, by using the stopcock to adjust the flow rate to ~0.75 ml/min.

22. Meanwhile, pre-equilibrate two 10-ml Zeba buffer exchange columns according to manufacturer's instructions with 4 × 5 ml exchange buffer plus Complete protease inhibitors (without EDTA).

23. Wash the SP-Sepharose column with 15 ml of SP wash buffer.

24. Elute the SP-Sepharose column with 8 ml of SP elution buffer (**Note 4**).

25. Apply 4 ml of eluate to each pre-equilibrated Zeba buffer exchange column. Centrifuge according to manufacturer's instructions, collect the buffer-exchanged samples, recombine, and freeze 0.5-ml aliquots at −70°C (**Fig. 1**) (**Notes 5–7**).

Fig. 1 Silver stain analysis of purified PrP^C substrate. Twelve percent SDS-PAGE showing (from left to right): (1) Page-ruler prestained protein ladder (MBI Fermentas cat. no. SM0671) (estimated molecular weights 170, 130, 100, 72, 55, 40, 33, 24, and 17 kDa); (2) eluate from 3F4 immunoaffinity column; (3) flow-through fraction of SP-Sepharose column; and (4–7) successive 2-ml eluate samples from SP-Sepharose column. All chromatographic fractions (lanes 2–7) were concentrated prior to electrophoresis by the method of Wessel and Flugge *(14)*

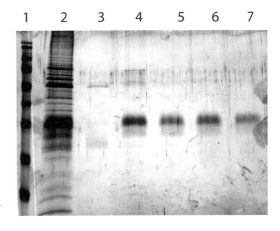

3.3. Preparation of PrP27-30 Template

1. Mix 100 µl of 10% scrapie brain homogenate, 895 µl of PBS, 1% Triton X-100, and 5 µl of 2 mg/ml proteinase K in a 1.5-ml micro-ultracentrifuge tube. Incubate the sample at 20°C in a Thermomixer, shaking at 300 rpm for 30 min (**Note 8**).
2. To stop the reaction, add 17 µl of 0.3 M PMSF to the 1-ml sample (final PMSF concentration 5 mM) and vortex.
3. Centrifuge at 100,000 g for 1 h.
4. Carefully remove and discard the supernatant.
5. Add 200 µl of ice-cold PBS, 1% Triton X-100 to the pellet.
6. Place sample in holder and sonicate for 2 × 30 s (with 1-min interval between bursts) by using Misonix 3000MPD with output setting 6.0 and filled with 350 ml of cold water.
7. Add 800 µl of ice-cold PBS, 1% Triton X-100 to sonicated sample and vortex.
8. Centrifuge at 100,000 g for 30 min.
9. Repeat **steps 4–8** two more times to wash the pellet thoroughly.
10. Vortex final sample and save 200-µl aliquots at −70°C (**Note 9**).

3.4. In Vitro *PrP^{Sc} Amplification*

1. Make a working solution of 60 µg/ml poly(A) RNA in RNase-free TE buffer.
2. Prepare a 1:2 dilution (equal volume) of PrP^C substrate in exchange buffer.
3. Prepare a cocktail containing 50 µl of diluted PrP^C substrate, 10-µl reaction buffer, and 30 µl of 60 µg/ml poly(A) RNA in TE, and 10 µl of PrP27-30 template. The final concentration of PrP^C in the cocktail is typically between 250 and 500 ng/ml.

Fig. 2 Western blot showing PrP^{Sc} amplification. Samples (from left to right): (1) reaction not subjected to proteinase K digestion, showing input PrP^{C} level; (2) time zero (frozen) reaction showing input PrP^{Sc} level; (3) amplification reaction without RNA; (4) amplification reaction with poly(A) RNA. Samples 2–4 were subjected to proteinase K digestion

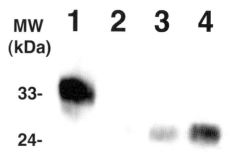

4. Incubate components in Thermomixer for 16 h at 37°C at 800 rpm. Freeze one sample at −70°C as a time zero control reaction to show the input PrP^{Sc} level.
5. After incubation, briefly centrifuge samples to consolidate the condensed liquid.
6. Add proteinase K to each sample at a final concentration of 50 μg/ml and vortex. For comparison, it is useful to prepare an extra sample that is not subjected to protease digestion. This sample shows the input PrP^{C} level.
7. Incubate in Thermomixer for 1 h at 37°C at 800 rpm.
8. After incubation, add equal volume (100 μl) of SDS loading buffer.
9. Using a preheated, metal tube block, immediately heat samples at 95°C for 10 min.
10. Briefly centrifuge samples to consolidate the condensed liquid and vortex (**Fig. 2**).

3.5. SDS-PAGE and Western Transfer

1. Assemble gel cassette by using 18- × 16-cm glass plates and 1.5-mm spacers. To avoid leaks, be sure that glass plates and spacers are completely flush at both the top and bottom of the cassette, to ensure that the bottom of the cassette does not leak.
2. Prepare 12% gel by mixing 18.35 ml of resolving gel mix, 9 ml of acrylamide solution, 2.5 ml of ammonium persulfate solution, and 30 μl of TEMED.
3. Quickly pipette the mixture into the gel cassette, leaving space for a stacking gel.
4. Carefully overlay with water-saturated butanol by using a plastic sample pipette to a level ~2 mm above the top of the gel mixture. Avoid disrupting the top of the gel mixture when applying the butanol.
5. After the gel polymerizes, turn the cassette upside down over paper towels to drain the butanol. Check to insure that the top of the gel is level.
6. Place a 15-well comb into the top of the cassette.

7. Prepare the stacking gel by mixing 8.2 ml of stacking gel mix, 1 ml of acrylamide solution, 1 ml of 10% ammonium persulfate, and 13 μl of TEMED.

8. Quickly pipette the mixture into the cassette, leaving ~5 mm space at the top of the cassette.

9. Prepare the running buffer by diluting 100 ml of the 10× Laemmli buffer with 900 ml of water in a measuring cylinder. Cover with parafilm and invert to mix.

10. Once the stacking gel has set, carefully remove the comb and fill the wells with running buffer.

11. Load 70 μl of each sample per well. Load 7 μl of prestained molecular weight marker.

12. Attach the upper chamber to the gel cassette and add running buffer to the upper and lower electrophoresis chambers.

13. Connect the apparatus to a power supply. The gel can be run at 180 V until the dye front reaches the bottom of the gel.

14. The samples that have been separated by SDS-PAGE are transferred to a charged, pre-equilibrated PVDF membrane by using semi-dry electrophoretic transfer as described in **steps 15–23**.

15. The gel unit is disconnected from the power supply and disassembled.

16. Using a ruler, gel spacer, or glass plate, remove and discard the stacking gel.

17. Use a ruler to measure the height and length of the resolving gel in centimeters. Soak the resolving gel in cathode buffer for 15 min.

18. Charge an appropriately sized PVDF membrane in methanol for 2 s, taking care not to wet the membrane before charging.

19. Rinse membrane in deionized water ~1–2 min.

20. Soak the PVDF membrane in anode buffer II for 15 min.

21. Carefully assemble the transfer sandwich as follows, by using a serological pipette to ensure that no air bubbles are trapped in the sandwich: (1) anode electrode plate, (2) two sheets of filter paper wetted in anode buffer I, (3) one sheet of filter paper wetted in anode buffer II, (4) equilibrated PVDF membrane, (5) resolving gel, (6) three sheets of filter paper wetted in cathode buffer, and (7) cathode electrode plat.

22. Finish assembling the trans-blot apparatus, connect to a power supply, and run on constant current at 2 mA/cm² gel surface area for 30 min.

23. After electrophoretic transfer, the low and average molecular weight prestained markers should transfer completely. A small amount (up to 20%) of high molecular weight prestained markers may be retained in the gel.

3.6. Protein Slot Blot

1. Mix 50 μl of each sample (previously boiled in SDS loading buffer) with 100 μl of 100% ethanol.

2. Charge a PVDF membrane in methanol for 2 s.

3. Equilibrate the PVDF membrane and two sheets of filter paper in water for ≥5 min.
4. Sandwich the water-equilibrated PVDF membrane backed by the two sheets of wet filter paper in the slot blot apparatus, following manufacturer's instructions. To insure a good seal, make sure that the membrane and filter paper do not overhang the slot blot gasket.
5. Apply samples to wells. Do not use corner wells.
6. Attach house vacuum gently (ideally, samples should take >1 min to pass through the membrane.
7. Continue to run the vacuum for 30 sec after the last sample passes through the membrane.
8. Disassemble the slot blot apparatus to recover the PVDF membrane. Use membrane marking pen to designate sample orientation. Proceed immediately to the blot development steps. Do not allow the membrane to dry out.

3.7. Blot Development and Analysis

1. Place the PVDF membrane (processed either by Western or slot blotting) in 3 M guanidine thiocyanate, 20 mM Tris, pH 8.0, at room temperature for 30 min to denature PrP^{Sc} molecules completely.
2. Quickly rinse membrane three times with TBST.
3. Block for 1 h with buffered nonfat milk either at room temperature or 4°C.
4. Wash membrane 3 × 10 min in TBST either at room temperature or 4°C.
5. Incubate membrane overnight with 3F4 primary antibody diluted 1:5000 in TBST (final concentration 0.2 μg/ml) at 4°C.
6. Wash membrane 3 × 10 min in TBST at 4°C.
7. Incubate membrane 1 h with HRP-labeled sheep anti-mouse IgG secondary antibody diluted 1:5000 in TBST (final concentration 0.1–0.2 μg/ml) at 4°C.
8. Wash membrane 3 × 10 min in TBST at 4°C.
9. Apply chemiluminescence substrate to membrane, according to manufacturer's instructions. Make sure that the entire membrane is covered with substrate.
10. Seal membrane in Saran wrap to prevent evaporation.
11. Immediately capture chemiluminescent signal with LAS-3000 photodocumentation system according to the owner's manual instructions. Use the following settings on Image Reader version 2.0 software: standard sensitivity, filter = 1: through, iris = F0.8S, and incremental exposures at 30-s intervals.
12. Save the longest exposure in which the signals are not saturated for quantitative analysis.
13. If desired, quantitate signals by using Image Gague version 4.22 software according to the owner's manual instructions.

Acknowledgements This work was supported by the National Institutes of Health grants R21 AI-058979 and R01 NS-046478 and the Burroughs Wellcome Fund.

4. Notes

1. During the purification of PrP^C, clean each column as soon as possible after its use.
2. To clean the protein A agarose column, wash sequentially with (1) 10 ml water to remove DOC, which precipitates under acidic conditions; (2) 8 ml of 0.2 M glycine, pH 2.5; (3) 15 ml of water; (4) 15 ml of PBS, 0.02% sodium azide. Store in PBS, 0.02% sodium azide.
3. To clean the 3F4 immunoaffinity column, wash sequentially with (1) 15 ml of water; (2) 15 ml of PBS, 0.02% sodium azide. Store in PBS, 0.02% sodium azide.
4. To clean the SP-Sepharose column, wash sequentially with (1) 8 ml of 2 M NaCl; (2) 10 ml of water; (3) 20 ml of 0.02% sodium azide. Store in 0.02% sodium azide.
5. Store all columns at 4°C and do not allow resins to dry out.
6. The expected yield of PrP^C from six hamster brains is ~16 µg PrP^C, and the expected concentration is ~2 µg/ml. If the concentration of PrP^C is higher or lower than expected, adjust the dilution of the substrate to be used in PrP^{Sc} amplification reactions accordingly.
7. The PrP^C purification protocol described here has only been used to purify hamster PrP^C, which is recognized by 3F4 antibody. Purification protocols for PrP^C molecules from other animal species would need to be developed empirically.
8. Handling of samples containing infectious prions should be carried out within a biosafety cabinet by using biosecurity level 2 preacautions. Infectious prion waste should be incinerated or inactivated by contact with 50% bleach for 1 h.
9. The concentration of PrP27-30 molecules varies from preparation to preparation. For each preparation, we recommend testing a range of PrP27-30 dilutions in PBS, 1% Triton X-100 (up to 1:8) to determine empirically the optimal dilution for PrP^{Sc} amplification.

References

1. Bessen, R. A., Kocisko, D. A., Raymond, G. J., Nandan, S., Lansbury, P. T., and Caughey, B. (1995) Non-genetic propagation of strain-specific properties of scrapie prion protein. *Nature* 375, 698–700.
2. Kocisko, D. A., Priola, S. A., Raymond, G. J., Chesebro, B., Lansbury, P. T., Jr., and Caughey, B. (1995) Species specificity in the cell-free conversion of prion protein to protease-resistant forms: a model for the scrapie species barrier. *Proc Natl Acad Sci U S A* 92, 3923–3927.
3. Kocisko, D. A., Come, J. H., Priola, S. A., Chesebro, B., Raymond, G. J., Lansbury, P. T., and Caughey, B. (1994) Cell-free formation of protease-resistant prion protein. *Nature* 370, 471–474.
4. Saborio, G. P., Permanne, B., and Soto, C. (2001) Sensitive detection of pathological prion protein by cyclic amplification of protein misfolding. *Nature* 411, 810–813.
5. Castilla, J., Saa, P., Hetz, C., and Soto, C. (2005) In vitro generation of infectious scrapie prions. *Cell* 121, 195–206.

6. Soto, C., Anderes, L., Suardi, S., Cardone, F., Castilla, J., Frossard, M. J., Peano, S., Saa, P., Limido, L., Carbonatto, M., Ironside, J., Torres, J. M., Pocchiari, M., and Tagliavini, F. (2005) Pre-symptomatic detection of prions by cyclic amplification of protein misfolding. *FEBS Lett* 579, 638–642.

7. Castilla, J., Saa, P., and Soto, C. (2005) Detection of prions in blood. *Nat Med* 11, 982–985.

8. Deleault, N. R., Geoghegan, J. C., Nishina, K., Kascsak, R., Williamson, R. A., and Supattapone, S. (2005) Protease-resistant prion protein amplification reconstituted with partially purified substrates and synthetic polyanions. *J Biol Chem* 280, 26873–26879.

9. Prusiner, S. B. (2000) Prion biology and diseases, Cold Spring Harbor Laboratory Press, Cold Spring Harbor, NY.

10. Prusiner, S. B. (1982) Novel proteinaceous infectious particles cause scrapie. *Science* 216, 136–144.

11. Prusiner, S. B. (1998) Prions. *Proc Natl Acad Sci U S A* 95, 13363–13383.

12. Orem, N. R., Geoghegan, J. C., Deleault, N. R., Kascsak, R., and Supattapone, S. (2006) Copper (II) ions potently inhibit purified PrPres amplification. *J Neurochem* 96, 1409–1415.

13. Laemmli, U. K. (1970) Cleavage of structural proteins during the assembly of the head of bacteriophage T4. *Nature* 227, 680–685.

14. Wessel, D., and Flugge, U. I. (1984) A method for the quantitative recovery of protein in dilute solution in the presence of detergents and lipids. *Anal Biochem* 138, 141–143.

Chapter 10
Expression and Purification of Full-Length Recombinant PrP of High Purity

Natallia Makarava and Ilia V. Baskakov

Summary Certain applications in the prion field require recombinant prion protein (PrP) of high purity and quality. Here, we report an experimental procedure for expression and purification of full-length mammalian prion protein. This protocol has been proved to yield PrP of extremely high purity that lacks PrP adducts, which are normally generated as a result of spontaneous oxidation or degradation.

Keywords Inclusion body; prion diseases; protein purification; recombinant prion protein.

1. Introduction

Recombinant prion protein (PrP) expressed in *Escherichia coli* has been used extensively in prion research for various applications. These applications include modeling of prion conversion in vitro, using PrP as an immunogen for generating anti-PrP antibody; developing antiprion therapeutic strategies that involve active immunization using PrP refolded in β-sheet–rich conformations; screening of anti-prion drugs by using in vitro conversion assays, and others. These applications require PrP of very high purity with minimal amounts of chemical modifications or degradation. Although several methods for purification and refolding of recombinant PrP have been described previously by different groups *(1–6)*, some of the previously developed protocols required a fusion of PrP to a histidine tags or they produced PrP of insufficient purity or PrP that was partially degraded. Inconsistent results in converting of PrP into β-sheet–rich conformations described in the past are attributed, at least in part, to differences in experimental protocols for expression and purification of PrP used by different laboratories. Here, we describe a reliable experimental protocol for expression of tag-free full-length recombinant PrP of high purity and with minimal amount of chemical modifications or degradation. This protocol yields 10 mg of PrP per liter of bacterial culture.

From: *Prion Protein Protocols.*
Methods in Molecular Biology, Vol. 459.
Edited by: A. F. Hill © Humana Press, Totowa, NJ

2. Materials

Unless otherwise noticed, all reagents are from Sigma-Aldrich (St. Louis, MO). All solutions are prepared with deionized water purified using Synergy 185 UV Ultrapure Water System (Millipore Corporation, Billerica, MA). Water and solutions for desalting and HPLC are degassed under vacuum. HPLC buffers are purged with helium.

2.1. Protein Expression

Whenever shaking at 37°C is required, we use an Innova 4300 incubator (New Brunswick Scientific, Edison, NJ) set at 200 rpm.

1. Plasmid DNA encoding mouse PrP 23-230 or Syrian hamster PrP 23-231 (*see* **Note 1**) in pET101/D-TOPO (Invitrogen, Carlsbad, CA).
2. Competent cells, BL21Star (DE3) One Shot *E. coli* (Invitrogen).
3. SOC medium is supplied with the competent cells (Invitrogen).
4. Luria-Bertani (LB) broth (Biosource, Rockville, MD).
5. Carbenicillin disodium salt (American Bioanalytical, Natick, MA): Dissolve at 100 mg/ml in water. Stored in aliquots at −20°C.
6. Two 2,800-ml baffled Pyrex flasks (Fisher Scientific, Hampton, NH).
7. TB medium composition for 1,200 ml (*see* **Note 2**): 14.4 g of Bacto tryptone (BD Biosciences, Sparks, MD), 28.8 g of Bacto yeast extract (BD Biosciences), 4.8 ml of glycerol (American Bioanalytical), and water to adjust 1,080 ml.

   a. TB medium needs to be autoclaved and then supplemented with 120 ml of filter-sterilized solution of 0.17 M KH_2PO_4, 0.72 M K_2HPO_4, and 100 μg/ml carbenicillin.

8. Isopropyl-β-D-thiogalactopyranoside (IPTG; American Bioanalytical): Dissolve in water at the concentration of 1 M. Store in aliquots at −20°C.

2.2. Isolation of Inclusion Bodies

1. Cell lysis buffer: 50 mM Tris-HCl, 1 mM EDTA, 100 mM NaCl, pH 8.0.
2. Phenylmethylsulfonyl fluoride (PMSF): Dissolve in acetonitrile at 9 mg/ml. Store at −20°C.
3. Lysozyme (American Bioanalytical) solution: Prepare at 10 mg/ml in lysis buffer. Store in aliquots at −20°C.
4. Deoxycholic acid (Alfa Aesar, Ward Hill, MA).
5. Deoxyribonuclease I (DNase I, type II): Dissolve in water at 2 mg/ml Store in aliquots at −20°C.

2.3. *Immobilized Metal Ion Affinity Chromatography (IMAC) and Oxidative Refolding*

1. Urea, 9 M aqueous solution. After urea is dissolved, 10 g/l mixed bed amberlite (MB-150) is added, and the solution is stirred further for at least 1 h. Before use, the solution is filtered using disposable filter units with polyethersulfone membrane (Nalge Nunc International, Rochester, NY). 9 M urea can be stored at −20°C.
2. Chelating Sepharose Fast Flow (GE Healthcare, Uppsala, Sweden).
3. Nickel sulfate ($NiSO_4$), 0.2 M solution.
4. Acidic buffer for elution of loosely bound ions from Sepharose: 0.02 M Na-acetate, 0.5 M NaCl, pH 3.0.
5. IMAC buffer A: 8 M urea, 0.1 M Na_2HPO_4, 10 mM Tris-HCl, 10 mM reduced glutathione, pH 8.0.
6. IMAC buffer B: 8 M urea, 0.1 M Na_2HPO_4, 10 mM Tris-HCl, 10 mM reduced glutathione, pH 4.5.
7. EGTA is prepared as 0.5 M solution, pH 8.0.
8. Desalting buffer: 6 M urea, 0.1 M Tris-HCl, pH 7.5.
9. 50 mM oxidized glutathione. Store in aliquots at −20°C.
10. Solutions for Sepharose regeneration and preservation: 2 M NaCl, 1 M NaOH, molecular grade ethanol.
11. XK chromatography column (GE Healthcare).
12. HiPrep 26/10 desalting column (GE Healthcare).

2.4. *HPLC*

1. HPLC buffer A: 0.1% trifluoroacetic acid (Supelco, Bellefonte, PA) in water.
2. HPLC buffer B: 0.1% trifluoroacetic acid in acetonitrile.
3. Protein C4 HPLC column: particle size 10 μm, i.d. 22 mm, length 250 mm; column guard: particle size 12 μm, cartridge 10 mm (Vydac, Hesperia, CA).
4. Disposable filter units with polyethersulfone membrane (Nalge Nunc International).

3. Methods

Researchers might face the following technical challenges during expression and purification of PrP:

1. Difficulties in achieving complete solubilization of PrP inclusion bodies.
2. Precipitation and irreversible binding of PrP to the IMAC matrix.
3. Copper-dependent self-cleavage of PrP.

4. Incomplete oxidative refolding of PrP.
5. Spontaneous formation of oxidative adducts.
6. Spontaneous methionine oxidation.

To minimize these problems and to achieve successful purification of PrP of high purity, the protocols described in **Subheadings 3.1.-3.6.** need to be followed closely.

3.1. Transformation of Bacterial Cells

Chemical transformation of BL21Star (DE3) One Shot *E. coli* is based on the protocol described in the pET Directional TOPO Expression kit instructional manual (Invitrogen).

1. Thaw one vial of cells on ice.
2. Add 1 µl (20 ng) of plasmid DNA into the vial of cells and mix by stirring gently with the pipette tip.
3. Incubate on ice for 30 min.
4. Heat shock the cells for 30 s at 42°C.
5. Immediately transfer the tube on ice.
6. Add 250 µl of room temperature SOC medium.
7. Tape the tube on its side to the bottom of incubator and shake at 37°C at 200 rpm for 30 min.
8. Add the entire transformation reaction into 50-ml centrifuge tube containing 10 ml of LB supplemented with 100 µg/ml carbenicillin.
9. Shake at 37°C at 200 rpm for 3–5 h.
10. Add the entire volume into the 500-ml flask containing 90 ml of LB supplemented with 100 µg/ml carbenicillin.
11. Shake overnight at 37°C at 200 rpm.

3.2. Induction of Protein Expression

1. Supplement 1,080 ml of autoclaved TB media with 120 ml of filter-sterilized solution of phosphates (0.17 M KH_2PO_4 and 0.72 M K_2HPO_4) and 100 µg/ml carbenicillin. Mix, and then save 1 ml of resulting mixture as an absorbance reference for cell growth monitoring.
2. Add 60 ml of overnight cell culture, mix, and divide equally between two baffled flasks (use sterile cylinder).
3. Incubate flasks with cell culture shaking at 37°C at 200 rpm until the absorbance at 600 nm reached 0.6. Dilute with fresh TB, if overgrown.
4. Induce expression by adding 1 mM IPTG.
5. Continue incubation for 4–5 h.

3.3. Cell Harvesting

1. To be able to determine cell pellet mass, weigh empty centrifuge bottles.
2. Divide bacterial culture between four 500-ml centrifuge bottles and centrifuge at 1,900 g for 10 min at 4°C.
3. Discard supernatant and calculate cell pellet mass.
4. At this point, cells can be stored overnight at −20°C.

3.4. Cell Lysis and Isolation of Inclusion Bodies

1. Thoroughly resuspend the pellet in lysis buffer (8.7 ml of buffer per each gram of bacterial pellet) by vortexing and pipetting up and down with a 25-ml pipette.
2. Freeze at −80°C and thaw the cells at least one time to ensure cell lysis. At −80°C, cells freeze in approx 10 min. Room temperature water bath is used to thaw the pellet quickly.
3. Pour cell lysate into a beaker, add 2 μl of PMSF and 20 μl of lysozyme per 1 ml of lysis buffer. Stir at room temperature for 20–40 min (see **Note 3**).
4. Add 1 mg/ml deoxycholic acid and stir for 20–30 min until the liquid becomes viscous.
5. Add DNase I to 5 μg/ml and stir for additional 30–45 min.
6. Divide the lysate between four 50-ml centrifuge tubes and centrifuge at 12,000 g for 30 min at 4°C. Decant the supernatant.
7. Thoroughly resuspend the pellet in 15 ml of lysis buffer by vortexing and pipetting up and down.
8. Repeat DNase I treatment: add DNase I to 5 μg/ml to each centrifuge tube and incubate on rotating platform for 20 min.
9. Centrifuge at 12,000 g for 30 min at 4°C. Decant the supernatant.
10. Dilute lysis buffer with water, 1:10. Thoroughly resuspend the pellet in 20 ml of diluted buffer. This step removes the excess of EDTA form the inclusion bodies to allow proper binding to Ni^{2+}-charged chromatography column.
11. Centrifuge at 12,000 g for 20 min at 4°C. Decant the supernatant. The resulting pellet contains recombinant PrP precipitated in form of inclusion bodies, which can be stored frozen at −80°C for at least 1 month.

3.5. IMAC Purification

IMAC purification is performed using a XK chromatography column (GE Healthcare), packed with 8 ml of Chelating Sepharose Fast Flow. Charging the Sepharose with Ni^{2+} ions and protein binding are performed in solution. The same Sepharose can be reused several times for purification of the same PrP variant (*see*

Note 4). Desalting column is stored at 4°C; however, it should be equilibrated to room temperature before use.

3.5.1. Preparing Sepharose

1. Transfer an aliquot of the Sepharose into a 50-ml centrifuge tube.
2. Let the Sepharose settle down by gravity and remove preservative solution.
3. To wash the Sepharose, add water to the top of the tube, cover the tube, and gently resuspend the Sepharose by inverting the tube several times. Let the Sepharose settle down (it takes approx 20 min); remove water.
4. Using the same procedure, wash with water again.
5. Remove water and charge the Sepharose by adding 2 ml of 0.2 M $NiSO_4$.
6. Wash the excess of ions with water twice.
7. To elute loosely bound ions, wash with acidic buffer, pH 3.0.
8. Wash two times with water.
9. Equilibrate the Sepharose: wash with IMAC buffer A two times. Keep the Sepharose under buffer until the protein is solubilized and ready for binding.

3.5.2. Protein Solubilization and Binding

1. Add 10 ml of IMAC buffer A to each tube of inclusion bodies (*see* **Notes 2** and **5**).
2. Thoroughly resuspend the pellet.
3. Incubate on rotating platform 1–1.5 h at room temperature.
4. Centrifuge at 12,000 g for 15 min at 4°C.
5. Remove equilibration buffer from the Ni^{2+}-charged Sepharose.
6. Add the supernatant containing solubilized recombinant PrP to the Sepharose.
7. Gently rotate the mixture of the Sepharose and the protein at room temperature, allowing 30–40 min for binding of the protein.

3.5.3. IMAC and Desalting

1. Secure empty chromatography column on a holder nearby chromatograph.
2. Close column outlet and load the mixture of Sepharose and protein solution.
3. Open the outlet and drain the excess of liquid from the column, collecting it as IMAC flow-through for the analysis of binding efficiency (**Figs. 1, 2**). Make sure not to drain the slurry completely: insert the adaptor and lock it above the Sepharose as soon as liquid front reaches the surface of the Sepharose.
4. Connect the column to the fast-performance liquid chromatography (FPLC) system (ÄKTA prime, GE Healthcare). Set flow rate to 2 ml/min and fraction size to 5 ml. Wash unbound proteins with IMAC buffer A until the UV readings from the chromatographer reach low plateau (**Fig. 1**; *see* **Notes 4** and **5**).

5. Switch to the IMAC buffer B, pH 4.5, to start elution of PrP. Fractions with protein are collected into borosilicate glass (13- × 100-mm tubes, Fisher Scientific), containing EGTA; the final concentration of EGTA in each tube should be 5 mM after fraction is collected. Typical profile of IMAC purification is shown in **Fig. 1**.

6. Combine fractions containing PrP (**Fig. 1**) in a 50-ml centrifuge tube. Typically, we collect 25 ml of protein solution and then proceed with desalting immediately.

7. To remove the Sepharose from the chromatography column, add water to the column, gently resuspend the slurry with the 25 ml pipette, and transfer the Sepharose with water to a new 50-ml tube for cleaning (*see* **Subheading 3.5.4.**).

8. Attach HiPrep 26/10 desalting column to the FPLC system, wash out storage solution, and equilibrate with desalting buffer: 6 M urea, 0.1 M Tris-HCl, pH 7.5 (*see* **Note 6**).

9. Desalting step is used to separate the protein from glutathione. Protein solution is loaded through the FPLC super loop. Because the total volume of the protein collected after IMAC exceeds column capacity (14 ml for HiPrep 26/10 desalting column), desalting is performed in two runs. First, 14 ml of the protein solution is loaded and desalted. Then, after the salt is washed out and the column is re-equilibrated once more, the rest of the protein (~11 ml) is run through the column (**Fig. 3**).

10. Wash desalting column immediately after the last run. Wash with water until the conductivity is at the baseline level. Disconnect the column form the FPLC system and reconnect it upside down. Wash with 0.2 M NaOH until the conductivity is on high plateau. Wash with water again. Finally, fill out the column with 20% ethanol, disconnect, close, and store at 4°C.

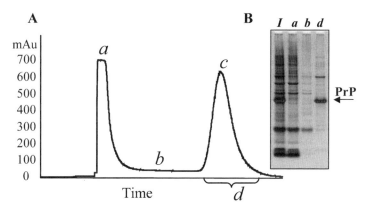

Fig. 1 Typical IMAC profile of mouse recombinant PrP purification. (**A**) IMAC profile. The peak of unbound proteins (stage *a*) typically reaches approx 700 mAu. After this peak drops to the baseline (<50 mAu, stage *b*), the IMAC buffer A (pH 8.0) is changed to the IMAC buffer B (pH 4.5). The PrP peak typically reaches approx 600 mAu (stage *c*). Fractions with UV values >50 mAu (stage *d*) are combined for subsequent purification. (**B**) Analysis of IMAC fractions in SDS-PAGE (12% bis-tris) following by silver staining. *I*, solubilized inclusion bodies; *a, b, d*, fractions collected at the stages *a, b,* and *d*, respectively

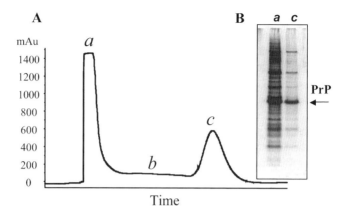

Fig. 2 Leakage of proteins from the IMAC column. (**A**) IMAC profile. Overloading of IMAC column results in higher then normal UV values of unbound proteins (>1,400 mAu, stage *a*) and subsequent baseline higher than 100 mAu (stage *b*). The height of the PrP peak remains at typical level of 600 mAu (stage *c*). (**B**) Analysis of IMAC fractions in SDS-PAGE (12% bis-tris) confirm the presence of high amount of PrP in the flow-trough (lane *a*)

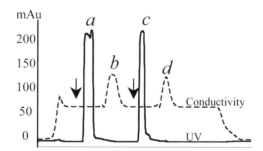

Fig. 3 Desalting profile. PrP collected after IMAC were divided into two parts and desalted using gel filtration chromatography. The arrows mark the points of loading of PrP onto the desalting column. During desalting, the protein (peaks *a* and *c*) is separated from glutathione (peaks *b* and *d*)

11. Combine the fractions containing recombinant PrP in a new 50-ml tube, mix. To estimate protein concentration (C), prepare 1:5 dilution of protein solution, measure absorbance at 280 nm, and calculate the concentration using the following equation: C (mg/ml) = A280 × 5 × 0.37 (for mouse PrP 23-230; 0.37 mg/ml = 1 o.e. at 280 nm). To minimize formation of dimers during oxidative refolding of PrP, dilute PrP solution with the desalting buffer (6 M urea, 0.1 M Tris-HCl, pH 7.5) to such extent that the concentration of PrP does not exceed 0.3 mg/ml.

12. Supplement the PrP solution with 5 mM EGTA and 0.2 mM oxidized glutathione, and gently rotate at room temperature overnight (*see* **Note 3**).

3.5.4. Cleaning Chelating Sepharose for Reusing

1. After the Sepharose is transferred from the chromatographer column to a 50-ml centrifuge tube, add water to the top of the tube, cover the tube, and gently resuspend the Sepharose by inverting the tube several times. Let the Sepharose settle down and then remove water.
2. Using the same procedure, wash with 2 M NaCl.
3. Wash with water three times.
4. Wash with 1 M NaOH.
5. Wash with water two times. When removing water after last wash, leave 2 ml above the Sepharose and add 2 ml of pure ethanol for preservation. Keep at 4°C.

3.6. HPLC

To perform C4 HPLC, we use Shimadzu (Columbia, MD) HPLC system operated with EZStart 7.3 SP1 software.

1. Prepare HPLC buffers A and B (see **Subheading 2.4.**), 1 liter each.
2. Degas buffers A and B for 15–20 min by stirring under vacuum, and then keep under constant purging of helium.
3. Before connecting the column, wash the tubing and pumps of HPLC system with running buffers A and B at 5 ml/min for 10 min.
4. Connect the C4 column, equilibrate with buffer A.
5. If visible precipitation occurs after overnight oxidation, filter protein solution by using disposable filter units with polyethersulfone membrane (Nalge Nunc International).
6. To reduce urea concentration, dilute the protein solution with HPLC buffer A (1:3 v/v) and load onto C4 column (see **Note 7**).
7. Wash unbound proteins from the C4 column with HPLC buffer A at the flow rate of 5 ml/min; monitor UV absorbance at 220 and 280 nm. When the baseline is reached, start the gradient (see **Note 8**).
8. Using HPLC profile as guidance, manually collect PrP fractions into borosilicate glass tubes (Fisher Scientific). α-PrP is eluted in major peak between 52.5 and 54.5 min (**Fig. 4**). Slow gradient separates correctly folded α-PrP from PrP adducts.
9. By the end of the run, wash the column with HPLC buffer A until the baseline of UV absorbance is reached (see **Note 9**).
10. The quality of the purified protein is checked by sodium dodecyl sulfate-polyacrylamide gel electrophoresis (SDS-PAGE) (**Fig. 4**) and by mass spectroscopy (see **Note 10**).
11. Freeze collected fractions at −80°C. Lyophilize (we use FreeZone 2.5 Plus freeze dry system from Labconco, Kansas City, MO). The protein is stored as lyophilized powder at −20°C. One purification (600 ml of bacterial culture) typically results in 6–8 mg of pure PrP.

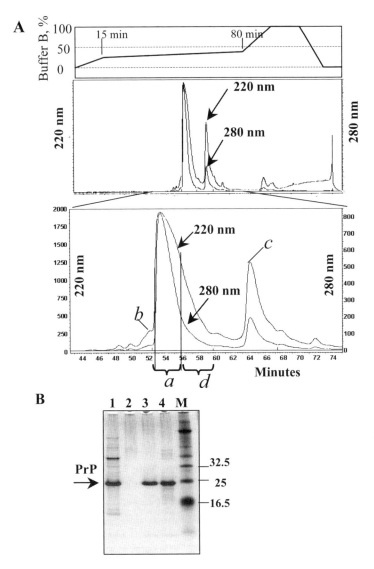

Fig. 4 Purification of PrP on C4 column. (**A**) Typical HPLC profile of elution of mouse recombinant PrP. Major peak contains pure PrP (peak *a*), it is separated from the peak containing PrP with oxidized methionines (peak *b*), from the peaks containing PrP with double glutathione adducts (peak *c*), and from other impurities. The right shoulder of the major HPLC peak (fractions *d* eluted at 56–60 min) is not collected; this shoulder may contain products of PrP degradation (see lane 4 in B). (**B**) Analysis of HPLC fractions in SDS-PAGE followed by silver staining: *lane 1*, PrP collected after IMAC and loaded onto C4 column; *lane 2*, HPLC flow-trough; *lane 3*, pure PrP collected from the major HPLC peak *a*; *lane 4*, right shoulder of the major peak (referred to as fraction *d*); and *lane M*, molecular marker. Lane 4 shows minor amounts of PrP degradation products that occurs due to self-cleavage (seen as a smear with mol. wt. <23 kDa). The extent of PrP degradation may vary from preparation to preparation (*see* **Note 3**)

4. Notes

1. Plasmids for expression of mouse PrP 23-230 or Syrian hamster PrP 23-231 were cloned from mouse of Syrian hamster cDNA, respectively, as described previously *(7, 8)*.
2. We have found convenient to grow bacteria in two flasks each containing 600 ml of TB media. This typically results in total of 12-g bacterial pellet used for isolation of inclusion bodies. To prevent overloading of HPLC column, only one half of the inclusion bodies should be used for IMAC. The other half is stored at −80°C.
3. Recombinant PrP is prone to copper-dependent self-cleavage *(9, 10)*. Performing cell lysis steps at lower temperature could slow down degradation; however, it would require prolonged incubation time. Therefore, the cell lysis steps are carried out at room temperature. EDTA and EGTA are added to the lysis buffer and during oxidative refolding, respectively, to minimize protein degradation.
4. Lack of Ni^{2+} ions on the Sepharose may cause the leakage of protein during wash step. For better performance, recharge the column with Ni^{2+} ions each time before reuse.
5. Loading the column with excessive amounts of protein (**Fig. 2**) or incomplete solubilization of inclusion bodies (**Fig. 5**) results in high UV readings during the wash with IMAC buffer A. To achieve proper solubilization of the protein, it is very important to prepare IMAC buffers A and B fresh, at the same day the IMAC purification is performed. Instead of keeping reduced glutathione in concentrated stock solution, we dissolved it directly in the IMAC buffers to the final concentration of 10 mM.

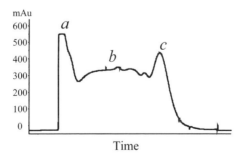

Fig. 5 IMAC profile of incompletely solubilized protein. Reduced glutathione degraded during preparation of stock solution (*see* **Note 5**). As a result, recombinant PrP was incompletely solubilized and leaked from the Sepharose during the wash step. The peak of unbound proteins (stage *a*) was <600 mAu; the UV values during the wash steps were >300 mAu (stage *b*) and the yield of eluted PrP was reduced (stage *c*)

6. Make sure to follow recommendations on pressure limit and buffer compatibility for the specific column. Our desalting column cannot tolerate urea at the concentrations above 6 M; therefore, desalting buffer contains less urea than IMAC buffers. Depending on solution viscosity, flow rate needs to be adjusted during the run to keep pressure under the limit.

7. Because after overnight oxidation PrP solution contains urea, loading the protein onto C4 column requires pressure above Shimadzu HPLC system limits. To avoid high pressure, we dilute PrP solution 1:3 with buffer A and use external HPLC pump 2010 from Varian (Walnut Creek, CA) to load the protein. We wash the pump with water and with 20 ml of buffer A before loading PrP. We load protein solution at flow rate 5 ml/min and finish by loading 20 ml of HPLC buffer A, to ensure that no protein is left in the pump. After loading is completed, we disconnect and thoroughly wash the loading pump with water.

8. In our gradient method, the percentage of HPLC buffer B (0.1% trifluoracetic acid in acetonitrile) grows from 0 to 25% during the first 15 min (at flow rate 5 ml/min). Then, a gentle gradient is applied (from 25 to 35% of buffer B in the next 65 min) to ensure efficient separation of PrP adducts (**Fig. 4**). Correctly folded, oxidized, α-helical form of PrP is eluted between 52.5 and 54.5 min. Then, the percentage of buffer B grows from 35 to 100% in 15 min; the column is washed with 100% buffer B for 15 min, and the gradient drops down to 0% within 10 min.

9. If the same C4 column is used for purification of different PrP variants, running a washing program with extended time of 100% acetonitrile will ensure the absence of cross-contamination. In our washing program, buffer B concentration grows from 0 to 100% within 20 min, remains constant 100% for 25 min, and then goes down to 0% during last 20 min. The run is finished by equilibration of the C4 column with buffer A for 25 min, which prepares the column for loading of new protein.

10. To perform mass spectroscopy, small amount of recombinant PrP (as a lyophilized powder) is dissolved in 20 μl of 1:1 water: methanol mixture containing 1% acetic acid. The solution is then analyzed on a Waters ZMD single quadrupole mass spectrometer (Waters, Milford, MA) operated in positive ion mode.

References

1. Zahn R, von Schroetter C, Wuthrich K. (1997) Human prion protein expression in *Escherichia coli* and purified by high affinity column refolding. FEBS Lett; 417:400–404.
2. Mehlhorn I, Groth D, Stöckel J, Moffat B, Reilly D, Yansura D, et al. (1996) High-level expression and characterization of a purified 142-residue polypeptide of the prion protein. Biochemistry; 35:5528–5537.
3. Rezaei H, Marc D, Choiset Y, Takahashi M, Hoa GHB, Haertle T, et al. (2000) High yield purification and physico-chemical properties of full-length recombinant allelic variants of sheep prion protein linked to scrapie susceptibility. Eur J Biochem; 267:2833–2839.

4. Hornemann S, Korth C, Oesch B, Riek R, Wide G, Wüthrich K, et al. (1997) Recombinant full-length murine prion protein, *m*PrP(23-231): purification and spectroscopic characterization. FEBS Lett; 413:277–281.

5. Jackson GS, Hil AF, Joseph C, Hosszu L, Power A, Waltho JP, et al. (1999) Multiple folding pathways for heterologously expressed human prion protein. Biochim Biophys Acta; 1431(1):1–13.

6. Yin SM, Zheng Y, Tien P. (2003) On-column purification and refolding of recombinant bovine prion protein: using its octarepeat sequences as a natural affinity tag. Protein Expr Purif; 32:104–109.

7. Bocharova OV, Breydo L, Parfenov AS, Salnikov VV, Baskakov IV. (2005) In vitro conversion of full length mammalian prion protein produces amyloid form with physical property of PrPSc. J Mol Biol; 346:645–659.

8. Breydo L, Bocharova OV, Makarava N, Salnikov VV, Anderson M, Baskakov IV.(2005) Methionine oxidation interferes with conversion of the prion protein into the fibrillar proteinase K-resistant conformation. Biochemistry; 44:15534–15543.

9. McMahon HEM, Mange A, Nishida N, Creminon C, Casanova D, Lehman S. (2001) Cleavage of the amino terminus of the prion protein by reactive oxygen species. J Biol Chem; 276:2286–2291.

10. Mange A, Beranger F, Peoc'h K, Onodera T, Frobert Y, Lehmann S. (2004) Alpha- and beta- cleavages of the amino-terminus of the cellular prion protein. Biol Cell; 96:125–132.

Chapter 11
Analysis of PrP Conformation Using Circular Dichroism

Sen Han and Andrew F. Hill

Summary The availability of recombinant prion proteins (recPrP) has been exploited as a model system to study PrP-mediated toxicity, conversion, and infectivity. According to the protein only hypothesis, the central event in the pathogenesis of prion diseases is the conversion of PrP^C to PrP^{Sc}. This involves a dramatic increase in β sheet conformation as PrP^C is converted to PrP^{Sc}, and it is widely believed that this conformational change affects the as-yet undefined function of PrP^C. Although there are many methods available to monitor for the changes in the structural makeup of PrP mutants and oligomers formed with respect to disease relevance, circular dichroism is one of the most popular methods used. In this chapter, we discuss the fundamental principles of circular dichroism and its current role and applications in prion disease research.

Keywords Circular dichroism; protein conformation; recombinant protein.

1. Fundamentals of Circular Dichroism (CD) in Peptides and Polypeptides

1.2. CD and Secondary Structure of Peptides and Proteins

CD spectroscopy measures differences in the absorption of left-handed polarized light versus right-handed polarized light that may arise due to structural asymmetry. Molecules exhibit asymmetry when they lack a center plane or an n-fold axis of symmetry. CD makes plane-polarized light elliptically polar due to this difference in absorption, forming the basis of the measurement of ellipticity during CD analysis. CD as exhibited by peptides and polypeptides are derived from the interaction of the amide bond (acting as the chromophore) with light and detected in the far UV region (190–250 nm). This results in the n-π* (forbidden transition) and π-π* transitions *(1–3)* that are detected by CD *(4, 5)* and that are dependent on the environmental and solvent conditions used to resuspend the peptide. The wavelength of n-π* transitions in water will blue shift compared with readings taken in organic solvents (the lone pair electrons in the ground "n" state are hydrogen bonded by water, hence requiring higher energy to promote electrons from the ground to excited state), whereas the wavelength of π-π^* transitions in water will red shift compared with wavelength readings taken in organic solvents (because the excited state is more polar than that of the ground state) *(6)*. Secondary structure analysis with CD is dependent on π-π^* orbital adsorptions of the amide bonds linking amino acids. These adsorption bands lie in the "vacuum ultraviolet" region (VUV) at wavelengths <200 nm, making them inaccessible in air due to adsorption of light by oxygen. This can be overcome by purging CD spectrometers with nitrogen gas to remove the oxygen.

Polypeptides usually exist as highly ordered structures and the transition dipoles of each amide bond display a tendency to interact with the transition dipole of the neighboring amide bond upon excitation, leading to an exchange of excitation energy between different molecules. This energy exchange will in turn split the transition dipoles into further transitions via "exciton" splitting. In the example of the excitation of two identical chromophores in an ordered array (as found in a polypeptide), the exchange of energy will result in two transitions; one transition with higher energy and the other transition with lower energy than that of the original chromophore in isolation. The placement of chromophores in ordered arrays also leads to the observation of hyper- or hypochromism of their optical transitions. It is this exciton splitting and hypo/hyperchroism observed in polypeptides upon excitation that allows for the identification and prediction of the secondary structure of characteristic features with CD spectroscopy. Examples include the observation of hypochromism with increasing α-helical content and hyperchromism with increasing β-sheet. A CD spectra of predominant random coil content will show a small positive π-π^* transition at 230 nm and a large single π-π^* at 195 nm. Conversely, α-helical CD profiles display a large negative n-π^* transition at 222 nm and a π-π^* transition split into two transitions (a negative band at 208 nm and a positive band at ~192 nm) due to exciton coupling *(7)*. For a CD spectrum rich with β-sheet content, the transition at π-π^* also will be split into two transitions: one transition at 218 nm for the negative band and the other transition at 195 nm for the positive band. The absence of regular structure results in zero CD intensity, whereas an ordered structure results in a spectrum that can contain both positive and negative signals. Based on these observations, it is possible to check for proper protein folding and to predict the dominant secondary structure content of the sample of interest from visualization of the CD spectrum obtained. Recommended working concentration of protein for prediction of secondary structure content from far-UV CD analysis can range from 0.5 to 50 µg/ml protein in any buffer that does not have high absorbance in the 190- to 250-nm region. It is, therefore, recommended to use water or buffers containing low salt concentrations as far as possible and to avoid buffers containing dithiothreitol, histidine, or imidazole. To obtain the best possible CD spectrum, samples should be prepared to the highest practicable purity and contain only the buffer required to maintain protein stability. Contamination from unfolded protein; particulate matter (light-scattering particles) that may contribute to noise must be avoided. Samples should ideally be filtered through a 0.22-µm filter or centrifuged to remove particulate contamination before analysis.

1.3. Near-UV CD and Tertiary Interactions

CD spectroscopy also can be used for deriving information about protein tertiary structure in the "near-UV" region (250–350 nm) where there is high sensitivity to aromatic amino acids and disulfide bonds. Signals in the 250- to 270-nm region are linked to phenylalanine residues, those obtained from 270 to 290 nm are attributable to tyrosine and tryptophan residues contribute to signals in the 280- to 300-nm range. The presence of disulfide bonds will lead to broadening and weakening of signals throughout the near-UV range. Furthermore, if a protein retains secondary structure but no defined three-dimensional structure (e.g., an incorrectly folded or "molten-globule" structure), the signals in the near-UV region will be nearly zero. Conversely, the presence of significant near-UV signals indicates that the protein is folded into a well-defined structure.

Near-UV CD is sensitive to subtle changes in tertiary interactions brought about by protein–protein interactions or solvent environmental variations. However, the signal strength from near-UV CD is much weaker than that from far-UV CD; hence, there is a need to use higher concentrations of sample for a good signal (recommended optical density 280 nm values between 0.5 and 1 A_{280} that may be in the range of 0.25–2 mg/ml concentration for most proteins).

1.4. Advantages and Limitations of CD

Although CD does not provide high-resolution analysis of protein structure as in the case of x-ray diffraction or nuclear magnetic resonance (NMR), it is acutely sensitive to changes in protein conformation. CD allows for preliminary analysis of any interaction or agent that perturbs the stability or structure of the sample protein as detected by conformational changes observed. The detection of these conformational changes can be used to check for differences in thermal stability of the protein at different physical or biochemical conditions of biological relevance. Such studies are habitually induced by thermal denaturation or chemical denaturants (solvent denaturants), such as guanidine or urea *(8)*. These chemicals induce the denaturation of the protein from a folded state to partially or completely unfolded states. The denaturation of proteins is accompanied by conformational changes consistent with a loss of ordered α and β content followed by the enhancement of random coil components These changes can be interpreted as thermodynamic expressions and used to elucidate the thermodynamics of protein folding through methods such as differential scanning microcalorimetry, CD, and intrinsic protein fluorescence. Although differential scanning microcalorimetry can provide for more highly accurate and complete thermodynamic characterization of the unfolding transition by determining the heat capacities of the native and denatured protein than CD, if executed accurately, it is highly dependent on the veracious determination of protein concentration. Protein concentration determination is hard to resolve with accuracy better than 2%, and such minute discrepancies may lead to gross errors with this method *(8)*. In contrast, CD does not have such stringent concentration accuracy requirements, and it can analyze equilibrium denaturation experiments with much less sample. Another shortcoming associated with calorimetry is that thermal denaturation of proteins are less well defined than those induced chemically, with the latter yielding more fully unfolded forms of the protein *(8)*.

Although one can derive secondary and tertiary structure interaction from other methods such as NMR, crystallography, or Fourier-transform infrared spectroscopy, there are advantages to the use of CD. CD analysis is typically nondestructive (this does not include denaturation studies), thereby allowing for downstream applications when other experiments or higher resolution studies are required. CD analysis also is rapid, and it does not require significant sample preparation or high sample concentration. Additionally, CD analysis provides for microsecond time resolution that is especially important for kinetics and thermodynamics studies *(8)*. CD analysis is not limited on the size of the macromolecule, and it is measurable in solution phase.

There are, however, limitations of CD analysis where one cannot account for specific residue contributions for secondary structure content or high-resolution tertiary information. The analysis of a CD spectrum is limited to the reliability of existing algorithms that are usually based on known reference sets. CD methodology also is hindered by incompatibilities with common aqueous buffer systems that absorb in the critical UV range of analysis. Phosphate, sulfate, acetate, and carbonate buffers are usually incompatible, and they can only be tolerated in diluted concentrations. Borate and ammonium salts are commonly used as pH buffers for CD analysis, whereas chloride and fluoride buffers are commonly used in far-UV CD because they absorb less in the far-UV region. Water is a commonly used solvent that does not affect the absorbance in the UV ranges. Shortening of the sample path-length (using a 0.1-mm cell path) is another way of obtaining CD spectra with less noise in the lower wavelength regions. Another consideration with CD analysis is the determination of the lowest wavelength data point to analyze. This is especially important if the data set is collected under conditions that are less than optimal or approaching lower wavelengths, resulting in a noisy CD spectrum. The intensity of the emitted signal drops significantly toward the lowest wavelengths in the range of a xenon lamp that is used in most conventional CD spectrometers. To compensate for the low signals, the detector will increase its sensitivity and resulting in an increased high-tension voltage (*htv*) which in turn causes readings to become unreliable and noisy. Data collected at high *htv* values is unreliable and should be avoided if the *htv* is >550 mV at 190 nm for sample or >500 mV for baseline. For the best CD spectra and for obtaining additional structural information at low wavelengths, synchrotron radiation CD (SRCD) *(9)* can be used. SRCD provides inherently polarized light at higher photo flux compared with short-arc xenon lamps used in conventional CD instruments. This allows for the recording of CD spectra down to 160 nm, enabling precise secondary structure analysis with shorter spectra collection times and an enhanced signal to noise.

2. Analysis of CD Spectra for Secondary Structure Predictions

A variety of methods are currently available for the analysis of CD spectra *(10–13)*. The majority of these methods involve using algorithms based on reference sets of known proteins (solved structurally via crystallography or NMR), enabling the prediction of the secondary structure of an unknown sample. These algorithms and other reference tools are readily available as Web-based applications. An example of such a service is DICHROWEB (http://www.cryst.bbk.ac.uk/cdweb/html/home.html). In this section, we present the basis of some of these methods and compare their strengths and weaknesses.

2.1. Linear Regression

The effects of mutations, ligands, and solvents on protein conformation can be analyzed with linear regression *(14)*. This method requires the use of standardized spectra sets derived from known proteins that had been evaluated with the least-squares methods and the convex constrain algorithm. Nonconstrained multilinear regression (MLR) is a variation of linear regression where the sum of the conformations is not constrained to equal 100%. Advantages of this method include being independent of protein concentration and having invariant standards. MLR, however, cannot accurately estimate for β-sheets and turns. In constrained least-squares fits, the sum of the contribution of each spectra are set to equal to 1. Although this allows for invariant database and provides better data fitting than MLR, it does not have optimal standards for β-turns.

2.1.1. Ridge Regression (Contin)

Contin is an algorithm that fits the CD spectra of unknown proteins by incorporating the spectra of a large data base of proteins with known conformations *(15)*. Using this method, the contribution of each reference spectrum is kept small unless it contributes to a good agreement between the theoretical best fit curve and the raw data. Although this method allows for a good estimate of β-turns, the downside is that the reference data sets used are different for each analysis.

2.1.2. Singular Value Decomposition (SVD)

SVD extracts curves with unique profiles from a set of spectra of proteins with known structures *(12)*. The basis curves are each characterized by a mixture of secondary structures, and they are then used to analyze the conformation of unknown proteins. Although SVD gives a good estimate of α-helical content, it is not suitable for fitting β-sheets and turns unless data are collected at low wavelengths of at least 184 nm.

2.1.3. Variable Selection (VARSLC) and CDSSTR

Starting with a large base of reference proteins of known spectra and secondary structure *(13, 16, 17)*, the number of reference proteins are subsequently removed systematically to create new data bases with a smaller number of standards. SVD analysis is then carried out to select for data sets that fit these criteria which are consequently averaged. A variation of VARSLC is CDSSTR (*Johnson's minimal basis-random selection* method for protein CD analysis) that implements the variable selection method by performing all the possible calculations using a fixed number of proteins from the reference set of choice before the SVD fit.

2.1.4. Self-Consistent Method (Selcon)

This is a rapid algorithm *(13, 17, 18)* that is a modification of VARSLC that is dependent on the references used. With the right references, it allows for accurate estimates of β-sheet and turns in proteins. This method, however, results in bad data fitting for polypeptides with high amounts of β-sheet content.

2.1.5. Neural Networks (K2D)

K2D *(19)* is one of the few neural networks available, and it is based on artificial intelligence programs that can "learn" to identify and correlate spectra trends in CD data to that of known references. This method is relatively independent of wavelength and results in relatively good data fits. To date, it seems that K2D gives higher accuracy to helical proteins as opposed to β-sheets and mixed protein.

2.2. CD in Prion Research

2.2.1. CD with PrP^C, PrP^{Sc}, PrP27-30, and PrP106-126

The central event in the prion hypothesis involves the conversion of the normal cellular prion protein (PrP^C), with a predominant α-helical content, into the abnormal infectious isoforms, denoted as PrP^{Sc}, with a marked increase in β-sheet content that also exhibits limited proteinase K (PK) resistance. As early as 1993 *(20)*, attempts to elucidate the secondary content differences in PrP^C, PrP^{Sc}, and PrP[27–30] from golden Syrian hamsters was undertaken with biophysical techniques, such as Fourier transform infrared spectroscopy (FTIR) and CD. FTIR analysis revealed PrP^C to be predominantly α-helical (42% α-helical, 3% ß-sheet), whereas PrP^{Sc} showed a marked increase in β-sheet content (30% α-helical, 43% β-sheet), and controlled digestion with proteases resulted in PrP[27–30] that showed 21% α-helical and 54% β-sheet). CD was used for secondary structure analysis of PrP^C, revealing a 36% α-helical content. PrP^{Sc} and PrP$^{27-30}$ were not analyzed due to insolubility. When PrP^{Sc} and PrP$^{27-30}$ were reconstituted as liposomes or applied as thin films, PrP^{Sc} was determined to contain 34% β-sheet, 20% α-helix, and 46% β-turns and random coil *(21)*. CD analysis of PrP$^{27-30}$ showed 43% β-sheet, no α-helical content, and 57% β-turn and random coil *(21)*. Thermal and solvent perturbations of hamster PrP$^{27-30}$ *(22)* demonstrated decreasing levels of infectivity with decreasing β-sheet content, suggesting that the β-sheet structure plays a key role in the physical stability and propagation properties of PrP^{Sc}.

Based on the Syrian hamster model and structural predictions *(23)*, a series of short PrP fragments (encompassing residues 109–122, 129–140, 178–191, and 202–218) that were predicted to be α-helical were actually rich in β-sheets when synthesized (as determined by FTIR). An interesting analogy with Gerstmann–Straussler–Scheinker syndrome (GSS) was made with residues 109–122 and 129–140 where truncated PrP58-150 is found to be the main constituent in GSS amyloid plaques.

Further experiments with 106–126 revealed neurotoxicity in murine hippocampal cells, and this has been analyzed using biophysical techniques *(24)*. CD analysis of PrP106-126 revealed high β-sheet content that is pH dependent. PrP106-126 in 200 mM phosphate buffer, pH 7.0, exists as a mixture of β-sheet and random coil. When dissolved in deionized water, it adopts a random coil conformation. The addition of 50% trifluoroethanol (TFE; enhances α-helical formation in residues prone to form helices) to PrP106-126 resulted in a pronounced change to double minimal at 206 and 220, indicative of predominant α-helical structure, whereas PrP106-126 retains a β-sheet–rich conformation at pH 5.0. Attempts to revert the β-conformation at pH 5.0 by changing conditions that favor adoption of α-helical conformation failed, indicating that once formed, the β-form is very stable and does not readily revert back to the α-helical conformation. PrP residues 106–126 are also resistant to limited PK treatment and aggregate upon dissolution. A scrambled control resulted in a dominant random coil content, protease sensitivity, in ability to aggregate at similar conditions indicative of sequence specificity for the β-sheet fold. Based on the similarities in neurotoxicity, high β-sheet content, and PK resistance of PrP106-126 and PrP^{Sc}, the former has since been widely used as a minimal model of toxicity in the study of prion diseases (see Chap. 6).

2.2.2. CD with Recombinant Prions

While the use of brain-derived material is the ideal and physiologically relevant model to work on, it is very hard to obtain, isolate, and purify. Although the PrP had been deemed indispensable for the pathogenesis of prion diseases, other components (e.g., RNA and metal interactions) have since been identified as important for the conversion process to occur *(25–29)*. Furthermore, the full-length PrP can be broken down into various different domains that can be either be neuroprotective or neurotoxic. It is important to dissect the individual components and functional domains of the PrP and their interactions with disease-associated cofactors. Because brain-derived material is likely to be a complex amalgam of these peripheral components in close association with the PrP, it will be difficult to determine the exact effects of each component on a molecular and functional level without isolating each of them. An alternative model system to address these issues and thus enabling better component analysis is through the use of bacterially expressed recombinant PrP (recPrP), a model system that allows for analysis of high yield of highly purified material while maintaining biological relevance.

Because the central event in prion diseases, according to the prion only hypothesis, involves the conversion of PrP^C of high helical content to PrP^{Sc} with increased β-sheet content, recombinant prion models have since been used to replicate that process through a variety of different experimental conditions. Various methodologies had been developed to obtain β-sheet–rich forms of recombinant PrP that may be a better representation of PrP^{Sc} *(30, 31)*. Many of these experiments involve the use of CD as a tool for monitoring the changes in secondary structure that may reflect the physiological events seen in the pathogenesis of prion disease. CD also has been used to monitor for the stability of the PrP under thermal and chemical denaturing conditions that may illuminate the mechanisms of conversion and refolding. In **Subheadings 2.2.3–2.2.5.**, we briefly discuss some examples of the use of CD with recPrP.

2.2.3. Comparison of an Array of PrP from Different Species

To date, the secondary structures of full-length recombinant PrP derived from various sources have been analyzed with CD, and they all reveal significant contributions of α-helices and random coils while having lower β-sheet content. Examples include recombinant full-length murine *(32)*, Syrian hamster, bovine 25-249 *(33)*, and human PrP with a high amount of α-helical content and low amount of β-sheet content *(34, 35)*. This α-helical content is mainly contributed by the carboxy-terminal region as shown with CD analysis comparisons with truncated fragments of 81–231, 90–231, and 121–231. They show an increase in random coil with increasing amino-terminal sequence compared with the 90–231 fragment *(34)*. The effects of membrane interactions with human recPrP (hurecPrP) simulated with liposomes also have been investigated for biological relevance. CD analysis of the human doppel protein 25–152 *(36)* sharing high sequence homology to human PrP (most of the N- and C-terminal regions are conserved, but doppel lacks the copper binding octapeptide repeat region and the predicted neurotoxic segment of 106–126) showed a maximum at 192 nm, and the characteristic double minimum at 208 and 222 nm, typical for proteins with a significant α-helical content. Comparative studies of PrP from different species are important, because they help elucidate why some species exhibit resistance to prion infection (e.g., the rabbit prion protein *[37, 38]*) and whether similar mechanisms can be adopted as therapeutics against prion disease.

2.2.4. PrP and Membrane Interactions

Although there is a wealth of data accumulated from the analysis of various forms of recombinant PrP from different sources relating structure, toxicity, and PrP conversion, most of these studies share a common assumption of protein behavior in solution. Few of these experiments take into account that the PrP does not exist as a free water-soluble protein but it is instead intimately associated with the surface of cellular membranes. This interaction or the lack of it may in turn be important to explain the change in structure of the protein during disease states. One such study attempted to investigate the effect of membrane interactions through model lipid membranes *(35)* that demonstrated that the N terminus of human PrP binds to acidic membranes strongly, whereas the C-terminal region destabilizes at low pH. This was revealed using intrinsic tryptophan fluorescence and CD analysis. Both full-length and truncated recombinant PrP were subjected to binding with different synthetic lipid membrane systems, and they displayed a change in conformation that was identified using CD. The conformational changes also were observed with acrylamide quenching and measurements of intrinsic tryptophan fluorescence. A series of work comparing the lipid binding interactions of recombinant α-PrP to that of β-PrP *(31, 39, 40)* with CD showed conformational changes upon lipid binding of β-PrP with a disruptive effect on the integrity of the lipid bilayer. These conformational changes also lead to aggregation and fibrillization events that relate well to disease phenotypes.

2.2.5. Influences of Polymorphisms and Amino Acid Substitutions in Regions of Postulated Importance or from Regions of Observed Mutations

An investigation on the effects of different amino acid substitutions related to known inherited human prion diseases on the secondary structure and thermal stability of the prion protein was carried out with hurecPrP[121–231] *(41)*. Monitoring the far-UV CD spectra revealed no differences between the valine and methionine polymorphs at position 129, a common polymorphism in human PrP. The study looked at the effects of eight mutations derived from known mutations in inherited TSE cases with murine PrP 121-231 relating to fatal familial insomnia (FFI) (D178N/M129), Creutzfeldt–Jakob disease (CJD) (D178N/V129, T183A, E200K, R208H, and V210I), and GSS (V180I and Q217R). Comparison of wild-type PrP to five of these variants indicated a destabilization of HuPrP[121–231], whereas the other variants have the same stability as the wild-type protein. This suggests that the destabilization of PrP[C] is neither a general mechanism critical to the formation of PrP[Sc] nor the basis of disease phenotypes in inherited TSEs. Another explanation for the lack of correlation between the mutations and disease may be due to a failure of the experimental conditions to be reflective of disease mechanisms and pathways.

Various epidemiological studies have determined that the polymorphism at codon 129 of human PrP (encoding either methionine or valine) is the major molecular determinant of susceptibility to prion diseases *(42–44)*. Both methionine and valine homozygotes are overrepresented, whereas heterozygotes are underrepresented *(45)*. Although these observations imply an important role played by the polymorphism, no difference had been observed on the in vitro thermodynamic stability of the recombinant PrP refolded into the α-helical form bearing either methionine or valine at codon 129 *(41)*. Structural studies, however, suggest hydrogen bonding between Asp178 and Try128 *(46)* that may provide a structural basis for the influence of the codon 129 polymorphism.

CD analysis of synthetic peptide 106–136 methionine and valine polymorphs of the human PrP *(47)* in different concentrations of TFE and pH ranges revealed the methionine polymorph to have a higher propensity for β-sheet formation and aggregation compared with the valine polymorph with increasing concentration of TFE or lower pH. Further work with the codon 129 polymorphs in a recombinant human PrP fragment of residues 90–231 (HuPrP$^{90-231}$Met^{129}/Val^{129}) indicated no significant differences in secondary structure, but they revealed HuPrP$^{90-231}$Met^{129} to have a higher propensity for oligomerization *(48)*. Another study probing into the roles of the polymorphism at position 129 in recombinant human PrP$^{91-231}$ *(49)*, with regard to thermodynamic stability through the use of CD failed to show any significant differences. The aforementioned studies implicate that polymorphism at position 129 does not confer susceptibility to disease by altering the structure or global stability of PrP^C and that their effects observed in disease may be mediated to be downstream in the disease mechanism after prion propagation or on different pathways not reflected by the experimental conditions used. Currently, there are five major PrP refolding and misfolding pathways postulated, studied, and identified through the use of recombinant PrP. These pathways include the conversion of denatured protein into α-monomer that had been commonly used as a representation of PrP^C *(50, 51)*, the conversion of denatured protein into a soluble β-monomer (β-PrP) *(30)*, the conversion of denatured protein into PK-resistant oligomeric states *(52)*, the conversion of the α-monomer into the amyloid form *(52)*, and the conversion of partially unfolded protein into mutimeric aggregate and eventually amyloid form *(53)*. Studies indicate that the pathway choice determines the eventual effect of the polymorphism at position 129 in recPrP. Adopting the fifth pathway as stated above, the presence of valine at residue 129 in HuPrP$^{90-231}$ was found to accelerate β-sheet conversion as detected by CD and amyloid formation *(53)*, whereas the presence of methionine at this residue when subjected to the conversion of denatured protein into PK-resistant oligomeric states resulted in a faster rate of oligomerization *(48)*. This finding is intriguing because sporadic CJD in patients homozygous for valine at codon 129 exhibit disease onset at a younger age (54–56 years), and the rapid amyloid formation and early adoption of β-sheet in the recombinant model may be reflective of valine 129 providing a better site of nucleation for the development of disease. The different behavior between the allomorphs when refolded/misfolded via different pathways is reflective of prion diseases as complex multifaceted disorders with multiple disease pathways. These results implicate a need to revisit existing findings from recPrP work and to compare findings when imposed with different refolding/unfolding pathways.

2.3. Basic CD Methodology

Spectra are collected on specialized CD spectrometers (such as the Jasco J-815), by using quartz cuvettes available from a variety of suppliers (such as Hellma). CD spectra are taken with lyophilized samples dissolved in Milli-Q–treated water diluted to final working concentration of 4.1–6 μM. Mean residue ellipticity MRE [θ] (measured in degrees cm^{2}dmol^{-1}residue^{-1}) is recorded between wavelengths 185 and 250 nm at 25°C. Analysis is performed via algorithms available on DICHROWEB (http://www.cryst.bbk.ac.uk/cdweb/html/home.html) by using the data obtained from the spectrometer.

   To analyze the thermodynamics of the PrP with CD, we first assume that the refolding process is a two-step process as suggested by previous studies (57). In this scenario, the protein can exist in two different states, namely, the folded (F) and unfolded (U) states. One can express the folding constant (K) at any experimental temperature (T) as demonstrated:

$$K=[F]/[U] = [F]/(1 - [F])   \qquad (1)$$

The free energy of folding can be conveyed by ΔG:

$$\Delta G = nRT\ln K   \qquad (2)$$

R is the universal gas constant, where R = 8.314472(15) J K^{-1} mol^{-1}.

   Denoting the fraction of folded protein at any temperature to be α;

$$\alpha=[F]/([F] + [U]) = K/(1 + K)   \qquad (3)$$

$$\alpha = (\theta_{obs} - \theta_U)/(\theta_F - \theta_U)   \qquad (4)$$

θ_{obs} is the observed ellipticity at any temperature (35). The ellipticity of the fully folded form and that of the unfolded form are denoted as θ_F and θ_U, respectively. The Gibbs–Helmholtz equation that describes folding as a function of temperature is used to fit the observed change in ellipticity as a function of temperature. The following equations are fit:

$$\Delta G = \Delta H(1 - T/T_M) - \Delta Cp((T_M - T) + T\ln(T/T_M))   \qquad (5)$$

$$K = \exp(-\Delta G/(RT))   \qquad (6)$$

$$\alpha = K/(1 + K)   \qquad (7)$$

$$[\theta]_{obs} = \alpha([\theta]_F - [\theta]_U) + [\theta]_U   \qquad (8)$$

T_M is the temperature where $\alpha = 0.5$ where there is an equal amount of folded and unfolded protein. Values of Δ_H, ΔC_p, T_M, $[\theta]_F$, and $[\theta]_U$ are estimated, and nonlinear least-squares analysis is used to best fit the raw data.

Alternatively one can derive the T_M and ΔH values from a nonlinear least-squares fit of the experimental data:

$$Y = (a_F + b_F{}^*T)/(1 + \exp((-\Delta H_M/T + \Delta H_M/T_M)/R)) + (a_U + b_U{}^*T){}^*(\exp((-\Delta H_m/T + \Delta H_M/T_M)/(R)/(1 + \exp((-\Delta H_M/T + \Delta H_M/T_M)/(R)))$$

Where a_F, b_F, a_U, and b_U represents the pre- and posttransitional baselines, ΔH_M and T_M as fitting parameters (58).

It should be noted that the derivation of T_M can be used to compare the thermal stability of different mutants to that of wild-type protein. A lower T_M indicates lower thermal stability, and it may be indicative of the mutation having a destabilizing effect on protein fold. ΔCp is the measure of heat capacity change from the folded to the unfolded state.

3. Concluding Comments

CD is a nondestructive method that can be used to rapidly determine the secondary structure of protein. It is an extremely convenient method of monitoring conformational changes accompanying mutations, folding, protein–protein interactions, and protein–ligand interactions. CD also can be used to follow the kinetics of folding and unfolding and determine differences in thermal stabilities brought about by mutations. These properties of CD have since been used in many prion studies to help monitor and elucidate the roles of different disease-associated epitopes or residues in the disease process that involves the conversion of PrP^C to PrP^{Sc}. In essence, CD can serve as semiquantitative methodology for preliminary data analysis that should be followed up with higher resolution methodologies such as X-ray diffraction, NMR, or calorimetry.

References

1. Peterson, D.L., and Simpson, W.T. (1957) Polarized electronic absorption spectrum of amides with assignments of transitions. *J Am Chem Soc* 79, 2375–2382.
2. Hunt, H.D., and Simpson, W.T. (1953) Spectra of simple amides in the vacuum ultraviolet. *J Am Chem Soc* 75, 4540–4543.
3. Sehellman, H. A., and Sehellman, C. (1964) Conformation of polypeptide chains. In: Neurath, H. and Hill, R.L. (eds.), Academic Press, New York.
4. Peggion, E., Fontana, A., and Cosani, A. (1969) Conformational studies on a modified poly-L-tryptophan: circular dichroism and optical rotatory dispersion of poly-2-(2-nitrophenyl-sulfenyl)-L-tryptophan and of random copolymers of L-tryptophan and 2-(2-nitrophenylsulfenyl)-L-tryptophan. *Biopolymers* 7, 517–526.

5. Greenfield, N.J., and Fasman, G.D. (1970) The circular dichroism of 3-methylpyrrolidin-2-one. *J Am Chem Soc* 92, 177–181.

6. Nielsen, E.B., and Schellman, J.A. (1967) The absorption spectra of simple amides and peptides. *J Phys Chem* 71, 2297–2304.

7. Greenfield, N.J., and Fasman, G.D. (1969) Optical activity of simple cyclic amides in solution. *Biopolymers* 7, 595–610.

8. Matouschek, A., Serrano, L., and Fersht, A.R. (1994) Analysis of protein folding by protein engineering. In: Mechanisms of protein folding (Pain, R.H., ed.), IRL Press, Oxford, UK.

9. Miles, A.J., and Wallace, B.A. (2006) Synchrotron radiation circular dichroism spectroscopy of proteins and applications in structural and functional genomics. *Chem Soc Rev* 35, 39–51.

10. Lees, J.G., Smith, B.R., Wien, F., Miles, A.J., and Wallace, B.A. (2004) CDtool–an integrated software package for circular dichroism spectroscopic data processing, analysis, and archiving. *Anal Biochem* 332, 285–289.

11. Greenfield, N.J. (2004) Analysis of circular dichroism data. *Methods Enzymol* 383, 282–317.

12. Hennessey, J.P., Jr., and Johnson, W.C., Jr. (1981) Information content in the circular dichroism of proteins. *Biochemistry* 20, 1085–1094.

13. Sreerama, N., and Woody, R.W. (1994) Protein secondary structure from circular dichroism spectroscopy. Combining variable selection principle and cluster analysis with neural network, ridge regression and self-consistent methods. *J Mol Biol* 242, 497–507.

14. Brahms, S., and Brahms, J. (1980) Determination of protein secondary structure in solution by vacuum ultraviolet circular dichroism. *J Mol Biol* 138, 149–178.

15. Provencher, S.W., and Glockner, J. (1981) Estimation of globular protein secondary structure from circular dichroism. *Biochemistry* 20, 33–37.

16. Manavalan, P., and Johnson, W.C., Jr. (1987) Variable selection method improves the prediction of protein secondary structure from circular dichroism spectra. *Anal Biochem* 167, 76–85.

17. Sreerama, N., and Woody, R.W. (1993) A self-consistent method for the analysis of protein secondary structure from circular dichroism. *Anal Biochem* 209, 32–44.

18. Sreerama, N., and Woody, R.W. (1994) Poly(pro)II helices in globular proteins: identification and circular dichroic analysis. *Biochemistry* 33, 10022–10025.

19. Andrade, M.A., Chacon, P., Merelo, J.J., and Moran, F. (1993) Evaluation of secondary structure of proteins from UV circular dichroism spectra using an unsupervised learning neural network. *Protein Eng* 6, 383–390.

20. Pan, K.M., Baldwin, M., Nguyen, J., Gasset, M., Serban, A., Groth, D., Mehlhorn, I., Huang, Z., Fletterick, R.J., Cohen, F.E., et al. (1993) Conversion of alpha-helices into beta-sheets features in the formation of the scrapie prion proteins. *Proc Natl Acad Sci U S A* 90, 10962–10966.

21. Safar, J., Roller, P.P., Gajdusek, D.C., and Gibbs, C.J., Jr. (1993) Conformational transitions, dissociation, and unfolding of scrapie amyloid (prion) protein. *J Biol Chem* 268, 20276–20284.

22. Safar, J., Roller, P.P., Gajdusek, D.C., and Gibbs, C.J., Jr. (1993) Thermal stability and conformational transitions of scrapie amyloid (prion) protein correlate with infectivity. *Protein Sci* 2, 2206–2216.

23. Gasset, M., Baldwin, M.A., Lloyd, D.H., Gabriel, J.M., Holtzman, D.M., Cohen, F., Fletterick, R., and Prusiner, S.B. (1992) Predicted alpha-helical regions of the prion protein when synthesized as peptides form amyloid. *Proc Natl Acad Sci U S A* 89, 10940–10944.

24. Selvaggini, C., De Gioia, L., Cantu, L., Ghibaudi, E., Diomede, L., Passerini, F., Forloni, G., Bugiani, O., Tagliavini, F., and Salmona, M. (1993) Molecular characteristics of a protease-resistant, amyloidogenic and neurotoxic peptide homologous to residues 106–126 of the prion protein. *Biochem Biophys Res Commun* 194, 1380–1386.

25. Gabus, C., Derrington, E., Leblanc, P., Chnaiderman, J., Dormont, D., Swietnicki, W., Morillas, M., Surewicz, W.K., Marc, D., Nandi, P., and Darlix, J.L. (2001) The prion protein has RNA binding and chaperoning properties characteristic of nucleocapsid protein NCP7 of HIV-1. *J Biol Chem* 276, 19301–19309.

26. Deleault, N.R., Lucassen, R.W., and Supattapone, S. (2003) RNA molecules stimulate prion protein conversion. *Nature* 425, 717–720.

27. Marc, D., Mercey, R., and Lantier, F. (2007) Scavenger, transducer, RNA chaperone? What ligands of the prion protein teach us about its function. *Cell Mol Life Sci* 64, 815–29.
28. Stockel, J., Safar, J., Wallace, A.C., Cohen, F.E., and Prusiner, S.B. (1998) Prion protein selectively binds copper(II) ions. *Biochemistry* 37, 7185–7193.
29. Brown, D.R., Hafiz, F., Glasssmith, L.L., Wong, B.S., Jones, I.M., Clive, C., and Haswell, S.J. (2000) Consequences of manganese replacement of copper for prion protein function and proteinase resistance. *EMBO J* 19, 1180–1186.
30. Jackson, G.S., Hosszu, L.L., Power, A., Hill, A.F., Kenney, J., Saibil, H., Craven, C.J., Waltho, J.P., Clarke, A.R., and Collinge, J. (1999) Reversible conversion of monomeric human prion protein between native and fibrilogenic conformations. *Science* 283, 1935–1937.
31. Kazlauskaite, J., Sanghera, N., Sylvester, I., Venien-Bryan, C., and Pinheiro, T.J. (2003) Structural changes of the prion protein in lipid membranes leading to aggregation and fibrilization. *Biochemistry* 42, 3295–3304.
32. Hornemann, S., Korth, C., Oesch, B., Riek, R., Wider, G., Wuthrich, K., and Glockshuber, R. (1997) Recombinant full-length murine prion protein, mPrP(23-231): purification and spectroscopic characterization. *FEBS Lett* 413, 277–281.
33. Negro, A., De Filippis, V., Skaper, S.D., James, P., and Sorgato, M.C. (1997) The complete mature bovine prion protein highly expressed in Escherichia coli: biochemical and structural studies. *FEBS Lett* 412, 359–364.
34. Zahn, R., von Schroetter, C., and Wuthrich, K. (1997) Human prion proteins expressed in Escherichia coli and purified by high-affinity column refolding. *FEBS Lett* 417, 400–404.
35. Morillas, M., Swietnicki, W., Gambetti, P., and Surewicz, W.K. (1999) Membrane environment alters the conformational structure of the recombinant human prion protein. *J Biol Chem* 274, 36859–36865.
36. Lu, K., Wang, W., Xie, Z., Wong, B.S., Li, R., Petersen, R.B., Sy, M.S., and Chen, S.G. (2000) Expression and structural characterization of the recombinant human doppel protein. *Biochemistry* 39, 13575–13583.
37. Loftus, B., and Rogers, M. (1997) Characterization of a prion protein (PrP) gene from rabbit; a species with apparent resistance to infection by prions. *Gene* 184, 215–219.
38. Vorberg, I., Groschup, M.H., Pfaff, E., and Priola, S.A. (2003) Multiple amino acid residues within the rabbit prion protein inhibit formation of its abnormal isoform. *J Virol* 77, 2003–2009.
39. Kazlauskaite, J., and Pinheiro, T.J. (2005) Aggregation and fibrilization of prions in lipid membranes. *Biochem Soc Symp* 72, 211–222.
40. Critchley, P., Kazlauskaite, J., Eason, R., and Pinheiro, T.J. (2004) Binding of prion proteins to lipid membranes. *Biochem Biophys Res Commun* 313, 559–567.
41. Liemann, S., and Glockshuber, R. (1999) Influence of amino acid substitutions related to inherited human prion diseases on the thermodynamic stability of the cellular prion protein. *Biochemistry* 38, 3258–3267.
42. Johnson, R.T., and Gibbs, C.J., Jr. (1998) Creutzfeldt-Jakob disease and related transmissible spongiform encephalopathies. *N Engl J Med* 339, 1994–2004.
43. Collinge, J. (2001) Prion diseases of humans and animals: their causes and molecular basis. *Annu Rev Neurosci* 24, 519–550.
44. Parchi, P., Zou, W., Wang, W., Brown, P., Capellari, S., Ghetti, B., Kopp, N., Schulz-Schaeffer, W.J., Kretzschmar, H.A., Head, M.W., Ironside, J.W., Gambetti, P., and Chen, S.G. (2000) Genetic influence on the structural variations of the abnormal prion protein. *Proc Natl Acad Sci U S A* 97, 10168–10172.
45. Palmer, M.S., Dryden, A.J., Hughes, J.T., and Collinge, J. (1991) Homozygous prion protein genotype predisposes to sporadic Creutzfeldt-Jakob disease. *Nature* 352, 340–342.
46. Riek, R., Wider, G., Billeter, M., Hornemann, S., Glockshuber, R., and Wuthrich, K. (1998) Prion protein NMR structure and familial human spongiform encephalopathies. *Proc Natl Acad Sci U S A* 95, 11667–11672.
47. Petchanikow, C., Saborio, G.P., Anderes, L., Frossard, M.J., Olmedo, M.I., and Soto, C. (2001) Biochemical and structural studies of the prion protein polymorphism. *FEBS Lett* 509, 451–456.

48. Tahiri-Alaoui, A., Gill, A.C., Disterer, P., and James, W. (2004) Methionine 129 variant of human prion protein oligomerizes more rapidly than the valine 129 variant: implications for disease susceptibility to Creutzfeldt-Jakob disease. *J Biol Chem* 279, 31390–31397.

49. Hosszu, L.L., Jackson, G.S., Trevitt, C.R., Jones, S., Batchelor, M., Bhelt, D., Prodromidou, K., Clarke, A.R., Waltho, J.P., and Collinge, J. (2004) The residue 129 polymorphism in human prion protein does not confer susceptibility to Creutzfeldt-Jakob disease by altering the structure or global stability of PrPC. *J Biol Chem* 279, 28515–28521.

50. Baskakov, I.V., Legname, G., Prusiner, S.B., and Cohen, F.E. (2001) Folding of prion protein to its native alpha-helical conformation is under kinetic control. *J Biol Chem* 276, 19687–19690.

51. Baskakov, I.V., Legname, G., Baldwin, M.A., Prusiner, S.B., and Cohen, F.E. (2002) Pathway complexity of prion protein assembly into amyloid. *J Biol Chem* 277, 21140–21148.

52. Tahiri-Alaoui, A., and James, W. (2005) Rapid formation of amyloid from alpha-monomeric recombinant human PrP in vitro. *Protein Sci* 14, 942–947.

53. Baskakov, I., Disterer, P., Breydo, L., Shaw, M., Gill, A., James, W., and Tahiri-Alaoui, A. (2005) The presence of valine at residue 129 in human prion protein accelerates amyloid formation. *FEBS Lett* 579, 2589–2596.

54. Parchi, P., Giese, A., Capellari, S., Brown, P., Schulz-Schaeffer, W., Windl, O., Zerr, I., Budka, H., Kopp, N., Piccardo, P., Poser, S., Rojiani, A., Streichemberger, N., Julien, J., Vital, C., Ghetti, B., Gambetti, P., and Kretzschmar, H. (1999) Classification of sporadic Creutzfeldt-Jakob disease based on molecular and phenotypic analysis of 300 subjects. *Ann Neurol* 46, 224–233.

55. Windl, O., Dempster, M., Estibeiro, J.P., Lathe, R., de Silva, R., Esmonde, T., Will, R., Springbett, A., Campbell, T.A., Sidle, K.C., Palmer, M.S., and Collinge, J. (1996) Genetic basis of Creutzfeldt-Jakob disease in the United Kingdom: a systematic analysis of predisposing mutations and allelic variation in the PRNP gene. *Hum Genet* 98, 259–264.

56. Alperovitch, A., Zerr, I., Pocchiari, M., Mitrova, E., de Pedro Cuesta, J., Hegyi, I., Collins, S., Kretzschmar, H., van Duijn, C., and Will, R.G. (1999) Codon 129 prion protein genotype and sporadic Creutzfeldt-Jakob disease. *Lancet* 353, 1673–1674.

57. Baskakov, I.V., Legname, G., Gryczynski, Z., and Prusiner, S.B. (2004) The peculiar nature of unfolding of the human prion protein. *Protein Sci* 13, 586–595.

58. Boer H., and Koivula A. (2003) The relationship between thermal stability and pH optimun studied with wild-type and mutant *Trichoderma reesei* cellobiohydrolase Cel7A. *Eur J Biochem* 270, 841–848.

Chapter 12
Effect of Copper on the De Novo Generation of Prion Protein Expressed in *Pichia pastoris*

Carina Treiber

Summary The prion protein (PrP) is the key protein implicated in diseases known as transmissible spongiform encephalopathies. PrP has been shown to be a metalloprotein that binds copper (Cu), and copper might have a role in the normal function of the protein. Conversely, PrP expression in yeast led us to suggest that the protein might be involved in the regulation of Cu homeostasis. In the presence of excess Cu in the growth medium, PrP expression limited the increase of the total number of Cu atoms per cell to a maximum of 14-fold compared with mock control cells, which showed a 52-fold increased intracellular Cu level. Conclusively, we suggest that PrP expression itself has a regulatory or buffering function for the cellular Cu level in yeast cells, most likely due to binding of Cu to the multiple Cu binding sites.

Keywords Copper homeostasis; inductively coupled mass spectrometry (ICP-MS); *Pichia pastoris*; prion protein (PrP); yeast expression system; scrapie.

1. Introduction

It is currently thought that the cellular isoform of the prion protein (PrP^C) may be involved in the regulation of intracellular metal ion homeostasis. For example, crude membranes and synaptosomal fractions from brains of mice lacking the expression of PrP^C have a major copper (Cu) deficiency *(1)*. It is discussed that an imbalance in brain trace elements, especially Cu deficiency, and a concomitant excess of manganese (Mn) might result in conditions that lead to the formation of the proteinase-resistant infectious isoform of the prion protein (PrP^{Sc}) *(2)*. Experiments in yeast, i.e., in the methylotrophic yeast *Pichia pastoris*, have clearly demonstrated the involvement of PrP in the regulation of intracellular metal ion homeostasis, and they have uncovered copper and more severely, manganese ions as in vivo risk factors for the conversion into PrP^{Sc} *(3)*.

Here, the use of the methylotrophic yeast *P. pastoris* is presented to be an appropriate and simple model system for heterologous expression of PrP and to investigate metal ion homeostasis in yeast. Antibodies commercially available were used

From: *Prion Protein Protocols.*
Methods in Molecular Biology, Vol. 459.
Edited by: A. F. Hill © Humana Press, Totowa, NJ

for Western blotting to recognize mouse PrP. Inductively coupled mass spectrometry (ICP-MS) is presented as a potent and sensitive method to analyze intracellular changes in metal homeostasis and based on the use state-of-the-art protocols for sample preparation.

2. Materials

2.1. *Cloning*

1. Yeast expression vector pPICZαA/B/C (3.6 kb) (Invitrogen, Carlsbad, CA).
2. Synthetic oligonucleotides (Sigma-Ark, Darmstadt, Germany): forward and reverse primer.
3. Pfu polymerase, Pfu buffer (Stratagene, La Jolle, CA).
4. dNTPs (GE Healthcare, Little Chalfont, Buckinghamshire, UK).
5. Agarose, electrophoresis grade (Invitrogen).
6. Ethidium bromide (Fluka, Buchs, Switzerland).

    a. Diluted in water to a stock solution of 10 mg/ml.

7. DNA buffer (6×): 0.25% (w/v) bromophenol blue, 60% (v/v) glycerol, 0.1 mM EDTA, pH 8.0.
8. 1-kb DNA Ladder (Invitrogen).
9. Tris borate-EDTA (TBE) buffer: 1.08% (w/w) Trizma base, 0.55% (w/w) boric acid, 0.09% (w/w) EDTA.
10. QIAquick gel extraction kit (QIAGEN GmbH, Hilden, Germany).
11. Quick Ligation™ kit (New England Biolabs, Ispwich, MA).
12. QIAquick PCR purification kit.
13. Electrocompetent *Escherichia coli*, TOP10 cells (Invitrogen).
14. low salt OR Luria-Bertani (LB) medium: 1% (w/v) Bacto-tryptone, 0.5% (w/v) yeast extract, 0.5% (w/v) NaCl, 16 g/l bacto agar, 25 μg/ml Zeocin™.
15. Nucleo Bond® PC 500 kit (Machery Nagel, Düren, Germany).
16. Restriction enzymes: EcoRI, NotI, SacI (New England Biolabs).
17. Sequencing primer: 3'-AOX1, 5'-AOX1 (Invitrogen).

2.2. *Yeast Culture*

1. *Pichia pastoris* strain SMD1168 (Invitrogen).
2. YPD medium: 1% (w/v) yeast extract, 1% (w/v) bacto-peptone, 2% (w/v) glucose.
3. YPDS selection plates: 1% (w/v) yeast extract, 2% (w/v) bacto-peptone, 1 M sorbitol, 2% (w/v) agarose, 100 μg/ml Zeocin (*see* **Note 1**).
4. YPDG: 1% (w/v) yeast extract, 2% (w/v) bacto-peptone, 2% (w/v) glucose, 15% (w/v) glycerol.

5. BMGY medium: 1% (w/v) yeast extract, 2% (w/v) bacto-peptone, 100 mM potassium phosphate, pH 6.0, 1.34 % (w/v) yeast nitrogen base, 4×10^{-5}% biotin, 1% (w/v) glycerol.

6. BMMY medium: 1% (w/v) yeast extract, 2% (w/v) bacto-peptone, 100 mM potassium phosphate, pH 6.0, 1.34% (w/v) yeast nitrogen base, 4×10^{-5}% biotin, 2% (v/v) methanol.

2.3. Protein Characterization

1. Breaking buffer: 50 mM potassium phosphate, pH 7.4, 1 mM EDTA, 5% glycerol, 5 μl of Complete protease inhibitor cocktail (Roche Diagnostics, Mannheim, Germany).

2. Acid-washed glass beads (0.5 mm) (Sigma Chemie, Deisenhofen, Germany).

3. 4× sodium dodecyl sulfate-polyacrylamide gel electrophoresis (SDS-PAGE) gel loading buffer: 4% (w/v) SDS, 40% (v/v) glycerol, 0.04% (w/v) bromophenol blue, 20% (v/v) mercaptoethanol, 250 mM Tris-HCl, pH 6.8.

4. 10–20% Tricine-gel (Anamed, Bensheim, Germany).

5. 10× running buffer: 0.1 M Tris-HCl, 0.1 M Tricine, 0.1% (w/v) SDS.

6. Molecular weight marker: Precision plus Protein Standard (Bio-Rad, Hercules, CA).

7. 1× transfer buffer: 0.19 M glycine, 25 mM Trizma base, 10% (v/v) methanol.

8. Blocking solution: 10% (w/v) milk powder in 1× phosphate-buffered saline (PBS) (137 mM NaCl, 2.7 mM KCl, 10 mM Na_2HPO_4, 2 mM KH_2PO_4).

9. Primary antibody: 3F4 (Sigma Chemie).

10. Secondary antibody: anti-mouse IgG horseradish peroxidase (Promega, Madison, WI).

11. Enhanced chemiluminescence (ECL) reagent A: 50 mg of luminol in 200 ml of 0.1 M Tris-HCl, pH 8.6.

    a. Store at 4°C.

12. ECL reagent B: 22 mg of *p*-hydrocycoumarinic acid in 20 ml of DMSO.

    a. Store at room temperature in the dark.

2.4. ICP-MS Analysis

1. ICP-MS 4500 Serie 300 Shield Torch System (Agilent Technologies, Palo Alto, CA) containing an ASX 500-autosampler (Cetac Technologies, Omaha, NE).

2. Argon 5.0 (Messer-Griesheim, Sulzbach, Germany).

3. Multi element 1 calibration standard 100.

4. Microwave: Ethos 900 (MLS) with temperature control ATC-CE, 400 mg/l (Merck, Darmstadt, Germany) and segment rotor MPR-600.

5. Rhodium calibration standard, 1,000 mg/l (Merck).
6. 65% HNO$_3$, suprapure (Merck).
7. Tune solution: 10 ng/nl Li, Y, Ce, Tl, Co; matrix 0.2% HNO$_3$ (Agilent Technologies).
8. 30% H$_2$O$_2$, (Merck).

3. Methods

In general, the methylotrophic yeast *P. pastoris* provides the advantage of high protein expression rates, and it possesses a high genetic stability. In comparison with *Saccharomyces cerevisiae*, *P. pastoris* provides a strongly reduced ethanolic fermentation, thereby inhibiting the accumulation of toxic ethanol concentrations despite high cell density. Additionally, hyperglycosylation is not as pronounced in *P. pastoris* as in *S. cerevisiae*. Also, *P. pastoris* has several advantages of expression in a eukaryotic system, such as protein processing, folding, and posttranslational modifications. Growth and maintenance is easy, fast, and cheap, with a generation time of 2–4 h. Another advantage of the *P. pastoris* strain SMD1168 is its protease deficiency, leading to high yields of recombinant protein (*4*). This system was already used for the expression of a set of functional active proteins as human lysosomal α-mannosidase, enterokinase, and invertase (*5*).

Especially for experiments concerning metal ion homeostasis in cells, yeast offer an appropriate system because most of what we know about Cu transport in eukaryotes is derived from yeast studies (*6–8*). Because of the high structural homologies of proteins in yeast and humans counterparts can reciprocally compensate the function of each other (*9*). Moreover, the yeast genome does not encode a direct structural homologue of PrP, making it an attractive model system for analysing the role of PrP in Cu homeostasis.

3.1. Cloning

3.1.1. Polymerase Chain Reaction (PCR) and DNA Preparation for Ligation

1. The PrP sequence to be inserted needs to be transferred in the yeast expression vectors pPICZαA/B/C (*see* **Note 2**). Thus, the designated sequence must be amplified by PCR.
2. The PCR reaction is done in a total volume of 50 µl. End concentrations of PCR mix are 20 pmol of synthetic oligonucleotides (*see* **Note 3**), 5 ng of DNA, 1x Pfu buffer, 1× dNTPs, and 5 U/µl Pfu polymerase. All pipetting is done on ice (*see* **Note 4**).

3. A Master Cycler Gradient (Eppendorf, Hamburg, Germany) or any other thermocycler can be used to run the following program: (1) 95°C for 2 min; (2) 95°C
   for 30 s; (3) 58°C for 30 s; (4) 72°C for 3 min; (5) go to step 2; 29×; (6) 72°C for
   7 min; and (7) hold 4°C (*see* **Note 5**).
4. PCR products are mixed with 6× DNA buffer and loaded adjacent to a 1-kb
   DNA ladder to a 1% agarose gel. Ethidium bromide is added to an end concentration of 1 µg/ml to the heated agarose solution (*see* **Note 6**). Running conditions are 120 V for 1 h in TBE buffer.
5. Bands at the right molecular weight are cut out of the gel, and they are prepurified by using a QIAquick gel extraction kit according to the manufacturer's
   instructions (QIAGEN).

3.1.2. DNA Ligation and DNA Preparation of Transformation

1. 50 ng of vector DNA (pPICZαA) is mixed with a threefold excess of amplified and
   prepurified PCR product and diluted with double-distilled water (ddH$_2$O) to 10 µl.
2. 10 µl of 2× Quick ligation buffer and 1 µl of Quick T4 DNA ligase is added after
   an incubation period at room temperature for 15 min. The reaction is blocked by
   short cooling on ice.
3. The QIAquick PCR purification kit is used to prepare DNA for transformation
   according to the manufacturer's instructions (QIAGEN).
4. Vector DNA (pPICZαA) and DNA to be cloned need to be incubated with
   restriction enzymes to allow ligation. One unit of restriction enzyme per micro-
   gram of DNA plus 100 µg/ml bovine serum albumin in combination with 1×
   buffer belonging to the enzyme is used. Incubation is done at optimum working
   temperature of the enzymes for 60–90 min. Reaction is stopped at 65°C for
   20 min. Here, EcoRI and NotI were used for cloning (*see* **Note 7**).

3.1.3. Electroporation and DNA Preparation

1. 10 pg–25 ng vector DNA (pPICZαA) and 40 µl of electrocompetent *E. coli* are
   mixed and incubated for 1 min on ice after transformation in a precooled electroporation cuvette. Electroporation is performed at 25 µF, 1.8 kV, and 200 Ω
   (*see* **Note 8**).
2. 1 ml of prewarmed low-salt LB medium is added and incubated under continuous shaking for 1 h at 37°C (*see* **Note 9**).
3. Aliquots of 50 µl are scratched out on low-salt LB plates containing Zeocin.
   Incubation is done overnight at 37°C.
4. Clones are picked and inoculated in 3 ml of low-salt LB medium containing
   Zeocin. Growth of bacteria was allowed for 8 h at 37°C at vigorous shaking after
   transfer of 1 ml of the preculture in 250 ml of low-salt LB medium containing
   Zeocin. A new incubation at 37°C for 16 h is accomplished.

5. Cells are centrifuged at $6,370 g$ at 4°C for 10 min, and DNA extraction is done using the Nucleo Bond PC 500 kit following the manufacturer's instructions (Machery Nagel).
6. Before transformation in yeast, vectors need to be linearized. This is done by restriction digestion (*see* **Subheading 3.1.2**) with SacI. DNA (20 μg) is needed for electrotransformation. It also is recommended to sequence the construct to check correct orientation of the gene to be expressed. 3'-AOX1 and 5'-AOX1-*Pichia* primers are available for this purpose.

3.2. Yeast Culture (P. pastoris)

3.2.1. Preparation of Electrocompetent Yeast and Electroporation

1. 5 ml of YPD medium is inoculated with SMD1168 under vigorously shaking at 30°C for 1–2 days. After transformation in 500 ml of YPD, shaking is repeated until an optical density $(OD)_{600 nm}$ of 1.0–1.6 is reached. Cells are harvested by centrifugation at $1,500 g$ for 10 min (*see* **Note 10**).
2. Sediments are first resuspended in 500 ml of sterile ddH$_2$O followed by 250 ml and centrifuged (*see* **step 1**). Resuspension is done first in 20 ml of 1 M sorbitol followed by 2 ml. Yeast cells can be stored up to 1 week at 4°C.
3. 5–20 μg of linearized DNA and 80 μl of electrocompetent yeast are mixed and incubated for 5 min on ice after transformation in a precooled electroporation cuvette. Electroporation is performed at 25 μF, 1.5 kV, and 400 Ω. As a negative control (mock), cells with just vector DNA missing any insert DNA also are transfected.
4. 1 ml of 1 M sorbitol is added and incubated for 1 h at 30°C.
5. Aliquots of 200 μl are scratched out on YPDS selection plates. Incubation is done over for 3–4 days at 30°C (*see* **Note 11**).

3.2.2. Long-term Storage of Yeast Clones

1. Cells cultivated and harvested as described in **Subheading 3.2.1.** are resuspended in 5 ml of YPDG to get an $OD_{600 nm, 1 cm}$ of 50–100. Aliquots to 50 μl are shock-frozen in liquid nitrogen and stored at −80°C.

3.2.3. Protein Expression in P. pastoris

1. Several transformed yeast clones are inoculated in 25 ml of BMGY medium and shacked at 30°C for 16 h.
2. Preculture is divided in three parts, and yeast cells are harvested at 4°C and $1,500 g$. Two of the sediments are used to prepare glycerol stocks for long-term storage (*see* **Subheading 3.2.2.**). The third cell sediment is used to inoculate BMMY medium

containing 2% methanol to an $OD_{600nm} = 1.0$. To investigate the effect of copper and manganese on intracellular metal content, induction medium is adjusted to 1 mM $Cu(II)Cl_2$ or 1 mM $Mn(II)Cl_2$, respectively, by using 1 M stock solutions.

3. To control expression levels, 1 ml of yeast cells culture is retained at different time points of the induction culture (recommended: 0, 6, 12, 24, 36, 48, 60, 72, 84, and 96 h). Samples are centrifuged for 2 min at 10,000 g. Sediments and supernatants are stored at −20°C for further analysis (*see* **Subheading 3.3.4.** and **Note 12**).

4. Glycerol stocks of yeast clones with high expression levels are cultured as described in **Subheading 3.2.3** and induction is terminated at the determined time point showing highest protein expression and slightest degradation.

3.2.4. Mechanical Disruption of Yeast Cells

1. Cell sediments corresponding to 1 ml of yeast culture are resuspended in 110 µl of breaking buffer, and an equal volume of acid-washed glass beads is added. For comparison of protein expression levels in different cultures, 8×10^7 cells were used, where $1\ OD_{600nm} = 5 \times 10^7$ cells/ml.

2. Samples were treated by eight cycles of vortexing for 30 s and alternating cooling for 30 s on ice.

3.2.5. SDS-PAGE and Western Blotting

Culture supernatant and lysed cell pellets were analyzed for protein expression. Instructions assume the use of the Anamed system, but they are easily adaptable to any other gel running system.

1. Precasted SDS gels are rinsed with H_2O and placed in the running chamber. Then, 1× running buffer is filled inside and outside the gels.

2. 10 µl of lysed cell pellets and 25 µl of supernatant are mixed with 4× SDS-PAGE gel loading buffer, boiled for 7 min, and loaded adjacent to a molecular weight standard on a 10–20% Tricine gel.

3. Gels are run for 60–90 min at 120 V.

4. To transfer the proteins from the gels to nitrocellulose tank, blotting can be done. The gel-nitrocellulose-sandwich is adjusted as follows: foam pieces are soaped in 1× transfer buffer and placed to each side of the transfer cassette. Two pieces of wetted 3 MM paper are placed to one side. On top of it, the gel is arranged and covered with the bright side of the nitrocellulose. After finishing the sandwich with two further pieces of 3 MM paper, air bubbles air eliminated by carefully rolling a glass pipette on top of the sandwich.

5. The cassette is placed in the transfer tank filled with 1× transfer buffer such that the nitrocellulose membrane is between the gel and the anode. Transfer is accomplished at 4°C at 380 mA for 3 h.

6. Once the transfer is complete, unspecific binding sites are blocked by incubation in blocking solution for 1 h followed by rinsing in 1× PBS three times for 10 min on a rocking platform.

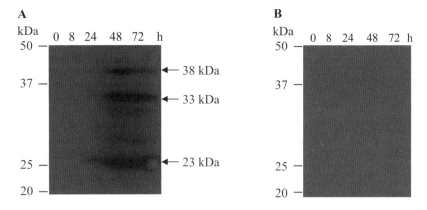

Fig. 1 Control of PrP expression in *P. pastoris* at different time points of induction. (**A**) Ten microliters of lysed cell pellets is separated in a 10–20% Tricine gel. Detection was done using the primary antibody 3F4. Results show that highest expression levels of PrP are reached after 48 h. The band migrating at 38 kDa is already weakened at 72 h of induction. (**B**) Twenty-five microliters of supernatant is separated in a 10–20% Tricine gel. Absolutely no signal could be detected after development with the primary antibody 3F4. This shows that PrP23-230 is not getting secreted into the culture medium. Arrows indicate PrP-specific bands at 38, 33, and 23 kDa

7. Incubation with first antibody (3F4) in a dilution of 1:10,000 in 1× PBS is performed for 1 h (*see* **Note 13**).
8. After three times rinsing for 10 min in 1× PBS, the secondary antibody is added in a dilution of 1:10,000 in 1× PBS.
9. The secondary antibody is discarded, and the membrane is washed three times with 1× PBS.
10. Detection of immunoreactivity is done by mixing 1 ml of ECL reagent A plus 1 ml of ECL reagent B with 0.3 µl of H_2O_2. This mixture is added immediately to the nitrocellulose, which is then rotated by hand for 1 min.
11. Nitrocellulose is transferred to X-ray film cassette and placed between two pieces of overhead foils avoiding air bubbles.
12. Exposure to ECL-Hyperfilms takes approximately a few minutes in the dark room until signal strength is strong enough for film development. An example is shown in **Figure 1**.

3.3. ICP-MS Analysis

3.3.1. Sample Preparation

1. After 48 h of induction, yeast cells of 15-ml culture are harvested (*see* **Subheadig 3.2.3**). To determine the amount of cells, cultures are diluted 1:1,000 in 1× PBS and counted in a Neubauer cell chamber.

2. Cell sediments for ICP-MS analysis are resuspended in 1 ml of H_2O, washed once with 0.5 M EDTA, pH 8.0, and then resuspended again in 1 ml of H_2O.

3. For microwave digestion, samples are resuspended in 1 ml of HNO_3 and transferred into 6-ml TFM inserts. Three TFM vials are placed in each sample holder and put into a polytetrafluoroethylene (PTFE) vessel. Then, 10 ml of a 1:1 mixture of H_2O/H_2O_2 is used as regeneration solution and filled into the internal space of the PTFE vessel. Closed PTFE vessels are placed into the MPR-rotor (medium pressure segment) and positioned into the microwave. Before digestion is started, the temperature sensor is attached to the rotor unit (*see* **Note 14**).

4. Microwave digestion is done using the following protocol: (1) 5 min, 250 W, 170°C; (2) 1 min, 0 W, 0°C; (3) 5 min, 250 W, 190°C; (4) 5 min, 450 W, 210°C; (5) 3 min, 550W, 210°C; and (6) 7 min, 300 W, 210°C.

5. After 1 h of annealing, the reaction container is opened, and samples are transferred in 15-ml reaction tube and diluted 1:10 with H_2O to adjust samples to an end concentration of 6.5% HNO_3.

6. To minimize metal contamination and preparation of sample vials for next applications vials are filled again with 1 ml of 65% HNO_3, and microwave digestion is repeated after rinsing in H_2O.

3.3.2. ICP-MS Measurement

1. Instrument parameters need to be checked before each run using a specific tune solution. Sample measurement can be started if all instrument specifications are correct. The purpose of this step is to achieve maximum count rate and minimum relative standard deviation for a specific atom mass unit in the selected tuning solution.

2. The apparatus is calibrated by using a 6.5% HNO_3 solution containing relevant elements at 1, 5, 10, 25, 50, 100, and 200 ppb, with Rh-103 as internal standard for all isotopes.

3. Measurements are done in peak-hopping mode, with spacing at 0.05 atomic mass units, three points per peak, three scans per replicate, two replicates per sample, and an integration time of 500 ms/point. The rate of plasma flow used is 15.8 l/min, with an auxiliary flow of 1.0 l/min and a blend gas flow rate of 0.1 l/min. The radio frequence power is 1.21 kW. The sample is introduced by using a crossflow nebulizer at a flow rate of 1.02 l/min. Samples are measured three times.

4. Raw data is further evaluated using a specific evaluation software (Agilent Technologies).

5. Data are normalized by the amount of yeast cells to compare the determined intracellular amount of metal ion per cell.

6. Statistical analysis is performed using three independent measurements by calculating the SEM. Statistical significance is determined by Student's *t* test.

An example of ICP-MS measurements with yeast cells is shown in **Figure 2**.

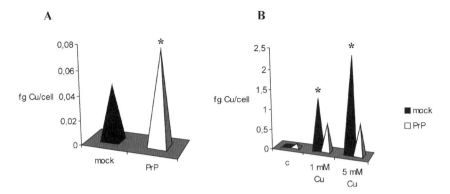

Fig. 2 Effect of PrP expression on intracellular copper concentration by ICP-MS. PrP and mock-transfected yeast cells were grown for 48 h in BMMY medium. (**A**) The increase of intracellular Cu is significant for PrP versus mock-transfected cells ($P < 0.00001$). (**B**) A statistically significant increase of intracellular copper was found between yeast cells grown in normal induction medium and yeast cells grown in copper-supplemented induction medium ($P < 0.0001$). When the cells were grown in the presence of 1 or 5 mM Cu, the increase of intracellular Cu was limited by PrP expression to 14-fold for both concentrations tested (white pyramids). In contrast, mock cells showed a 52-fold increased intracellular Cu level upon treatment with 1 mM Cu, which further increased to 94-fold when cells were grown in medium supplemented with 5 mM Cu (black pyramids)

Acknowledgements The author would like to thank G. Multhaup for his advice and encouraging discussion and M. Nuendel for technical assistance.

4. Notes

1. The use of sorbitol in selection plates is optional. But inclusion stabilizes electroporated cells, because they seem to be somewhat osmotically sensitive.
2. Vectors pPICZαA/B/C differ in their multiple cloning sites and in their reading frames, and they need to be chosen depending on the insert to be cloned.
3. Oligonucleotides should contain equal amounts of G/C and A/T ratios, and they should be between 12 and 50 bp. Melting temperature (Tm) of oligonucleotides is evaluated by the formula $Tm = 4°C (G + C) + 2°C (A + T)$. Annealing temperature should be chosen 4°C below Tm.
4. To improve yield of PCR, 1–2.5 mM $MgCl_2$ can be added to the master mix. Enhancement of PCR specificity is often reached by formamide (<5%) or dimethyl sulfoxide (<10%).
5. The lid of the thermocycler is prewarmed to avoid sample condensation. The length of the polymerization step depends on the length of the DNA fragment to be amplified (~1 min/1,000 bp). Denaturation should be as short as possible to avoid premature inactivation of the polymerase.
6. Ethidium bromide is carcinogen, and it should be handled only with gloves.

7. It is important to ensure that the insert does not contain the restriction sites used for digestion.
8. Electroporation is the most efficient methodology, and it is the method of choice for big plasmids, but chemical transformation also can be done.
9. pPICZα vectors contain the Zeocin resistance (*Sh ble*) gene to allow selection of the plasmid by using Zeocin. Low-salt LB medium (0.5%, w/v, NaCl) is necessary because high-salt LB medium (1%, w/v, NaCl) and acidity and basicity would inactivate Zeocin.
10. The strain SMD1168 contains the AOX1 gene, and it is therefore able to metabolize methanol-inducing expression of the recombinant protein.
11. Zeocin is only required for initial screening and selection of recombinant clones. It is not necessary to maintain selected clones in medium containing Zeocin.
12. All pPICZα vectors contain a *S. cerevisiae* α-factor secretion signal. Recombinant proteins are expressed as fusions to this N-terminal peptide to allow directed secreting of recombinant protein. The α-factor signal sequence is processed first at the specific Kex2 cleavage site. Also, remaining Glu-Ala repeats are further cleaved by the STE3 gene product to result in native recombinant protein containing only very few extrinsic amino acids.
13. The antibody 3F4 recognizes the epitope 109–112 of mouse PrP sequence. pPICZα vectors offer the opportunity to express recombinant proteins not only fused to the N-terminal signal peptide but also to C-terminal peptides containing the c-myc epitope and the polyhistidine tag to allow detection with commercially available tag-specific antibodies. If expression of protein without the C-terminal peptides is desired, a stop codon needs to be inserted. Antibodies can be saved for subsequent experiments by addition of 0.02% final concentration of sodium azide and storage at 4°C.
14. Only suprapure reagents and Milli-Q water should be used to avoid external metal contaminations. The amount of 65% HNO_3 used for microwave digestion is variable, but it needs to be kept as low as possible to ensure the metal concentration of the sample in the detection level of the ICP-MS device. The end concentration of 6.5% HNO_3 in the sample finally analyzed by ICP-MS is fixed.

References

1. Brown, D. R., Qin, K., Herms, J. W., Madlung, A., Manson, J., Strome, R., Fraser, P. E., Kruck, T., von Bohlen, A., Schulz-Schaeffer, W., Giese, A., Westaway, D., and Kretzschmar, H. (1997) The cellular prion protein binds copper in vivo. *Nature* **390,** 684–687.
2. Thackray, A. M., Knight, R., Haswell, S. J., Bujdoso, R., and Brown, D. R. (2002) Metal imbalance and compromised antioxidant function are early changes in prion disease. *Biochem J* **362,** 253–258.
3. Treiber, C., Simons, A., and Multhaup, G. (2006) Effect of copper and manganese on the de novo generation of protease-resistant prion protein in yeast cells. *Biochemistry* **45,** 6674–6680.

4. White, C. E., Hunter, M. J., Meininger, D. P., White, L. R., and Komives, E. A. (1995) Large-scale expression, purification and characterization of small fragments of thrombomodulin: the roles of the sixth domain and of methionine 388. *Protein Eng* **8,** 1177–1187.

5. Liao, Y. F., Lal, A., and Moremen, K. W. (1996) Cloning, expression, purification, and characterization of the human broad specificity lysosomal acid alpha-mannosidase. *J Biol Chem* **271,** 28348–28358.

6. Huffman, D. L., and O'Halloran, T. V. (2001) Function, structure, and mechanism of intracellular copper trafficking proteins. *Annu Rev Biochem* **70,** 677–701.

7. Labbe, S., and Thiele, D. J. (1999) Pipes and wiring: the regulation of copper uptake and distribution in yeast. *Trends Microbiol* **7,** 500–505.

8. Thiele, D. J. (2003) Integrating trace element metabolism from the cell to the whole organism. *J Nutr* **133,** 1579S–1580S.

9. Pena, M. M., Lee, J., and Thiele, D. J. (1999) A delicate balance: homeostatic control of copper uptake and distribution. *J Nutr* **129,** 1251–1260.

Chapter 13
Biophysical Investigations of the Prion Protein Using Electron Paramagnetic Resonance

Simon C. Drew and Kevin J. Barnham

Summary The binding of paramagnetic metal ions is thought to be an essential function of the prion protein and lends itself to interrogation by electron paramagnetic resonance (EPR), which probes the local coordination environment of bound metal ions to provide details of the metal-binding affinity, stoichiometry, and the symmetry and identity of its ligating atoms. It is also capable of identifying reactive oxygen/nitrogen species and peptide-derived radicals, in addition to monitoring protein-membrane dynamics and conformation by using site-directed spin labeling. An overview of the EPR technique as applied to the prion protein is given, key results are summarized, and some future experimental avenues are outlined.

Keywords Magnetic resonance; metalloprotein; site-directed spin labeling; spin-trapping.

1. Introduction

Electron paramagnetic resonance (EPR) spectroscopy, also known as electron spin resonance, measures the resonant absorption of microwave radiation by paramagnetic materials in the presence of a uniform magnetic field. It is a particularly useful technique for probing the local structural environment of a paramagnetic centre by characterizing the interaction of the magnetic moment produced by its unpaired electrons with the local electric and magnetic fields produced by its environment. In biological magnetic resonance, these unpaired electrons are associated with transition metal ions and free radicals.

In proteins, paramagnetic centers may be encountered as naturally stable cofactors (involving metal ions such as Cu^{2+}, Fe^{3+}, and Mn^{2+}), transiently generated radicals (such as reactive oxygen species and protein-derived Tyr, His, and Trp radicals), or more stable nitroxide radicals associated with spin labels and spin traps. It frequently occurs that metalloproteins are too large for detailed structural analysis by nuclear magnetic resonance (NMR), or that single crystals are not

From: *Prion Protein Protocols.*
Methods in Molecular Biology, Vol. 459.
Edited by: A. F. Hill © Humana Press, Totowa, NJ

available for structure solution by X-ray crystallography. Even when crystallographic structures are available, one must still relate this geometric structure to the electronic structure and function at a fundamental quantum mechanical level. In these instances, EPR techniques can fill the gaps to provide more detailed information about a protein's active site(s).

Every paramagnetic transition metal binding site in a protein has a characteristic EPR spectrum. The prion protein (PrP) is known to have at least five binding sites (**Fig. 1**); hence, EPR is a useful probe of this metal binding. Histidine-containing fragments of the PrP are reported to have an affinity for various transition metal ions in the order: $Pd^{2+} \gg Cu^{2+} \gg Ni^{2+} \geq Zn^{2+} > Cd^{2+} \sim Co^{2+} > Mn^{2+}$ *(1)*. Binding by Pd^{2+}, Zn^{2+} and Cd^{2+} cannot be monitored by EPR because these ions are not paramagnetic, while the apparently low affinity of PrP for paramagnetic ions other than Cu^{2+} has meant that EPR studies to date have almost exclusively examined Cu^{2+} coordination. However, it is worth noting that a complete and reversible copper displacement in PrP by Co^{2+}, Ni^{2+}, and Mn^{2+} has been reported in bovine brain homogenate under reductive conditions, the uptake of Cu^{2+} being restored in an oxidizing medium *(2)*. The observation that incorporation of Mn^{2+} imparts protease resistance to cellular PrP *(3)* left plenty of scope for EPR investigation of Mn^{2+}-bound PrP, although relevant studies to date are scarce *(4, 5)*, perhaps due in part to the difficulty in extracting useful structural information from Mn^{2+}-bound protein EPR spectra.

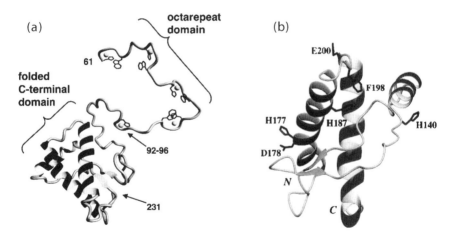

Fig. 1 (**A**) Three-dimensional rendering of PrP(61–231) with Cu^{2+} ions included in the N terminus. The Cu^{2+} binding to the PrP(92–96) segment is postulated based upon the results of Burns et al. *(43)*; however, His111 also is thought to be involved in Cu^{2+} binding in the unstructured 91–115 region, possibly acting as a coligand with His96 *(47)* (*see* **Subheading 4.2.**). (Reprinted with permission from Burns et al. *(43)*. Copyright (2003) American Chemical Society). (**B**) Three-dimensional structure of the folded C-terminal domain of PrP (PrP(121–231)) showing the location of the potential Cu^{2+} binding sites around His140, His177, and His187, together with the location of point mutations associated with Gerstmann-Sträussler–Scheinker disease (F198S), Creutzfeld–Jakob disease (E200K), and fatal familial insomnia (D178N). (Reprinted with permission from Cereghetti et al. *(48)*. Copyright (2003) American Chemical Society)

The body of work using EPR spectroscopy to probe PrP and its related model compounds is substantial. Although not a "how-to" guide, the present chapter discusses the contributions the technique has made to the study of mammalian prion disease, with a bias toward citing the most recent and relevant contributions; the references herein can be pursued for further detail. Unless otherwise specified, the abbreviation PrP refers to the native α conformation (cellular form) of the prion protein and residues are numbered according to the human PrP sequence.

2. Materials

2.1. EPR Spectrometer

Several texts and monographs cover the principles of EPR theory and practice (6–10) and an excellent review of the EPR of Cu^{2+} in biological systems is given by Boas et al. (11, 12). We, therefore, briefly summarize only the key aspects. A summary of the instrumentation is described in **Fig. 2.**

In the simplest case of a spin 1/2 system such as Cu^{2+}, a relatively large external magnetic field of magnitude B interacts with the (spin) magnetic moment of the single unpaired electron of each Cu^{2+} ion and splits its "spin up" and "spin

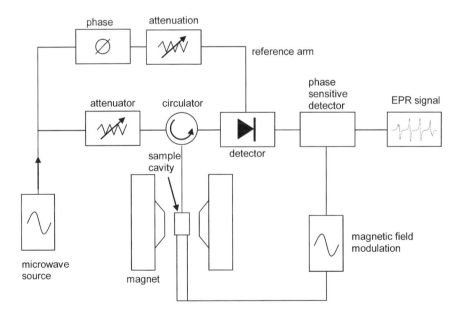

Fig. 2 Block diagram of a basic CW-EPR spectrometer. A simplified block diagram of a pulsed EPR spectrometer can be found in **ref. 6**

down" states by an amount $g\beta B$. If one applies an alternating magnetic field (microwave radiation) perpendicular to the static field, some of the lower energy electrons in the spin down state will absorb energy from the microwave radiation and make a transition to the higher energy state. This resonance condition occurs when

$$hv = g\beta B \tag{1}$$

where h is Plank's constant, v is the microwave frequency, β is the Bohr magneton, and g is the spectroscopic "g-factor." An EPR spectrum is a measure of the sample's microwave absorption and can be acquired in practice by either varying the strength of the applied magnetic field ($\sim10^2$–10^4 Gauss) at a fixed microwave frequency (typically 1–100 GHz) or by varying the frequency at a fixed magnetic field. For technical reasons, most EPR spectroscopy is performed under the former condition (*6*).

There are broadly two types of EPR spectroscopic experiment, depending on whether continuous (CW-EPR *[6]*) or pulsed (pulsed-EPR *[9]*) microwave radiation is used. The former involves constant application of continuous low-power (approximately milliWatts) microwaves, whilst the magnetic field strength is varied, whereas the latter uses short ($\sim$10-ns), high power (approximately kiloWatt) microwave pulses at one or more discrete magnetic fields. Pulsed-EPR encompasses Fourier-transform (FT) EPR (the analog of FT-NMR), in addition to other time-domain techniques such as electron spin echo envelope modulation (ESEEM) spectroscopy (*13*) and pulsed electron-nuclear double resonance (ENDOR) and pulsed electron-nuclear triple resonance (*9*).

Because relatively few unpaired electrons are excited to their higher energy state, the microwave absorption is rather weak. A signal detection system known as phase-sensitive detection (PSD) is, therefore, used in a CW-EPR spectrometer to extract the signal of interest from the background noise level. Due to the way PSD is implemented, most published CW-EPR spectra are presented as the first derivative of the absorption spectrum (i.e., the rate of change of the absorption as a function of magnetic field), which gives EPR spectra their rather unique appearance.

2.2. Sample Preparation

2.2.1. Peptide Concentration

The minimum Cu^{2+} concentration required for X-band ($\sim$9 GHz) CW-EPR is $\sim$20 μM, whereas at S-band ($\sim$2–4 GHz), where sensitivity is generally lower, $\sim$0.5–1 mM is desirable. The minimum peptide concentration required is determined by the desired molar ratio of Cu^{2+} added. The concentration of rec PrP commonly used for routine X-band EPR is $\sim$100 μM.

2.2.2. Sample Containment

For frozen solution EPR, typically 3-mm-i.d. quartz tubes (Wilmad-LabGlass, Buena, NJ). Purge tubes with dry N_2 gas before sealing to avoid liquid oxygen at low temperatures. Plastic caps supplied with tubes should be discarded for low-temperature EPR, because these caps can form hazardous projectiles during sample transfer; seal with parafilm. Any tube should be checked for baseline signal as a matter of routine.

For metal-binding or spin trapping/labeling experiments in the liquid phase, ordinary 3-mm-i.d. tubes cannot be used for solvents with high dielectric constant (e.g., water) due to the strong absorption of microwave radiation at X-band frequencies. Instead, either a quartz flat cell, one or more ~50-μl capillary tubes, or folded lengths of Teflon tubing can be used. Ordinary glass melting point tubes will also suffice, because trace Fe^{3+} contamination does not interfere with the narrow spectral region of interest around $g \sim 2$.

2.2.3. Sample Volume

For frozen solution spectra in 3-mm-i.d. tubes, typically 100–300 μl, or ~25–50 μl/ glass capillary for solution spectra. There is generally no upper limit to sample volume; however, there exists a limited region over which the length of the tube is exposed to a uniform microwave magnetic field within the cavity/resonator. For Cu^{2+} quantitation, (*see* **Subheading 3.5.**), however, it can be helpful to work with sample (and standard sample) volumes that exceed the active cavity length to avoid differences in sample positioning and cavity filling factor.

2.2.4. Sample Temperature

For CW-EPR of frozen solutions, the working temperature is typically in the range 100–150 K by using a liquid nitrogen flow system. The precise working temperature is not critical, because the signal intensity will not vary significantly over this temperature range. Pulsed-EPR experiments rely on the observation of an electron spin echo, which, for transition metals such as Cu^{2+}, requires the use of a liquid helium flow system. Samples for use at low temperatures should be snap-frozen first in liquid nitrogen (outside the magnet) to ensure a truly randomly oriented system is obtained.

Glycerol is commonly added at 10–20% (v/v) as a "cryoprotectant," preventing the separation of aqueous solutions into purified ice and concentrated solute phases upon freezing. The latter reduces resolution due to dipolar Cu^{2+} interactions. Even where no appreciable improvement in CW spectral resolution is observed, glycerol may still be essential for acceptable resolution in time-domain (pulsed-EPR) experiments by lengthening phase-memory times *(42)*. Ethylene glycol is also sometimes used. To ensure there is no interference with the structure of the metal binding site, it is advisable to compare spectra from samples prepared with and without the cyroprotectant.

2.2.5. Metal Addition

Isotopically enriched ^{63}Cu or ^{65}Cu should be used in preference to natural abundance to avoid loss of spectral resolution due to overlapping hyperfine spectra from each isotope. Consideration should be given to whether to refold PrP in the presence of Cu^{2+} *(14)* or to add the metal after refolding. At pH 5.6, Schweiger et al. *(41)* reported no substantial difference in the EPR spectrum of mPrP(23–231) refolded in the presence of Cu^{2+} and that obtained after metal addition to the refolded apoprotein. However, this finding may not be applicable under all conditions.

3. Methods

3.1. Spin Hamiltonian and EPR Spectrum

The interpretation of an EPR spectrum from first principles is complicated *(15)*, so it is conventional to parameterize it in terms of an empirical spin Hamiltonian (H) that accounts for the energy of the various interactions which contribute to the spectrum:

$$H = \beta_e \, \mathbf{B} \cdot \mathbf{g} \cdot \mathbf{S} + \sum_{\substack{k=\text{metal,} \\ \text{ligand}}} \mathbf{S} \cdot {}^k\mathbf{A} \cdot {}^k\mathbf{I} + \mathbf{S} \cdot \mathbf{D} \cdot \mathbf{S} + {}^k\mathbf{I} \cdot {}^k\mathbf{Q} \cdot {}^k\mathbf{I} - {}^k\gamma \mathbf{B} \cdot {}^k\mathbf{I} \qquad (2)$$

where $\mathbf{S}$ and $\mathbf{I}$ are the electron and nuclear vector spin operators, $\mathbf{g}$ and $\mathbf{A}$ are the 3 × 3 electron Zeeman and hyperfine coupling matrices, $\mathbf{D}$ and $\mathbf{Q}$ are the second rank zero-field splitting and nuclear quadrupole tensors, γ the nuclear gyromagnetic ratio, β is the Bohr magneton, and $\mathbf{B}$ the applied magnetic field vector. The summation incorporates the hyperfine, quadrupole, and nuclear Zeeman interactions of the paramagnetic metal ion (or radical) in addition to the superhyperfine interactions with the surrounding nuclei. Distance information pertaining to the kth nucleus is contained in the $^k\mathbf{A}$ matrix, whose anisotropic terms vary with the internuclear distance r from the Cu^{2+} center as $1/r^6$. The form of the spin Hamiltonian parameters in **Eq. 2** yields information about oxidation state and ligand symmetry (geometric structure), and it can be related to the spatial and energetic properties of the molecular orbitals (electronic structure) at a fundamental quantum mechanical level *(15)*.

For metal centers with more than one unpaired electron (i.e., $S > 1/2$), such as high-spin Fe^{3+} and Mn^{2+} (both S = 5/2), the fine-structure term $\mathbf{D}$ accounts for interactions between the unpaired electrons centered on a single ion. It is zero for S = 1/2 centers such as Cu^{2+}. It is also possible to detect resonances due to dipolar coupling and exchange between spatially separated paramagnetic centers, such as Cu..Cu dimers or between metal centers and radicals. These interactions can be accounted for by inclusion of an additional term of the form $\mathbf{S}_1 \cdot \mathbf{D}_{dd} \cdot \mathbf{S}_2$ in the spin Hamiltonian, where $\mathbf{S}_1$ and $\mathbf{S}_2$ correspond to the electron spins of the two distinct paramagnetic centers. Characteristic broadening becomes evident in the $g \sim 2$ region of the spectrum in addition to a weak but frequently observable feature at "half-field" near $g \sim 4$, which is diagnostic of binuclear Cu^{2+} centers with a very specific geometric relationship *(6, 11, 12)*. Analogously to the hyperfine matrix, the dipole-dipole matrix $\mathbf{D}_{dd}$ depends on the interspin distance and simulation of the dipolar-coupled spectrum can yield the distance between the two paramagnetic centers up to approx 8 Å apart. In addition to magnetic dipolar coupling, quantum mechanical exchange interactions arising from the overlap of electronic wave functions can lead to further complications of the EPR spectrum *(6, 12)*, which are not discussed here. However, provided second order exchange effects are negligible, a reliable interspin distance can still be obtained. We also note that for essentially dipolar-coupled Cu^{2+} pairs in the absence of strong antiferromagnetic exchange, the triplet state $S = 1$ is not energetically separated from the singlet state $S = 0$, and, therefore, the dimeric EPR spectrum cannot be "frozen out" at low temperatures.

The effects of the nuclear quadrupole interaction $\mathbf{Q}$ and the nuclear Zeeman interaction (last term of **Eq. 2**) usually remain unresolved in CW-EPR. However, they can be resolved for ligand nuclei using high-resolution pulsed EPR techniques such as ESEEM and ENDOR (*vide infra*). Note that the interactions $\mathbf{g}$, $\mathbf{A}$, $\mathbf{D}$, and $\mathbf{Q}$ are represented by 3×3 matrices because they are generally anisotropic, i.e., their magnitude varies depending on the molecular orientation with respect to the applied magnetic field.

The resolution of various interactions is dependent upon the choice of microwave frequency. The most common microwave frequency used is X-band (9–10 GHz). Higher frequencies (e.g., Q-band, ~35 GHz) yield better resolution of g due to a larger Zeeman splitting, whereas lower frequencies (e.g., S-band, 2–4 GHz) afford better resolution of hyperfine interactions $^k\mathbf{A}$ due to a reduction in the effects of mechanisms such as "strain broadening" (*[6]*, **Fig. 4**).

3.2. EPR Spectra of Cu^{2+} Binding Proteins

For the case of copper that is essentially tetragonally coordinated to PrP, one is dealing with an $S = \frac{1}{2}$ system due to the $3d^9$ electronic configuration of Cu^{2+}, and one unpaired electron resides in a $d_{x^2-y^2}$-based molecular orbital *(6)*, where x and y are directed along the metal-ligand bonds. Spectra of Cu^{2+} bound to PrP are characteristic of "type II" copper complexes *(6, 11, 12, 16)*, which are relatively high symmetry and approximately axial, so the g and ^{Cu}A matrices are both diagonal with principal components $g_{zz} = g_{\parallel}$ and $g_{xx} \approx g_{yy} = g_{\perp}$ (similarly for $A_{\parallel}$ and $A_{\perp}$). Here, the "$\parallel$" and "$\perp$" notation signifies the magnitude of the anisotropic parameter when the magnetic field is aligned parallel or perpendicular to the symmetry axis of the respective interaction. In general, the orientation dependence of g and A parameters is given by *(6, 7, 8)*

$$g^2 = g_{\perp}^2 \cos^2 \theta + g_{\parallel}^2 \sin^2 \theta \tag{3}$$

$$A^2 = \frac{1}{g^2}\left(g_\perp^2 A_\perp^2 \cos^2\theta + g_\parallel^2 A_\parallel^2 \sin^2\theta\right)\tag{4}$$

where θ denotes the angle the magnetic field makes with the "$\parallel$" principal direction of g and A. An intuitive picture of the physical origin of g anisotropy in metalloproteins is given by Palmer *(17)*.

Because single crystals are rarely available, frozen solution spectra are usually required. In this instance, all molecular orientations are possible and an orientationally averaged spectrum results, in which instance prominent features corresponding to the principal values $g_\parallel$, $g_\perp$, $A_\parallel$, and $A_\perp$ can all be simultaneously visible. To a first approximation, $A_\parallel$ and $g_\parallel$ are obtained by measuring the average splitting between each of the $2(^{Cu}I) + 1 = 4$ low-field hyperfine peaks, and their centre of gravity, respectively (**Fig. 3**). More precise values can be obtained using numerical simulations involving exact quantum mechanical solution of **Eq. 2** *(6)*. Here, trial spin Hamiltonian parameters are input, the simulated spectrum is compared with the experiment, and the parameters are iteratively refined until satisfactory agreement

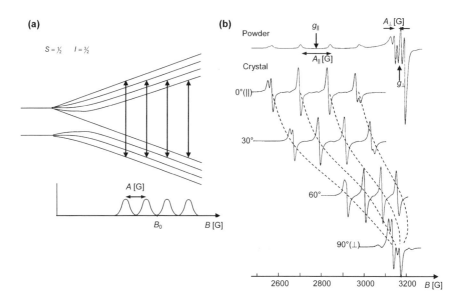

Fig. 3 (A) Energy levels of an $S = 1/2$, $I = 3/2$ system; in the simplest case of an isotropic hyperfine interaction, showing the allowed transitions (*upper*) and the resultant microwave absorption spectrum (*upper*). (B) Typical powder and single crystal EPR spectra of a representative Cu^{2+} complex [Cu^{2+} doped into $CaCd(CH_3COO)_4 \cdot 6H_2O$] with anisotropic g and ^{Cu}A. The powder-like spectrum is obtained from either a polycrystalline solid or a frozen solution and represents a weighted average of all possible molecular orientations with respect to the magnetic field B. First approximations to the principal g and A values can be obtained directly from the spectrum as indicated. Dashed lines schematically depict the anisotropic behavior described by **Eq. 4**. In this example, overlapping hyperfine spectra of the ^{63}Cu and ^{65}Cu isotopes are visible due to a 7% difference in their nuclear magnetic moments. No superhyperfine structure is present due to the absence of ligands with non-zero nuclear spin. (Adapted with permission from Pilbrow *(6)*. Copyright JR Pilbrow 1990)

is achieved. A listing of many of the programs available for spectral simulation and manipulation can be found in the NIEHS EPR Software Database http://tools.niehs. nih.gov/stdb/esdb.cfm.

A word about the choice of units for expressing hyperfine splittings is in order. The hyperfine matrix is a measure of the electron-nuclear interaction energy; hence, it is preferable to use energy units such as wavenumbers (cm^{-1}) or megahertz ($10^{-4}cm^{-1} = 2.9979\,MHz$) rather than field units such as Gauss or Tesla ($1\,T = 10^4\,G$). Using field units is appealing because $A_{\parallel}[G]$ and and $A_{\perp}[G]$ then be estimated directly from the CW-EPR spectrum as indicated in **Fig. 3**; however, the values obtained depend upon the corresponding g-value. For example, $A_{\parallel}[G] = A_{\parallel}[cm^{-1}]$. $hc/(g_{\parallel}\beta_e)$, where h is Plank's constant, $c = 2.9979 \times 10^{10}\,cm.s^{-1}$ and $\beta_e = 9.274 \times 10^{-28}$ $J.G^{-1}$. However, a hyperfine interaction expressed in energy units is field-independent, which makes comparison of principal A values between systems possessing different principal g-values more transparent.

3.3. Determination of Coordination Environment from g and ^{Cu}A

Characterization of the pH sensitivity of Cu^{2+} binding may help in determining its physiological function. It has been suggested that Cu^{2+} stimulates endocytosis of PrP (18); hence, PrP may release this copper in the low pH environment of endosomes (19). As expected for any Cu^{2+} protein, the Cu^{2+} binding of PrP will be highly pH-dependent, as reflected by significant shifts in $g_{\parallel}$ and $A_{\parallel}$ due to changes in the ligand environment of Cu^{2+}. In addition, as the pH is lowered, aqueous copper makes a progressively larger contribution to the EPR spectrum, and it is identifiable by spin Hamiltonian parameters characteristic of planar coordination by four oxygen atoms (16). Control spectra of aqueous Cu^{2+} at low pH can be run to confirm this contribution.

The Blumberg–Peisach plots (16) provide an empirical relationship between $g_{\parallel}$ and $A_{\parallel}$, the number and type of equatorially coordinated atoms (typically O, N, and S) and the overall charge of the metal–ligand complex. For type II (non-blue) Cu^{2+} proteins, the plots are strictly valid only for tetragonal Cu^{2+} coordination, and, at best, they provide a rough guide due to the substantial overlap of regions delineating certain coordination modes (e.g., often "3N1O," "2N2O," and "1N3O" can all be possibilities). Tetrahedral distortion of otherwise tetragonal Cu^{2+} is usually associated with increases in $g_{\parallel}$ and decreases in $A_{\parallel}$, and the degree of distortion can be implied from the ratio of $g_{\parallel}/A_{\parallel}$ (19). More definitive assignments can often be made by resolving superhyperfine splittings in the CW-EPR spectrum and by using pulsed-EPR techniques to resolve superhyperfine, nuclear quadrupole, and nuclear Zeeman interactions with neighboring ligand nuclei (see **Subheading** 3.4). Note that in the original work (16), the Blumberg–Peisach relations were plotted with $A_{\parallel}$ versus $g_{\parallel}$ for a range of natural and artificial copper proteins by expressing $A_{\parallel}$ in the rather uncommon units of millikaiser (mK). The Blumberg–Peisach plots can be converted into more familiar units by noting that $1\,mK = 0.1 \times 10^{-4}\,cm^{-1}$, as done in the text of Pilbrow (6).

Coupling of the paramagnetic center to N equivalent ligand nuclei with spin I generates $2NI + 1$ additional spectral lines *(8)*, and their resolution in the CW-EPR spectrum can provide information about the number of ligands coordinated to Cu^{2+}. Naturally abundant sulfur and oxygen contain $< 1\%$ ^{33}S ($I = 3/2$) and $< 0.1\%$ ^{17}O ($I = 5/2$), respectively; hence, coordinating O and S atoms make no discernible contribution to the superhyperfine spectrum. However, directly coordinated nitrogen atoms ($> 99\%$ ^{14}N, $I = 1$) have nearly isotropic couplings with the unpaired electron of Cu^{2+} with a magnitude in the vicinity of $10–15 \times 10^{-4}$ cm^{-1}. At X-band, it is frequently possible to observe these superhyperfine splittings (the ^{k}A interactions of **Eq. 2**) in the $g_{\perp}$ region of the spectrum, especially if isotopically enriched copper is used to reduce inhomogeneous broadening of the spectral features. In reality, ligand nuclei are frequently inequivalent, and $^{Cu}A_{\perp}$ splittings of a similar magnitude may overlap with ligand superhyperfine structure. Hence, spectral simulations are required to unambiguously determine the number and type of ligand nuclei coordinating the metal binding site. Reduced strain broadening at S-band frequencies also can help to resolve superhyperfine structure in the $g_{\parallel}$ region where $^{Cu}A_{\parallel}$ does not interfere (**Fig. 4**); however, sensitivity is diminished at lower frequency, and, therefore, higher peptide concentrations (~1 mM) are required.

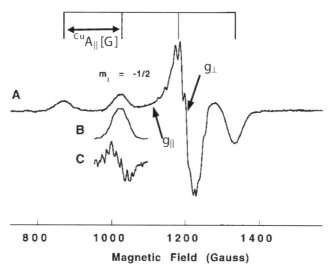

Fig. 4 S-band EPR (3.5 GHz) of ^{63}Cu binding to the octarepeat (PHGGGWGQ) segment. At S-band, the "parallel" and "perpendicular" features partially overlap. However the reduced strain broadening at this frequency narrows one of the ^{63}Cu hyperfine lines ($m_1 = -1/2$) enough to resolve the interactions of the unpaired electron with ligand nuclei. *Trace A* is the full field scan; *trace B* is the $m_1 = -1/2$ line scanned with high resolution and signal averaging; *trace C* is the derivative of B (i.e., the second derivative of the absorption lineshape) emphasizing the superhyperfine structure (Adapted with permission from Aronoff-Spencer et al. *(35)*. Copyright (2000) American Chemical Society)

When the number and type of donor atoms can be determined from the CW-EPR spectrum, site-specific labeling (e.g., ^{15}N, $I = 1/2$ and ^{17}O, $I = 5/2$) may be required to ascertain from which residues they originate. The change in nuclear spin will produce a change in the number of superhyperfine lines only if that residue ligates the Cu^{2+} ion. For example, Burns et al. *(36)* used a combination of site-specific ^{15}N labeling and S-band CW-EPR to show that the fundamental octarepeat HGGGW segment involves coordination by amide nitrogens from the first two Gly residues.

3.4. Determination of Coordination Environment Using Pulsed-EPR

In cases where many overlapping contributions arise from the anisotropic nature of the various interactions, or when there is more than one paramagnetic species causing additional overlap, the spectral lines are often so broadened that it is not possible to resolve the hyperfine coupling between nuclei and the electron spin. In such cases, multidimensional pulsed-EPR extends the capabilities and removes many of the limitations of CW-EPR, even though it can be considerably more complicated to carry out and analyze. Due to its reliance on the observation of an electron spin echo *(9)*, which is usually only visible at low temperatures, use of liquid helium as a cryogen also is required. Furthermore, simulation of pulsed-EPR spectra involves more complicated mathematical techniques than does simulation of CW-EPR spectra *(9)*. A detailed description of the underlying principles of pulsed-EPR techniques is beyond the scope of this review is only briefly mentioned below.

ESEEM *(9, 13)* and ENDOR *(9)* techniques allow the extraction of the ligand superhyperfine parameters often obscured in CW-EPR. They can provide spectral resolution of hyperfine interactions comparable to that seen in NMR. The ESEEM experiment is particularly sensitive to extremely weak couplings from remote nuclei not directly coordinated to the paramagnetic centre, whereas the ENDOR experiment is more sensitive to strongly coupled (directly coordinated) nuclei. Hence, for Cu^{2+} coordination by histidine, the remote amino nitrogen of the imidazole ring is observable using ESEEM, whereas the directly coordinated imino nitrogen is observable using ENDOR. Two-dimensional correlation experiments can be applied, such as the hyperfine sublevel correlation (HYSCORE) experiment *(9, 13)*, yielding further information regarding the number and type of nuclei present and their distance and relative orientation from the paramagnetic centre.

At X-band, it frequently occurs that a special situation known as "exact cancellation" occurs in an ESEEM experiment, which leads to the so-called "3-narrow-plus-a-broad-line" spectrum *(9, 13)*, which is characteristic of the coupling from the remote imidazole N of a coordinated His residue (**Fig. 5**). If there is coordination from more than one His residue, several combination frequencies occur between the narrow and broad peaks. With sufficient spectral resolution, the number of coordinated His residues can be determined, although the combination peaks are not always readily apparent and can even occur as "holes" in the broad background *(20)*.

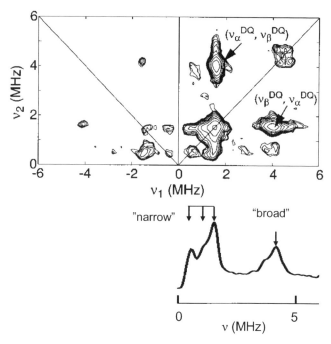

Fig. 5 X-band HYSCORE spectrum (*top*) and three-pulse ESEEM spectrum (*bottom*) of 100µM PrP(23–231) fully loaded with $CuCl_2$ at pH 5.6 and a magnetic field corresponding to $g_\perp$. At this pH, the octarepeat domain is not expected to bind Cu^{2+}. Peaks (cross-peaks) in the one-dimensional (two-dimensional) experiment are characteristic of a single His imidazole coordinating to the Cu^{2+} ion (Adapted with permission from van Doorslaer et al. *(42))*. Copyright (2001) American Chemical Society)

It often occurs that remote amide nitrogen nuclei in the right distance range also may contribute to ESEEM and HYSCORE spectra; these may create peaks in regions where combination lines from multiple His coordination might be expected, thereby potentially complicating interpretation (*see* **Subheading 4.1.**).

3.5. Spin Quantitation and Stoichiometry

The intensity of the microwave absorption is directly proportional to the number of paramagnetic centers. CW-EPR detects the first derivative of the microwave absorption as a function of magnetic field; hence, double integration of the EPR spectrum enables a measure of the number of metal ions present. Comparison with a spectrum of a standard sample of accurately known concentration and volume, run under identical conditions, yields the absolute number of metal ions bound (**Fig. 6**). In conjunction with an accurately known peptide concentration, this further enables the stoichiometry of metal binding to be established but usually with

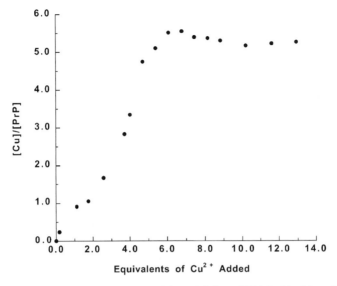

Fig. 6 Integrated EPR intensity as a function of titrated Cu^{2+} at pH 7.4 for 21 μM rec SHaPrP(29–231) at 85 K. (Reprinted with permission from Burns et al. *(43)*. Copyright (2003) American Chemical Society)

no better than approx 10% uncertainty *(12)*. Combined with the uncertainty in spectrophotometric determination of peptide concentrations, the uncertainty in metal-peptide stoichiometry by using EPR can be approx 20% *(43)*. At pH ≥ 7.0, aquo Cu^{2+} precipitates as EPR silent $[Cu(OH)]_n$; therefore, it does not contribute to the doubly-integrated spectrum. It is difficult to distinguish protein-bound EPR-silent Cu^{2+} (such as an exchange-coupled dimer *[6, 12]*) from that not bound by the protein (hydroxo-bridged Cu^{2+}); hence, determination of metal binding stoichiometry can be potentially ambiguous in some instances.

3.6. Radical Spectra, Spin Trapping

EPR spectra of radicals can be detected at room temperature and in frozen solution. Although it is sometimes possible to observe radicals directly in an EPR experiment, in most cases they are short-lived, and they decay within the timeframe of an EPR experiment. Spin trapping *(21–23)* involves the reaction of an unstable free radical with nitroso or nitrone compounds, which react with the free radical to form a more stable radical adduct (aminoxyl radical). Lifetimes are typically minutes to hours.

The resulting nitroxide moieties exhibit a hyperfine interaction between the spin density of the unpaired electron and the nitrogen nucleus of the N–O group of the trap. Anisotropy of *g* is very small and generally unresolved at X-band. For rapidly tumbling species in solution, such as small peptides/complexes, the hyperfine splitting is averaged to yield an isotropic 3-line spectrum due to the $I = 1$ nuclear spin of ^{14}N. The magnitude of the hyperfine splitting is affected by the electron withdrawing nature of the neighboring atoms and the polarity of the solvent. This may be primarily determined by the spin trap itself, as occurs for the nitrone spin traps, or be more characteristic of the atoms of the primary radical as for the nitroso traps *(21, 23)*. The commonly used nitrone spin traps include phenyl-tert-butylnitrone, α-(4-pyridyl 1-oxide)-*N-tert*-butylnitrone, and 5,5-dimethyl-1 pyrroline *N*-oxide (DMPO). Nitroso traps include 2-methyl-2-nitrosopropane. In addition, several newer spin traps have been developed with improved stability and specificity for reactive oxygen species *(24)*, in addition to lipophilic spin traps for detecting lipid peroxidation *(25)*.

Allsop and coworkers investigated Cu^{2+}-dependent H_2O_2 generation of PrP(106–126) and PrP(121–231) in which ferrous iron was added to reduce hydrogen peroxide to ˙OH, which was subsequently trapped by DMPO *(26–28)*. The characteristic 4-line EPR signal from the DMPO/˙OH adduct is quenched in the presence of metal-ion chelators and catalase, indicating that Fenton chemistry involving H_2O_2 takes place. To date, however, there seem to have been no studies of genuine protein-centered radicals, although these are purported to be formed in amyloid β (Aβ) *(29)*, the aggregating peptide implicated in the pathogenesis of Alzheimer's disease. Based upon sequence and structural similarities between PrP and the amyloid β peptide, protein-based tyrosyl radicals also have been postulated to be formed by the prion protein *(36)*.

3.7. Site-directed Spin Labeling (SDSL)

This technique enables analysis of secondary structure and mobility and relies on the covalent attachment of a nitroxide side chain to sulfhydryl group of cysteine residues via a disulfide bond *(31)*. Site-directed mutagenesis is used to introduce cysteine residues at the location where the nitroxide group is to be inserted. The linewidth of the solution EPR spectrum then provides a measure of immobilization of the nitroxide label, and, as a consequence, local protein mobility. SDSL inevitably involves mutations to insert Cys residues and may, in principle, impose unwanted effects on peptide conformation.

An early study by Millhauser, Prusiner, and coworkers *(32)* applied SDSL to a modified SHaPrP(113–120) sequence AGAACAGA to monitor the kinetics of aggregation over a period of 3.5h. The width of the central line of the motionally averaged 3-line hyperfine spectrum remained constant, indicating the absence of oligomerization. In contrast, the integrated EPR spectrum showed a decrease in the sharp 3-line spectrum (assigned to the monomer) and a concomitant increase in an underlying broad spectral component (assigned to the aggregate) arising from exchange-coupled spins in proximity. Subsequent electron microscopy analysis was able to show that the aggregate consisted of both amorphous aggregate and amyloid species.

More recently, Inanami and coworkers *(33, 34)* attached a methane thiosulfonate spin label to mPrP(23–231) with cysteine mutations in Helix 1 and Helix 2. From the differences in EPR linewidths of the nitroxide, they concluded that the region containing the fifth Cu^{2+} binding site around His96 is more flexible than Helix 1 and Helix 2. A linewidth study between pH 4.0 and 7.8 also revealed a dependence of the mobility of Helix 2 on pH and points to its possible role in the pH-dependent conformational change involved in PrP^{Sc} conversion.

4. EPR Studies of Metal Binding Domains of PrP

It is difficult to present a coherent overview of the EPR spectroscopic evidence of the way full-length recombinant PrP behaves in vitro because the issue is highly debated. Although there exists some agreement between different studies, the take-home message is that, much like the Aβ peptide in the Alzheimer's field, Cu^{2+} binding seems to depend critically upon sequence length, sample history, and preparation; buffer conditions; metal binding stoichiometry; and to some degree, even the spectroscopic technique used.

Several EPR studies of the highly-conserved octarepeat (PHGGGWGQ) region of PrP have been carried out by Millhauser and coworkers *(35–39)*. Using various length PrP constructs, they examined the dominant Cu^{2+} interactions with SHaPrP(23–28, 57–91) at pH 7.4, and they observed three distinct coordination modes *(38)* (**Table 1**). The binding mode at low Cu occupancy (≤ 1.0 Eq) involved coordination by the His residues from three to four octarepeats ("component 3"). At intermediate copper occupancy, a second binding mode ("component 2") also coordinated Cu^{2+} via a deprotonated amide nitrogen from His, an imidazole nitrogen from within a single octarepeat, and possibly an axial imidazole nitrogen from a neighboring octarepeat. At full copper occupancy the coordination switched to a mode ("component 1") whereby each HGGGW segment bound a single Cu^{2+} via the His imidazole nitrogen, the deprotonated amide nitrogens of the next two Gly residues, the amide carbonyl oxygen from the second Gly, and an axially coordinated water that hydrogen bonds with the indole NH of the Trp residue. This coordination environment at full occupancy agreed with an X-ray crystallographic structure of the HGGGW segment *(35, 37)*. Dipolar-coupled Cu^{2+} signals also occurred in the CW-EPR spectrum, as confirmed by the presence of a half-field signal at $g \sim 4$ due to the interaction of Cu^{2+} ions separated by 3.5–6.0 Å and thought to be bound by neighboring octarepeats. The affinity for Cu^{2+} binding was observed to decrease by nearly 5 orders of magnitude upon increasing the number of binding equivalents from 1 to 4 *(39)*. This behavior suggested negative cooperativity between binding modes 1 to 3, and it was in direct contrast to many earlier results obtained by different methods where positive cooperativity between the octarepeats was thought to exist *(37, 39* and references therein).

The Cu^{2+} binding modes within the octarepeats region at intermediate Cu^{2+} stoichiometries remain controversial, especially with regard to comparison of results

Table 1.1 Summary of selected spin Hamiltonian parameters of various synthetic and recombinant PrP constructs. Where indicated, values in parentheses represent uncertainty (±) in the last decimal place for g-values and in the absolute uncertainty in A values. Parameters from **ref. 43** represent the dominant spectral feature only. See original references for precise conditions.

System	$g_\parallel$	$A_\parallel$ (MHz)[a]	pH	Reference
Complex 1				
rec mPrP(121–231)	2.332 (5)	452 (10)	3–6	41
rec mPrP(23–231)			4, 6	41
rec mPrP(23–231) D178N			4, 6	48
rec mPrP(23–231) E200K			4, 6	48
Complex 2				
rec mPrP(121–231)	2.295 (5)	457 (10)	3–8	41
rec mPrP(23–231)			4, 6	41
rec mPrP(23–231) D178N			4, 6, 7.4	48
rec mPrP(23–231) E200K			7.4	48
rec mPrP(23–231) H140S			7.4	48
Complex 3				
rec mPrP(121–231)	2.230 (5)	495 (10)	7–8	41
mPrP(58–91)			7.4	41
rec mPrP(23–231)			7.4	41
rec mPrP(23–231) F198S			7.4	48
rec mPrP(23–231) H140S			7.4	48
rec mPrP(23–231) H187S [#]			7.4	48
Complex 4				
mPrP(58–91)	2.270 (5)	520 (10)	6, 7.4	41
Component 1 (cf. complex 3)				
SHaPrP(23–28, 57–91)	2.24 (1)	492 (6)	7.4	38
Component 2 (cf. complex 4)				
SHaPrP(23–28,57–91)	2.27 (1)	530 (6)	7.4	38
Component 3				
SHaPrP(23–28,57–91)	2.25 (1)	576 (6)	7.4	38
Species I				
rechPrP(23–231)	2.249	535[b]	7.0	40
hPrP(60–91)				
Species II				
rechPrP(23–231)	2.264	599[b]	7.0	40
hPrP(60–91)			7.5	40
hPrP(60–67)			7.1	40
rec mPrP(23–231) F198S	2.232 (5)	480 (10)	4	48
rec mPrP(23–231) F198S	2.295 (5)	510 (10)	6	48
PrP(Ac180-193NH$_2$)	2.230 (2)	516 (6)	7.3	50
(cf. complex 3)				
recSHaPrP(29–231)	2.22 (1)	541 (6)	7.47 (6)	43
recSHaPrP(90–231)	2.22 (1)	541 (6)	7.47 (6)	43
PrP(90–101)	2.21 (1)	588 (6)	7.47 (6)	43
PrP(90–116)	2.22 (1)	544 (6)	7.47 (6)	43
PrP(92–96)	2.21	588 (6)	7.47 (6)	43
PrP(94–96)	2.23	562 (6)	7.47 (6)	43
PrP(106–116)	2.23	518 (6)	7.47 (6)	43
PrP(106–126)	2.17	484	7.3	45
PrP(106–126)	2.20	584	7.8	46

(continued)

Table 1 (continued)

System	$g_{\parallel}$	$A_{\parallel}$ (MHz)[a]	pH	Reference
PrP(106–115)	2.27	459	6.5, 7.4	47
PrP(106–115)	2.22	509	7.4, 9.3	47

[a]Hyperfine data is given for ^{63}Cu in MHz, where $A_{\parallel}[MHz]/2.9979 = A_{\parallel}[10^{-4}cm^{-1}] = 10^{4}(g_{\parallel}\beta_{e}/hc) \times A_{\parallel}[G]$, h is Plank's constant, $c = 2.9979 \times 10^{10}$ cm.s^{-1}, and $\beta_{e} = 9.274 \times 10^{-28}$ J.G^{-1}.
[b]Hyperfine parameters in the original reference are quoted in units of mT, but it is assumed here that the intended units were in fact Gauss (1 G = 0.1 mT), otherwise their magnitudes seem non-sensical. The authors *(48)* recommend to interpret data cautiously due to tendency of H187S to aggregate.

obtained from truncated (*vide supra*) and full-length PrP. For example, a combined EPR, ENDOR, and EXAFS study by Parak and coworkers *(40)* of hPrP(23–231) and of PHGGGWGQ and hPrP(60–91) has produced a different model of Cu^{2+} binding at intermediate metal occupancy, in which they attributed the differences to the influence of the inclusion of the full PrP sequence. Near physiological pH, they observed only two binding modes ("species I" and "species II") in both hPrP(60–91) (1.0–4.0 Eq Cu^{2+}) and hPrP(23–231) (approx 2.7 Eq Cu^{2+}) that do not correlate closely with components 1–3 identified by Millhauser and coworkers (*[38, 39]*, **Table 1**) *(38, 39)* (**Table 1**). Species II, which was dominant in hPrP(23–231) with 2.7 Eq Cu^{2+}, was modeled assuming two Cu^{2+} binding modes within the octarepeats domain. The first binding mode involves imidazole nitrogen coordination from His61 and His69 together with the deprotonated amides of Gly70 and Gly71 and an analogous binding mode involving two His and two Gly was postulated for the second two octarepeats. This model is very different from models of components 1–3 found in SHaPrP(23–28, 57–91) at pH 7.4. Moreover, they found no evidence of dipolar coupling between Cu^{2+} centers in hPrP(60–91) with 4.0 Eq Cu^{2+}, nor of binding by His96 or His111. These findings suggest a profound influence of the buffer conditions, sequence, and sample preparation history on the nature of metal binding by the prion protein in vitro.

Van Doorslaer and coworkers have used CW-EPR *(41, 48)*, ESEEM, and ENDOR *(42)* to examine mPrP(23–231) and mPrP(121–231), and they identified three distinct Cu^{2+} coordination types between pH 3.0 and 8.0 (**Table 1**). "Complex 1" was evident in the pH 3.0–6.0 range, "complex 2" in the pH 3.0–8.0 range, and "complex 3" at pH ≥7.0. Species with complex 3 parameters also were observed for mPrP(58–91) *(48)*, with evidence for the involvement of direct coordination by backbone nitrogens *(42)*, which correlates well with the component 1 EPR data for SHaPrP(23–28, 57–91) fully loaded with Cu^{2+} at pH 7.5 (**Table 1**). Only two components in mPrP(58–91) were reported at pH 7.4 when 2 Eq Cu^{2+} were added, one of which was complex 3 and another signal due to "complex 4" (**Table 1**). These findings are fully consistent with the appearance of only component 1 and component 2 for SHaPrP(23–28, 57–91) in the presence of 2 Eq Cu^{2+} *(39)*.

The observation of EPR parameters resembling complex 3 for both mPrP (58–91) and mPrP(121–231), suggested binding sites in the both the N and C termini with near-identical EPR parameters. Up to pH 6.0, the same CW-EPR spectra were observed for both mPrP(23–231) and mPrP(121–231), but not mPrP(58–91), suggesting that complex 2 and (one of the) complex 3 species are associated with the C-terminal part of the protein. Furthermore, it suggested preferential Cu^{2+} binding to the C-terminal region of full-length mPrP at pH <6.0. Similar behavior was noted by the Millhauser group, who observed that SHaPrP(23–28, 57–91) and SHaPrP(29–231) released the majority of its Cu^{2+} below pH 6.0 *(43)*. HYSCORE and ESEEM spectra of mPrP(23–231) fully loaded with $CuCl_2$ indicated Cu^{2+} coordination by a single His imidazole group at pH 5.6, implying that the C-terminal binding involves His coordination (**Fig. 5**).

Both the Millhauser and Van Doorslaer groups indicated that at physiological pH, PrP(90–231) can bind 2 Eq Cu^{2+}. Two contributions were clearly evident from the CW-EPR spectrum of mPrP(90–231) at pH 7.4 (complex 2 and complex 3) *(48)* and a binding of 2 Eq Cu^{2+} to SHaPrP(90–231) and SHaPrP(90–116) (**Table 1**) also was reported, but the second equivalent was only taken up at severalfold excess *(43)*. Similarly, SHaPrP(29–231) at pH 7.4 (**Fig. 6**) was able to bind 6 Eq Cu^{2+}, with the last equivalent once again only being taken up in the presence of excess Cu^{2+} *(43)*, but here they deemed that the second site was of such low affinity that it was not biologically significant and did not pursue it further. The first equivalent of Cu^{2+} bound by SHaPrP(90–231) was assigned to binding by the GGGTH segment corresponding to PrP(92–96), with K_d values in the micromolar range *(43, 44)*. Similar to the octarepeat domain, this binding site was found to lose its affinity for Cu^{2+} binding below pH 6.0. Cu^{2+} uptake of full-length PrP was suggested to occur sequentially, with His96 site taking up the first copper ion followed by uptake in the adjacent octarepeat domain *(43)*. The recent finding that the octarepeat domain binds its first equivalent of Cu^{2+} with very high affinity ($K_d \sim 0.1\,nM$) *(39)* seems inconsistent with this picture, although as usual caution should be exercised when comparing results from different length constructs in isolation.

Evidence exists to support both independent Cu^{2+} binding of His96 and His111, and a binding mode in which both His residues coordinate a single Cu^{2+} ion. Binding by His111 has been shown for the neurotoxic fragment PrP(106–126). Jobling et al. *(45)* studied PrP(106–126) with 2 Eq Cu^{2+} at pH 7.3, suggesting 2N1S1O ligation involving His111, Met109, or Met112 and the N terminus Lys106. Acetylating the N terminus was shown to significantly reduce Cu^{2+} binding, peptide aggregation and neurotoxicity. Belosi et al. *(46)*, in contrast, found no evidence of Met coordination at physiological pH (Table 1). Jones et al. *(47)* examined the Cu^{2+} binding of PrP(91–115) with associated H111A and H96A mutants; they provided evidence of a high-affinity binding site that involves ligation by both His96 and His111 and cited a dissociation constant $K_d \sim$ nanomolar range for PrP(91–115).

To determine which His residues are important for Cu^{2+} binding in the C-terminal domain, the van Doorslaer group examined H140S, H177S, and H187S mutations of the isolated C-terminal domain of mPrP(121–231) by using CW-EPR *(48)*.

Additionally, the pathogenic mutations D178N, F198S, and E200K mPrP(23–231) associated with inherited forms of prion diseases were studied (**Fig. 1**). The D178N mutant aggregated at physiological pH even in the absence of added Cu^{2+}. Although CD spectroscopy revealed no differences in secondary structure compared with wild-type mPrP, the F198S and E200K variants showed different Cu^{2+} binding (**Table 1**), suggesting the presence of a Cu^{2+} binding site in the vicinity of residue 200.

At pH 4, the H140S, H177S, and H187S mutants showed no evidence of Cu^{2+} binding and the spectrum was identical to that of aqueous Cu^{2+}. Complementary CD spectra showed the histidine mutants did not properly fold at this pH. At pH 6.0, all three mutants exhibited a marked tendency to aggregate at 100 µM in the presence of 1 Eq Cu^{2+}. At pH 7.4, H140S was soluble, H187S transiently soluble, and H177S completely insoluble. No evidence of the involvement of His140 in Cu^{2+} binding was found, and no conclusions could be drawn regarding His177 due to insolubility.

Rizzarelli and coworkers *(49, 50)* recently studied the terminally blocked IKQHT segment corresponding to PrP(184–188) and PrP(180–193). Although H187S seemed to have EPR parameters similar to complex 3 (**Table 1**), caution was recommended *(42)* when interpreting this result due to possible aggregation effects. With this in mind, a correspondence was suggested between the EPR parameters for PrP(184–188) and complex 3 observed by van Doorslaer et al. *(42)* between pH 7.0 and 8.0 in PrP(121–231) and PrP(23–231) (**Table 1**), and they postulated this coordination mode may therefore correspond to the binding site anchored around His187. They also suggested this region can bind a single Cu^{2+} more tightly than both the octarepeat region and the peptide fragment PrP(106–126). However, caution should be exercised when comparing isolated peptide fragments.

5. Summary and Outlook

An increasing number of multidisciplinary studies incorporating EPR spectroscopy continue to seek a better understanding of the manner and physiological significance of metal binding to PrP. Although there is still some disagreement as to the coordination mode, affinity, and cooperativity of each site, up to seven putative binding sites have now been identified in natively folded recombinant PrP, four within the octarepeat region, one or two at His96 and His111 in the amyloidogenic region, and another in the C-terminal region around His187.

The overwhelming body of EPR research has focused on metal binding of the isolated PrP. Physiologically, PrP is derived from a cell surface glycoprotein; hence, the interaction of PrP with the membrane surface cannot be neglected. Parak and coworkers *(51)* have provided evidence that in the presence of dodecylphosphocholine (DPC) micelles, the otherwise unstructured octarepeat region of PrP is restricted to form well-defined conformations. This was seen from their EPR spectra of a 1:1 Cu^{2+}:octapeptide complex in aqueous (100 mM KCl) and micellular (DPC/100mM KCl) environments at neutral pH. A better resolved spectrum was obtained in the lipid environment, which may reflect a reduction in strain broadening (*see* **Subheading 3.1.**) and hence a more well-defined binding site, possibly due to a structuring effect of the membrane on the N-terminal region. Conversely, this ordering effect was not as apparent for octarepeat peptides suspended in other aqueous environments containing DPC micelles *(40)*. A more realistic model is one which includes anchoring of PrP to the membrane surface. A glycosylphosphatidylionositol anchor mimetic has recently been synthesized, enabling the study of lipid-anchored PrP constructs *(52)*. No EPR studies to date have examined the effect of such membrane anchoring.

With the use of synthetic lipid environments as membrane mimetics comes the possibility of observing lipid radicals that may be generated as part of pathological processes that involve membrane insertion. Lipophilic spin traps that identify radicals localized at specific depths within the membrane are now being developed *(25)*, offering the possibility of obtaining spatial and molecular information about radicals formed within biological membranes.

In addition to EPR spin trapping, it is also possible to use fluorogenic spin traps. These consist of an EPR-active nitroxide functional group covalently linked to a fluorophore whose fluorescence yield is effectively quenched through an intermolecular electron exchange interaction. Radical trapping by such compounds reduces the nitroxide to a diamagnetic (EPR silent) hydroxylamine, thereby restoring the fluorescence *(53, 54)*. The advantage of fluorescence-based radical trapping is the ability to conduct microtiter plate assays, enabling high-throughput screening of various experimental conditions. The drawback, however, is the loss of chemical information that identifies the nature of the radical trapped. EPR spectroscopy using the same spin traps cannot be used concurrently to provide such information, because the adduct is EPR silent.

Yet another technique derived from EPR radical trapping method is immuno-spin trapping *(55)*. Here, DMPO traps protein-centered radicals, forming a transiently stable adduct. Within a few minutes, it decays to an EPR-silent nitrone that is then detected on a Western blot or with enzyme-linked immunosorbent assay by using an anti-DMPO antibody. The method requires much lower protein concentrations than EPR; hence, spin trapping in cell-based assays becomes possible without the need for unrealistic sample concentrations and without the loss of specificity encountered in fluorogenic spin trapping. This principle can be transferred to other spin traps; however, only an anti-DMPO antibody is currently available.

The disease-related isomer of the PrP is rich in β-sheet structure (β-PrP) and is thought to consist of the C1 and/or C2 fragments after cleavage around residues His111/Met112 and Glu90, respectively *(56)*. Cleavage at the C2 site is particularly associated with disease, and it has been linked to an H_2O_2-induced copper-catalyzed oxidative mechanism mediated by hydroxyl radical-like species. The presence and potential significance of metal binding by soluble monomeric β-PrP in the disease pathway is yet to be established. Given the existence of metal binding sites within C-terminally truncated PrP, EPR studies of recombinant β-PrP refolded in the presence of metal ions therefore represent a novel target. Ultimately, we seek to study

Cu^{2+} binding of PrP isolated from mammalian cell systems, to see whether Cu^{2+} incorporated into the protein by the cells themselves is bound in the same manner as found in recombinant PrP.

References

1. Jószai V, Zoltán N, Ősz K, Sanna D, Di Natale G, La Mendola D, Pappalardo G, Rizzarelli E, Sóvágó I (2006) Transition metal complexes of terminally protected peptides containing histidyl residues. J. Inorg. Biochem. 100: 1399–1409.
2. Deloncle R, Guillard O, Bind JL, Delaval J, Fleury N, Mauco G, Lesage G (2006) Free radical generation of protease-resistant prion after substitution of manganese for copper in bovine brain homogenate. Neurotoxicology 27: 437–444.
3. Brown DR, Hafiz F, Glassmith LL, Wong B-S, Jones IM, Clive C, Haswell S (2000) Consequences of manganese replacement of copper for prion protein function and proteinase resistance. EMBO J. 19: 1180–1186.
4. Gaggelli E, Bernardi F, Molteni F, Pogni R, Valensin D, Valensin G, Remelli M, Luczkowski M, Kozlowski H (2005) J. Am. Chem. Soc. 127: 996–1006.
5. Sutoh Y, Nishida Y (2005) Formation of a Mn(IV) species in the reaction mixture of a manganese(II) complex and an aliphatic aldehyde. Synthesis and Reactivity in Inorganic, Metal-Organic and Nano-Metal Chemistry 35: 575–577.
6. Pilbrow JR (1990) Transition Ion Electron Paramagnetic Resonance, Oxford, UK: Clarendon Press.
7. Abragam A, Bleaney B (1970) Electron Paramagnetic Resonance of Transition Ions, Oxford, UK: Clarendon Press.
8. Pake GE, Estle TL (1973) The Physical Principles of Electron Paramagnetic Resonance, New York: WA Benjamin.
9. Schweiger A, Jeschke G (2001) Principles of Pulse Electron Paramagnetic Resonance, Oxford, UK: Oxford University Press.
10. Berliner LJ (ed) (1976) Spin Labeling–Theory and Applications, New York: Academic Press.
11. Boas JF, Pilbrow JR, Smith TD (1978) ESR of copper in biological systems. In: Biological Magnetic Resonance, Vol. 1, Berliner LJ and Reuben J (eds), New York: Plenum Press, 277–342.
12. Boas JF (1984) Electron paramagnetic resonance of copper proteins. In: Copper Proteins and Copper Enzymes, Vol. 1, Lontie R (ed), Boca Raton, FL: CRC Press, 5–62.
13. Deligiannakis Y, Louloudi M, Hadjiliadis N (2000) Electron spin echo envelope modulation (ESEEM) spectroscopy as a tool to investigate the coordination environment of metal centers. Coord Chem Rev 204: 1–112.
14. Wong B-S, Vénien-Bryan C̄, Williamson RA, Burton DR, Gambetti P, Sy M-S, Brown DR, Jones IM (2000) Copper refolding of prion protein. Biochem. Biophys. Res. Commun. 276: 1217–1224.
15. Kaupp M, Bühl M, Malkin VG (Eds) (2004) Calculation of NMR and EPR Parameters: Theory and Applications, Wiley-VCH, Germany: Weinheim.
16. Peisach J, Blumberg WE (1974) Structural implications derived from the analysis of electron paramagnetic resonance spectra of natural and artificial copper proteins. Arch. Biochem. Biophys. 165: 691–708.
17. Palmer G (1985) The electron paramagnetic resonance of metalloproteins. Biochem. Soc. Trans. 13: 548, 560.
18. Pauly PC, Harris DA (1998) Copper stimulates endocytosis of the prion protein. J. Biol. Chem. 273: 33107–33110.

19. Sakaguchi U, Addison AW (1979) Spectroscopic and redox studies of some copper(II) com-plexes with biomimetic donor atoms: implications for protein copper centres. J. Chem. Soc. Dalton Trans. 4: 600–608.
20. Jeschke G (1996) PhD thesis, Swiss Federal Institute of Technology, section 6.3.2.
21. Janzen EG (1971) Spin trapping. Acc. Chem. Res. 4: 31–40.
22. Mason RP, Buettner GR (2003) Spin-trapping methods for detecting superoxide and hydroxyl free radicals in vitro and in vivo. In: Critical Reviews of Oxidative Stress and Aging: Advances in Basic Science, Diagnostics and Intervention, Cutker RG and Rodriguez H (eds), Hackensack, NJ: World Scientific, 27–38.
23. Kennedy CH, Maples KR, Mason RP (1990) In vivo detection of free radical metabolites. Pure Appl. Chem. 62: 95–299.
24. Shi H, Timmins G, Monske M, Burdick A, Kalyanaraman B, Liu Y, Clément J-L, Burchiel S, Liu KJ (2005) Evaluation of spin trapping agents and trapping conditions for detection of cell-generated reactive oxygen species. Arch. Biochem. Biophys. 437: 59–68.
25. Haya A, Burkittb MJ, Jones CM, Hartley RC (2005) Development of a new EPR spin trap DOD-8C for the trapping of lipid radicals at a predetermined depth within biological mem-branes. Arch. Biochem. Biophys. 435: 336–346.
26. Turnbull S, Tabner BJ, Brown DR, Allsop D (2003) Copper-dependent generation of hydro-gen peroxide from the toxic prion protein fragment PrP106–126. Neurosci. Lett. 336: 159–162.
27. Turnbull S, Tabner BJ, Brown DR, Allsop D (2003) Generation of hydrogen peroxide from mutant forms of the prion protein fragment PrP121–231. Biochemistry 42: 7675–7681.
28. Tabner BJ, Turnbull S, Fullwood NJ, German M, and Allsop D (2005) The production of hydrogen peroxide during early-stage protein aggregation: a common pathological mechanism in different neurodegenerative diseases? Biochem. Soc. Trans. 33: 548–550.
29. Barnham KJ, Haeffner F, Ciccotosto GD, Curtain CC, Tew D, Carrington D, Mavros C, Beyreuther C, Carrington R, Masters CL, Cherny RA, Cappai R, Bush AI (2004) Tyrosine gated electron transfer is key to the toxic mechanism of Alzheimer's disease β-amyloid. FASEB J. 2004; 18: 1427–1429.
30. Barnham KJ, Cappai R, Beyreuther K, Masters CL, Hill AF (2006) Delineating common molecular mechanisms in Alzheimer's and prion diseases. Trends Biochem. Sci. 31: 465–472.
31. Berliner L (1983) The spin-label approach to labeling membrane protein sulfhydryl groups. Ann. N Y Acad. Sci. 414: 153–161.
32. Lundberg KM, Stenland CJ, Cohen FE, Prusiner SB, Millhauser GL (1997) Kinetics and mechanism of amyloid formation by the prion protein H1 peptide as determined by time-dependent ESR. Chem. Biol. 4: 345–355.
33. Inanami O, Hashida S, Iizuka D, Horiuchi M, Hiraoka W, Shimoyama Y, Nakamura H, Inagaki F, Kuwabara M. (2005) Conformational change in full-length mouse prion: a site-directed spin-labeling study. Biochem. Biophys. Res. Commun. 335: 785–792.
34. Watanabe Y, Inanami O, Horiuchi M, Hiraoka W, Shimoyama Y, Inagaki F, Kuwabara M (2006) Identification of pH-sensitive regions in the mouse prion by the cysteine-scanning spin-labeling ESR technique. Biochem. Biophys. Res. Commun. 350: 549–556.
35. Aronoff-Spencer E, Burns CS, Avdievich NI, Gerfen GJ, Peisach J, Antholine WE, Ball HL, Cohen FE, Prusiner SB, Millhauser GL (2000) Identification of the Cu²⁺ binding sites in the N-terminal domain of the prion protein by EPR and CD spectroscopy. Biochemistry 39: 13760–13771.
36. Burns CS, Aronoff-Spencer E, Dunham CM, Lario P, Avdievich NI, Antholine WE, Olmstead MM, Vrielink A, Gerfen GJ, Peisach J, Scott WG, Millhauser GL (2002) Molecular features of the copper binding sites in the octarepeat domain of the prion protein. Biochemistry 41: 3991–4001.
37. Millhauser GL (2004) Copper binding in the prion protein. Acc. Chem. Res. 37: 79–85.
38. Chattopadhyay M, Walter ED, Newell DJ, Jackson PJ, Aronoff-Spencer E, Peisach J, Gerfen GJ, Bennett B, Antholine WE, Millhauser GL (2005) The octarepeat domain of the prion

protein binds Cu(II) with three distinct coordination modes at pH 7.4. J. Am. Chem. Soc. 127: 12647–12656.

39. Walter ED, Chattopadhyay M, Millhauser GL (2006) The affinity of copper binding to the prion protein octarepeat domain: evidence for negative cooperativity. Biochemistry 45: 13083–13092.

40. del Pino P, Weiss A, Bertsch U, Renner C, Mentler M, Grantner K, Fiorino F, Meyer-Klaucke W, Moroder L, Kretzschmar HA, Parak FG (2007) The configuration of the Cu²⁺ binding region in full-length human prion protein. Eur. Biophys. J. 36: 239–252.

41. Cereghetti GM, Schweiger A, Glockshuber R, Van Doorslaer SV (2001) Electron paramagnetic resonance evidence for binding of the Cu²⁺ to the C-terminal domain of the murine prion protein. Biophys. J. 81: 516–525.

42. Van Doorslaer SV, Cereghetti GM, Glockshuber R, Schweiger A (2001) Unraveling the Cu²⁺ binding sites in the C-terminal domain of the murine prion protein: a pulse EPR and ENDOR study. J. Phys. Chem. B 105: 1631–1639.

43. Burns CS, Aronoff-Spencer E, Legname G, Prusiner S, Antholine WE, Gerfen GJ, Peisach J, Millhauser GL. (2003) Copper coordination in the full-length prion protein. Biochemistry 42: 6794–6803.

44. Hureau C, Charlet L, Dorlet P, Gonnet F, Spadini L, Anxolabéhère-Mallart E, Girerd J-J (2006) A spectroscopic and voltammetric study of the pH-dependent Cu(II) coordination to the peptide GGGTH: relevance to the fifth Cu(II) site in the prion protein. J. Biol. Inorg. Chem. 11: 735–744.

45. Jobling MF, Huang X, Stewart LR, Barnham KJ, Curtain C,Volitakis I, Perugini M, White AR, Cherny RA, Masters CL, Barrow CJ, Collins SJ, Bush AI, Cappai R (2001) Copper and zinc binding modulates the aggregation and neurotoxic properties of the prion peptide PrP106–126. Biochemistry 40: 8073–8084.

46. Belosi B, Gaggelli E, Guerrini R, Kozłowski H, Łuczkowski M, Mancini FM, Remelli M, Valensin D, Valensin G. (2004) Copper binding to the neurotoxic peptide PrP₁₀₆–₁₂₆: thermodynamic and structural studies. Chem. Biol. Chem. 5: 349–359.

47. Jones CE, Abdelraheim SR, Brown DR, Viles JH (2004) Preferential Cu²⁺ coordination by His96 and His111 induces β sheet formation in the unstructured amyloidogenic region of the prion protein. J. Biol. Chem. 279: 32018–32027.

48. Cereghetti GM, Schweiger A, Glockshuber R, Van Doorslaer S (2003) Stability and Cu(II) binding of prion protein variants related to inherited human prion diseases. Biophys. J. 84: 1985–1997.

49. Grasso D, Grasso G, Guantieri V, Impellizzeri G, La Rosa C, Milardi D, Micera G, Õsz K, Pappalardo G, Rizzarelli E, Sanna D, Sóvágó I (2006) Environmental effects on a prion's Helix II domain: copper(II) and membrane interactions with PrP180–193 and its analogues. Chem. Eur. J. 12: 537–547.

50. Brown DR, Guantieri V, Grasso G, Impellizzeri G, Pappalardo G, Rizzarelli E (2004) Copper(II) complexes of peptide fragments of the prion protein. Conformation changes induced by copper(II) and the binding motif in C-terminal protein region J. Inorg. Biochem. 98: 133–143.

51. Renner C, Fiori S, Fiorino F, Landgraf D, Deluca D, Mentler M, Grantner K, Parak FG, Kretzschmar H, Moroder L (2004) Micellar environments induce structuring of the N-terminal tail of the prion protein. Biopolymers 73: 421–433.

52. Hicks MR, Gill AC, Bath IK, Rullay AK, Sylvester ID, Crout DH,Pinheiro TJT (2006) Synthesis and structural characterization of a mimetic membrane-anchored prion protein. FEBS J. 273: 1285–1299.

53. Blough NV, Simpson DJ (1998) Chemically mediated fluorescence yield switching in nitroxide-fluorophore adducts: optical sensors of radical/redox reactions. J. Am. Chem. Soc. 110: 1915–1917.

54. Jiang J, Borisenko GG, Osipov A, Martin I, Chen R, Shvedova AA, Sorokin A, Tyurina YY, Potapovich A, Tyurin VA, Graham SH, KaganVE (2004) Arachidonic acid-induced carbon-centered radicals and phospholipid peroxidation in cyclo-oxygenase-2-transfected PC12 cells. J. Neurochem. 90: 1036–1049.

55. Ramirez DC, Gomez Mejiba SE, Mason RP (2005) Copper-catalyzed protein oxidation and its modulation by carbon dioxide. J. Biol. Chem. 280: 27402–27411.
56. Abdelraheim SR, Královicová S, Brown DR (2006) Hydrogen peroxide cleavage of the prion protein generates a fragment able to initiate polymerisation of full length prion protein. Int. J. Biochem. Cell Biol. 38: 1429–1440.

Chapter 14
Molecular Diagnosis of Human Prion Disease

Jonathan D. F. Wadsworth, Caroline Powell, Jonathan A. Beck, Susan Joiner, Jacqueline M. Linehan, Sebastian Brandner, Simon Mead, and John Collinge

Summary Human prion diseases are associated with a range of clinical presentations, and they are classified by both clinicopathological syndrome and etiology, with subclassification according to molecular criteria. Here, we describe procedures that are used within the MRC Prion Unit to determine a molecular diagnosis of human prion disease. Sequencing of the *PRNP* open reading frame to establish the presence of pathogenic mutations is described, together with detailed methods for immunoblot or immunohistochemical determination of the presence of abnormal prion protein in brain or peripheral tissues.

Keywords Bovine spongiform encephalopathy; Creutzfeldt–Jakob disease; fatal familial insomnia; Gerstmann–Sträussler–Scheinker disease; kuru; prion; prion disease; prion protein; transmissible spongiform encephalopathy; variant Creutzfeldt–Jakob disease.

1. Introduction

Prion diseases are fatal neurodegenerative disorders that include Creutzfeldt–Jakob disease (CJD), Gerstmann–Sträussler–Scheinker disease (GSS), fatal familial insomnia (FFI), kuru, and variant CJD (vCJD) in humans *(1–3)*. Their central feature is the posttranslational conversion of host-encoded, cellular prion protein (PrP^C), to an abnormal isoform, designated PrP^{Sc} *(1–3)*. Human prion diseases are biologically unique in that the disease process can be triggered through inherited germline mutations in the human prion protein gene (*PRNP*), infection (by inoculation, or in some cases by dietary exposure) with prion-infected tissue or by rare sporadic events that generate PrP^{Sc} *(1–5)*. Substantial evidence indicates that an abnormal PrP isoform is the principal, if not the sole, component of the transmissible infectious agent, or prion *(1, 2, 6)*. The existence of multiple strains or isolates of prions, has been difficult to accommodate within a protein only model of prion propagation; however, considerable experimental evidence suggests that prion strain diversity is encoded within PrP itself and that phenotypic diversity in human

prion diseases relates to differing physicochemical properties of abnormal PrP isoforms *(7–15)*. Furthermore, the propagation of distinct abnormal PrP isoforms may
be determined by the host genome *(16, 17)*.

Human prion diseases are associated with a range of clinical presentations, and they
are classified by both clinicopathological syndrome and etiology, with subclassification according to molecular criteria *(3, 18, 19)*. Approximately 85% of cases
occur sporadically as Creutzfeldt–Jakob disease (sporadic CJD) at a rate of one to
two cases per million population per year across the world, with an equal incidence
in men and women *(2, 4, 20, 21)*. Approximately 15% of human prion disease is
associated with autosomal dominant pathogenic mutations in *PRNP*, and to date
more than 30 mutations have been described *(2, 4, 22, 23)*. These include one to
nine octapeptide repeat insertions (OPRI) within the octapeptide repeat region
between codons 51 and 91, a two-octapeptide repeat deletion (OPRD), and various
point mutations causing missense or stop substitutions **(Fig. 1)**.

Although human prion diseases are transmissible diseases, acquired forms have,
until recently, been confined to rare and unusual situations. The most frequent
causes of iatrogenic CJD occurring through medical procedures have arisen as a

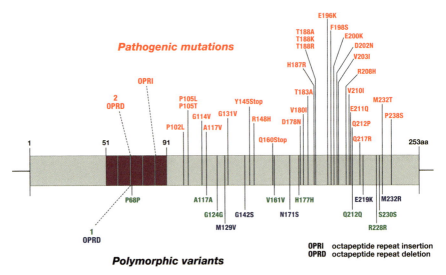

Fig. 1 Pathogenic mutations and polymorphisms in human prion protein. The pathogenic mutations associated with human prion disease are shown above the human PrP coding sequence. These
consist of one to nine OPRI within the octapeptide repeat region between codons 51 and 91, a two
OPRD, and various point mutations causing missense or stop codon substitutions. Some of these
changes have been observed in individual patients only and should be considered as possible pathogenic mutations that require confirmation. Point mutations are designated by the wild-type amino
acid preceding the codon number, followed by the mutant residue, using single-letter amino acid
nomenclature. Polymorphic variants are shown below the PrP coding sequence (synonymous
changes, *green*; nonsynonymous changes, *blue*). Codon 129 and 219 polymorphisms have profound susceptibility, disease-modifying effects, or both. Deletion of one octapeptide repeat is not
associated with prion disease in humans

result of treatment with growth hormone derived from human cadavers or implantation of dura mater grafts *(24, 25)*. Less frequent incidences of iatrogenic human prion disease have resulted from transmission of CJD prions during corneal transplantation, contaminated electroencephalographic electrode implantation, and surgical operations using contaminated instruments or apparatus *(24, 25)*. The most well-known example of acquired prion disease in humans is kuru, transmitted by cannibalism among the Fore and neighboring linguistic groups of the Eastern Highlands in Papua New Guinea *(19, 26–28)*. Remarkably, kuru demonstrates that incubation periods of infection with human prions can exceed 50 years *(28)*. The appearance of vCJD in the United Kingdom from 1995 onward *(29)*, and the experimental confirmation that this is caused by the same prion strain as that causing bovine spongiform encephalopathy (BSE) in cattle *(8, 17, 30, 31)*, has led to widespread concern that exposure to the epidemic of BSE poses a distinct and conceivably a severe threat to public health in the United Kingdom and other countries *(3, 32)*. The extremely prolonged and variable incubation periods seen with prion diseases when crossing a species barrier means that it will be some years before the parameters of any human epidemic can be predicted with confidence *(28, 32–35)*. In the meantime, significant numbers in the population may be incubating this disease, with risk of secondary transmission via blood transfusion, blood products, tissue and organ transplantation, and other iatrogenic routes *(32, 36–39)*.

Polymorphism at residue 129 of human PrP (encoding either methionine [M] or valine [V]) powerfully affects susceptibility to human prion diseases *(21, 27, 40–43)*. About 38% of northern Europeans are homozygous for the more frequent methionine allele, 51% are heterozygous, and 11% homozygous for valine. Homozygosity at *PRNP* codon 129 predisposes to the development of sporadic and acquired CJD *(21, 27, 40–43)*, and most sporadic CJD occurs in individuals homozygous for this polymorphism (**Fig. 2**). This susceptibility factor is also relevant in the acquired forms of CJD, most strikingly in vCJD where all clinical cases studied so far have been homozygous for codon 129 methionine of *PRNP (3, 39, 44)*.

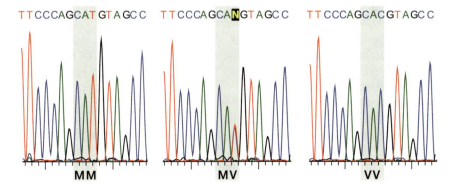

Fig. 2 *PRNP* codon 129 polymorphism. Electropherogram traces that illustrate all *PRNP* codon 129 genotypes in the reverse DNA strand orientation

The clinical presentation of human prion disease varies enormously, and there is considerable overlap observed between individuals with different disease etiologies *(3, 4, 23)* and even in family members with the same pathogenic *PRNP* mutation *(22, 23, 45–49)*. Progressive dementia, cerebellar ataxia, pyramidal signs, chorea, myoclonus, extrapyramidal features, pseudobulbar signs, seizures, and amyotrophic features can be seen in variable combinations. Criteria used for diagnosis of human prion disease are available *(3, 50)*, and definite diagnosis of sporadic and acquired prion disease relies upon neuropathological examination and the demonstration of abnormal PrP deposition in the central nervous system by either immunoblotting or immunohistochemistry. In the appropriate clinical setting with family history, identification of a pathogenic *PRNP* mutation provides diagnosis of inherited prion disease and subclassification according to mutation; *PRNP* analysis is also used for presymptomatic genetic testing in affected families *(3, 19, 23, 51)*. Because of the extensive phenotypic variability seen in inherited prion disease and its ability to mimic other neurodegenerative conditions, notably Alzheimer's disease, *PRNP* analysis should be considered in all patients with undiagnosed dementing and ataxic disorders *(3, 23)*.

The brains of patients with prion disease frequently show no recognizable abnormalities on gross examination at necropsy; however, microscopic examination of the brain at either necropsy or in ante-mortem biopsy specimens typically reveals characteristic histopathologic changes, consisting of neuronal vacuolation and degeneration, which gives the cerebral gray matter a microvacuolated or "spongiform" appearance accompanied by a reactive proliferation of astroglial cells *(52, 53)*. Although spongiform degeneration is frequently detected, it is not an obligatory neuropathologic feature of prion disease; astrocytic gliosis, although not specific to the prion diseases, is more constantly seen. The lack of an inflammatory response is also an important characteristic. Demonstration of abnormal PrP immunoreactivity, or more specifically biochemical detection of PrP[Sc] in brain material by immunoblotting techniques is diagnostic of prion disease and some forms of prion disease are characterized by deposition of amyloid plaques composed of insoluble aggregates of PrP *(52, 53)*. Amyloid plaques are a notable feature of kuru and GSS *(52, 54)*, but they are less frequently found in the brains of patients with sporadic CJD, which typically show a diffuse pattern of abnormal PrP deposition *(12, 52)*. The histopathological features of vCJD are remarkably consistent and distinguish it from other human prion diseases, with large numbers of PrP-positive amyloid plaques that differ in morphology from the plaques seen in kuru and GSS in that the surrounding tissue takes on a microvacuolated appearance, giving the plaques a florid appearance *(29, 55)*. The tissue distribution of PrP[Sc] in vCJD differs strikingly from that in classical CJD with uniform and prominent involvement of lymphoreticular tissues *(36, 56–60)*. Depending upon the density of lymphoid follicles, PrP[Sc] concentrations in vCJD peripheral tissues can vary enormously, with levels relative to brain as high as 10% in tonsil *(36, 56)* or as low as 0.002% in rectum *(36, 61)*. Tonsil biopsy is used for diagnosis of vCJD, and to date it has shown 100% sensitivity and specificity for diagnosis of vCJD at an early clinical stage *(3, 36, 39, 56, 62)*.

In this chapter, we describe the procedures that are currently used within the MRC Prion Unit to provide a molecular diagnosis of human prion disease. Methods for sequencing the *PRNP* open reading frame to establish the presence of pathogenic mutations and to determine *PRNP* polymorphic codon 129 status are described together with procedures used for immunoblot or immunohistochemical determination of the presence of abnormal PrP in brain or peripheral tissues.

2. Materials

2.1. Molecular Genetics

1. BACC2 DNA extraction kit from Nucleon Biosciences.
2. TE buffer: 10 mM Tris and 1 mM EDTA, pH 8.0.
3. MegaMix Blue (Microzone, Haywards Heath, UK).
4. HyperLadder IV (Bioline, London, UK).
5. MicroClean (Microzone).
6. BigDye version 1.1 Cycle Sequencing kit (Applied Biosystems, Foster City, CA).
7. BetterBuffer (Microzone).
8. 0.5 M EDTA, pH 8.0, diluted fourfold in water.
9. Hi-Di formamide (Applied Biosystems).
10. Performance-optimized polymer 7 (Applied Biosystems).
11. DdeI restriction endonuclease, including NEBuffer 3 (New England Biolabs, Ipswich, MA).
12. MetaSieve agarose (Flowgen, Ashby, Leicestershire, UK).
13. PflFI restriction endonuclease, including NEBuffer 4 (New England Biolabs).
14. BsaI restriction endonuclease, including NEBuffer 3 (New England Biolabs).
15. TOPO TA cloning kit for sequencing (Invitrogen, Paisley, UK).
16. Luria-Bertani (LB) broth.

2.2. Immunoblotting

1. Dulbecco's sterile phosphate-buffered saline (PBS) lacking Ca^{2+} and Mg^{2+} ions.
2. Duall tissue grinders (Anachem Ltd., Luton, Bedfordshire, UK).
3. Proteinase K (specific enzymatic activity ~30 Anson units/g) prepared as a stock solution of 1 mg/ml in water.
4. Sodium dodecyl sulfate (SDS) sample buffer.

   a. A stock concentrate of 2× SDS sample buffer [142 mM Tris, 22.72% (v/v) glycerol, 4.54% (w/v) SDS, and 0.022% (w/v) bromphenol blue] is prepared in water and titrated to pH 6.8 with HCl.

   b. This solution requires adjustment with reducing agent and proteinase K inhibitor immediately before use to produce 2× working SDS sample buffer.
   c. For preparation of 0.5 ml of 2× working SDS sample buffer, mix the following: 440 µl of stock concentrate of 2× SDS sample buffer plus 20 µl of 2-mercapto-ethanol plus 40 µl of 100 mM 4-(2-aminoethyl)-benzene sulfonyl fluoride prepared in water.
   d. This produces 2× working SDS sample buffer of the following final composition: 125 mM Tris-HCl, 20% (v/v) glycerol, pH 6.8, containing 4% (w/v) SDS, 4% (v/v) 2-mercaptoethanol, 8 mM 4-(2-aminoethyl)benzenesulfonyl fluoride, and 0.02% (w/v) bromphenol blue.
5. 16% Tris-glycine SDS-polyacrylamide gel electrophoresis (PAGE) gels (Invitrogen).
6. Seeblue prestained molecular mass markers (Invitrogen).
7. SDS-PAGE electrophoresis buffer: 100 ml of 10× Tris-glycine, SDS concentrate [0.25 M Tris, 1.92 M glycine, 1% (w/v) SDS] plus 900 ml water.
8. Immobilon P transfer membrane.
9. Electroblotting buffer: 100 ml of 10× Tris-glycine concentrate (0.20 M Tris and 1.50 M glycine), 700 ml of water, and 200 ml of methanol.
10. PBST: 100 ml of 10× PBS concentrate (low in phosphate) (BDH, Poole, Dorset, UK), 900 ml of water, and 0.5 ml of Tween-20.
11. Anti-PrP monoclonal antibody 3F4 (Signet Laboratories, Dedham, MA).
12. Goat anti-mouse IgG (fab-specific) alkaline phosphatase conjugate (absorbed with human serum proteins) (Sigma Chemical, Poole, Dorset, UK).
13. CDP-star chemiluminescent substrate (Tropix, Bedford, MA).
14. Kodak Biomax MR film (Eastman Kodak, Rochester, NY).
15. AttoPhos chemifluorescent substrate (Promega, Madison, WI): Mix 36 mg of Attophos substrate in 60 ml of Attophos buffer. Store as 3 ml aliquots at −20°C.
16. Sodium lauroylsarcosine (Calbiochem, San Diego, CA).
17. Benzonase (Benzon nuclease purity 1 [25 U/µl], Merck, Darmstadt, Germany).
18. Sodium phosphotungstic acid stock solution.

   a. Stock solution is 4% (w/v) sodium phosphotungstic acid containing 170 mM $MgCl_2$ prepared in water, pH 7.4.
   b. For preparation of 10 ml of stock solution, add 0.4 g of sodium phosphotungstic acid and 0.35 g $MgCl_2 \cdot 6H_2O$ to a 50 ml polypropylene tube and make to ~9 ml with water.
   c. The pH of this solution is acidic and needs to be titrated with 5 M NaOH to pH 7.4 before adjusting to a final volume of 10 ml with water.
   d. On addition of NaOH, immediate formation of insoluble $MgOH_2$ occurs that will redissolve on vortexing.
   e. Addition of 5 M NaOH followed by vortexing and measurement of pH needs to be done repetitively.
   f. For 10 ml of stock solution, addition of 360 µl of 5 M NaOH will generate pH 7.4.

2.3. Immunohistochemistry

2.3.1. Procurement

1. 10% buffered formal-saline.
2. Biopsy cassettes (R. A. Lamb, Eastbourne, UK).

2.3.2. Prion Deactivation with Formic Acid

1. Biopsy cassettes (R. A. Lamb).
2. 98% formic acid.
3. 2 M sodium hydroxide: 80 g of sodium hydroxide pellets in water to 1 liter.
4. 10% buffered formal-saline.

2.3.3. Tissue Processing

1. 10% buffered formal-saline.
2. Industrial methylated spirits (J.M. Loveridge Ltd., Southampton, UK), diluted in water to desired concentration.
3. Xylene.
4. Pure paraffin wax (R A Lamb).

2.3.4. Tissue Sectioning

1. Microtome (Leica, Wetzlar, Germany).
2. SuperFrost microscope slides (VWR, West Chester, PA).

2.3.5. Tissue Staining

1. Xylene.
2. Absolute ethanol diluted in water to desired concentration.
3. Harris hematoxylin (BDH).
4. Acid alcohol: 1% HCl in absolute ethanol.
5. Eosin Y solution 0.5%, aqueous (VWR).
6. Pertex mounting medium (Cox Scientific Ltd, Kettering, UK).
7. Benchmark staining machine (Ventana Medical Systems, Illkirch CEDEX, France).
8. Protease 1 (Ventana Medical Systems).
9. Rabbit anti-glial fibrillary protein (Dako UK Ltd., Ely, Cambridgeshire, UK); antibody diluent (Ventana Medical Systems).
10. IViewDAB detection kit (Ventana Medical Systems), containing an inhibitor solution (3% hydrogen peroxide) (4 min), universal biotinylated secondary

antibody (10 min), streptavidin-horseradish peroxidase solution (10 min), 3,3-diaminobenzidine and hydrogen peroxide (20 min), copper solution (4 min).

11. Hematoxylin (Ventana Medical Systems).
12. Bluing reagent (Ventana Medical Systems).
13. Tris-EDTA-citrate buffer, pH 7.8: 2.1 mM Tris, 1.3 mM EDTA, and 1.1 mM sodium citrate.
14. 98% formic acid.
15. 10 mM sodium citrate buffer, pH 6.0: solution A: 10.5 g of citric acid in 500 ml of deionized water and solution B: 29.41 g of sodium citrate in 1000 ml of water.

  a. Add 18 ml of solution A to 82 ml of solution B, and then adjust to 1 liter final volume with water.

16. Anti-PrP monoclonal antibody ICSM35 (D-Gen Ltd., London, UK).
17. Protease 3 (Ventana Medical Systems).
18. Superblock (Pierce Chemical, Rockford, IL).
19. Prepared with methanol GPR and 30% hydrogen peroxide.
20. Tris-buffered saline (TBS): 50 mM Tris, 145 mM NaCl, pH 7.6. For 1 liter of 10× stock, add 60.55 g of Trizma base and 85 g of sodium chloride to 800 ml of water. Adjust to pH 7.6 with 32.00 ml of concentrated HCl. Make to 1 liter final volume with water. Dilute 10× TBS to 1× concentration with water.
21. 4 M guanidine thiocyanate: Add 472.64 g guanidine thiocyanate in water to a final volume of 1 liter.
22. Normal rabbit serum (Dako UK Ltd.).
23. Biotinylated rabbit anti-mouse immunoglobulins (Dako UK Ltd.).
24. Strept ABComplex/HRP Duet kit (Dako UK Ltd.).
25. 3,3-diaminobenzidine tetrachloride: Use at a final concentration of 25 mg in 100 ml of 1× TBS.

3. Methods

Numbers in parentheses refer to the materials listed in **Subheading 2**.

3.1. Molecular Genetics

3.1.1. Isolation of Genomic DNA from Blood

1. All procedures are performed within a class 1 microbiological safety cabinet situated within an ACDP level II containment laboratory with strict adherence to local rules of safe working practice. Informed consent for the analysis of a sample must be established before investigation. This may be obtained from the patient (or next of kin or physician in the circumstance of incapacity). For predictive testing, we expect evidence of appropriate genetic counseling before analysis.

2. Genomic DNA is extracted from whole anticoagulated blood (typically from a 5-ml EDTA tube) by using the Nucleon BACC2 DNA extraction kit (*see* **Subheading 2.1., item 1**) following the manufacturer's instructions. DNA concentrations are determined using a Nanodrop ND-1000 spectrophotometer, and adjusted to 200–250 ng/µl in TE buffer (*see* **Subheading 2.1., item 2**). Concentrations are remeasured before dilution of DNA in TE buffer to a final concentration of 20 ng/µl and storage at 4°C.

3.1.2. Sequencing of PRNP Open Reading Frame

3.1.2.1. PCR of *PRNP* Open Reading Frame

1. Prepare a premix of MegaMix Blue (*see* **Subheading 2.1., item 3**) containing primers at 500 µM sufficient for 25-µl reactions on a 96-well plate. PCR primers used to amplify the open reading frame are 5'-CTA TGC ACT CAT TCA TTA TGC-3' (forward) and 5'-GTT TTC CAG TGC CCA TCA GTG-3' (reverse).
2. Add 1 µl of 20 ng/µl genomic DNA.
3. Thermal cycling is performed on an MJ Research (Watertown, MA) Tetrad 1 PCR machine or similar using the following cycling parameters:

    a. 94 °C for 5 min.
    b. 94 °C for 30 s.
    c. 55 °C for 40 s.
    d. 72 °C for 45 s.
    e. repeat steps b–d 34 times.
    f. 72°C for 5 min.

4. Assess polymerase chain reaction (PCR) by electrophoresis of 5 µl of product on a 2% ethidium bromide-stained agarose gel with 5 µl of HyperLadder IV (*see* **Subheading 2.1., item 4**) size standard. The gel is viewed using a Gel Doc 1000 transilluminator (Bio-Rad, Hemel Hempstead, UK) and Quantity One 4.5.1 software or similar.

3.1.2.2. PCR Product Cleanup

1. An equal volume of MicroClean (*see* **Subheading 2.1., item 5**) is added to the PCR product and mixed well by pipetting or vortexing.
2. The mixture is left at room temperature for 15 min.
3. The plate is centrifuged at 2,000–4,000 g for 40 min at room temperature.
4. The supernatant is removed by centrifuging the plate at 40 g for 30 s in an inverted position on tissue paper by using centrifuge plate holders.
5. Remove the plate to the bench and leave to air dry for 5 min.
6. Resuspend the cleaned PCR product in 40 µl of water.

3.1.2.3. Sequencing Reactions

1. For each sequencing reaction, prepare a premix of 1 μl of BigDye (*see* **Subheading 2.1., item 6**) 5 μl of BetterBuffer (*see* **Subheading 2.1., item 7**), 0.75 μl of sequencing primer at 5 pmol/μl, 2.5 ng of cleaned PCR product, and water to a final volume of 15 μl. The amount of PCR product is estimated using visual comparison with known amounts of HyperLadder IV size standard.
2. Sequencing primers are 5'-GAC GTT CTC CTC TTC ATT TT-3' (forward 1), 5'-CCG AGT AAG CCA AAA ACC AAC-3' (forward 2), and 5'-CAC CAC CAC TAA AAG GGC TGC-3' (reverse 1), 5'-TTC ACG ATA GTA ACG GTC C-3' (reverse 2).
3. Sequencing reactions are thermal cycled on an MJ Research Tetrad 1 PCR machine or similar using the following cycling parameters:

   a. 96°C for 1 min.
   b. 96°C for 10 s.
   c. 50°C for 5 s.
   d. 60°C for 4 min.
   e. Repeat steps b–d 24 times.
   f. Hold at 15°C until ready to purify.

3.1.2.4. Sequencing Product Cleanup

1. To each sequencing reaction, add 3.75 μl of 0.125 M EDTA, pH 8.0 (*see* **Subheading 2.1., item 8**).
2. Add 45 μl of 100% ethanol to each reaction and mix by pipetting or vortexing.
3. Leave reactions at room temperature for 15 min.
4. Centrifuge the plate at 3,000 g for 30 min at 4°C.
5. The supernatant is removed by centrifuging the plate at 185 g for 1 min in an inverted position on tissue paper using centrifuge plate holders.
6. Add 50 μl of 70% ethanol in water.
7. Centrifuge the plate at 1,650 g for 15 min at 4°C.
8. The supernatant is removed by centrifuging the plate at 185 g for 1 min in an inverted position on tissue paper by using centrifuge plate holders.
9. Place the plate on the PCR block held at 37°C for 5 min to remove final traces of ethanol.

3.1.2.5. Electrophoresis

1. Add 10 μl of Hi-Di formamide loading solution (*see* **Subheading 2.1., item 9**) and vortex the plate for 30 s.
2. Denature the samples by placing on the PCR block held at 95°C for 5 min and then immediately transfer to ice.

3. Standard run conditions are applied to electrophoresis of sequencing products on an Applied Biosystems 3130xl, using polymer POP7 (*see* **Subheading 2.1., item 10**), 50-cm arrays, and a standard run module with a sample injection time of 15 s.

3.1.2.6. Data Analysis

1. Data analysis is performed using Applied Biosystems Seqscape software version 2.5.
2. Analysis filter settings are adjusted to allow assembly of poor data due to insertions or deletions (maximum mixed bases 95%, maximum Ns 95%, minimum clear length bp of 1, and minimum sample score of 1).
3. Poor data or failed reactions are removed from projects by visual inspection of data.

3.1.3. PCR Size Fractionation to Investigate Insertion or Deletion Variants

1. Prepare a premix of Mega Mix Blue (*see* **Subheading 2.1., item 3**) containing primers at 500 μM sufficient for 25-μl reactions on a 96-well plate. PCR primers are 5'-GAC CTG GGC CTC TGC AAG AAG CGC-3' (forward) and 5'-GGC ACT TCC CAG CAT GTA GCC G-3' (reverse).
2. Add 1 μl of 20 ng/μl genomic DNA.
3. Thermal cycling is performed on an MJ Research Tetrad 1 PCR machine or similar using the following cycling parameters:

    a. 94°C for 5 min.
    b. 94°C for 30 s.
    c. 65°C for 30 s.
    d. 72°C for 1 min.
    e. Repeat steps b–d 34 times.
    f. 72°C for 5 min.

4. Assess PCR by electrophoresis of 5 μl of product on a 2% ethidium bromide-stained agarose gel with 5 μl of HyperLadder IV (*see* **Subheading 2.1., item 4**) size standard. The gel is viewed using a Bio-Rad Gel Doc 1000 transilluminator and Quantity One 4.5.1 software or similar.
5. 1 OPRD and 6 OPRI controls are run on each gel (**Fig. 3**).

3.1.4. Mutation Confirmation

A second assay is performed to confirm the presence or absence of missense or stop mutations when a predictive genetic test is being carried out. PCR size fractionation, as described above, is sufficient in addition to sequencing when testing for insertion mutants in a predictive setting; however, unexpected or unknown insertion mutations may require cloning to confirm exact base pair composition.

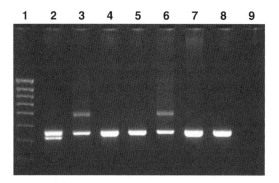

Fig. 3 Analysis of *PRNP* OPRI mutations. Image from agarose gel electrophoresis of *PRNP* amplicons illustrating the presence of an insertional mutation. Lane 1, HyperLadder IV; lane 2, 1-OPRD control; lane 3, 6 OPRI control; lane 4–8, patient samples: lane 6 demonstrates amplification of a heterozygous insertional mutation of 144 base pairs (6-OPRI mutation positive), lanes 4, 5, 7, and 8 are wild-type alleles only; lane 9, no-template control

Examples of confirmatory assays used to detect the more common *PRNP* missense mutations and cloning methodology are described in **Subheading 3.1.4.1.**

3.1.4.1. Confirmation of P102L

1. Prepare a premix of MegaMix Blue (*see* **Subheading 2.1., item 3**) containing primers at 500 μM sufficient for 25-μl reactions on a 96-well plate. PCR primers are 5'-GAC CTG GGC CTC TGC AAG AAG CGC-3' (forward) and 5'-GGC ACT TCC CAG CAT GTA GCC G-3' (reverse).
2. Add 1 μl of genomic DNA at 20 ng/of μl.
3. Thermal cycling is performed on an MJ Research Tetrad 1 PCR machine or similar using the following cycling parameters:
   a. 94°C for 5 min.
   b. 94°C for 30 s.
   c. 65°C for 30 s.
   d. 72°C for 1 min.
   e. Repeat steps b–d 34 times.
   f. 72°C for 5 min.
4. Prepare restriction endonuclease reaction by adding 10 μl of PCR product, 1 μl of DdeI (*see* **Subheading 2.1., item 11**), 2.5 μl of 10× NEBuffer 3 (*see* **Subheading 2.1., item 11**), and 11.5 μl of H_2O.
5. Incubate reaction at 37°C for 3 h.
6. Electrophorese 10 μl of digested PCR product on a 3% 2:1 MetaSieve agarose (*see* **Subheading 2.1., item 12**) ethidium bromide-stained gel using 5 μl of HyperLadder IV (*see* **Subheading 2.1., item 4**) size standard. The gel is viewed using a Bio-Rad Gel Doc 1000 transilluminator and Quantity One 4.5.1 software.
7. Digested positive and negative controls are run on each gel to visualize mutant DNA (95-, 101-, and 152-bp) and wild-type DNA (101- and 247-bp) fragment patterns.

3.1.4.2. D178N

1. Prepare a premix of MegaMix Blue (*see* **Subheading 2.1., item 3**) containing primers at 500 µM sufficient for 25-µl reactions on a 96-well plate. PCR primers are 5'-CTA TGC ACT CAT TCA TTA TGC-3' (forward) and 5'-GTT TTC CAG TGC CCA TCA GTG-3' (reverse).
2. Add 1 µl of genomic DNA at 20 ng/µl.
3. Thermal cycling is performed on an MJ Research Tetrad 1 PCR machine or similar using the following cycling parameters:

    a. 94°C for 5 min.
    b. 94°C for 30 s.
    c. 55°C for 40 s.
    d. 72°C for 45 s.
    e. Repeat steps b–d 34 times.
    f. 72°C for 5 min.

4. Prepare restriction endonuclease reaction by adding 10 µl of PCR product, 1 µl of PflFI (*see* **Subheading 2.1., item 13**), 2.5 µl of 10× NEBuffer 4 , (*see* **Subheading 2.1., item 13**) and 11.5 µl of H_2O.
5. Incubate reaction at 37°C for 3 h.
6. Electrophorese 10 µl of digested PCR product on a 2% ethidium bromide-stained agarose gel by using 5 µl of HyperLadder IV (*see* **Subheading 2.1., item 4**) size standard. The gel is viewed using a Bio-Rad Gel Doc 1000 transilluminator and Quantity One 4.5.1 software.
7. Digested positive and negative controls are run on each gel to visualize mutant DNA (1015-bp) and wild-type DNA (386- and 629-bp) fragment patterns.

3.1.4.3. E200K

1. Prepare a premix of MegaMix Blue (*see* **Subheading 2.1., item 3**) containing primers at 500 µM sufficient for 25-µl reactions on a 96-well plate. PCR primers are 5'-CTA TGC ACT CAT TCA TTA TGC-3' (forward) and 5'-GTT TTC CAG TGC CCA TCA GTG-3' (reverse).
2. Add 1 µl of genomic DNA at 20 ng/µl.
3. Thermal cycling is performed on an MJ Research Tetrad 1 PCR machine or similar using the following cycling parameters:

    a. 94°C for 5 min.
    b. 94°C for 30 s.
    c. 55°C for 40 s.
    d. 72°C for 45 s.
    e. Repeat steps b–d 34 times.
    f. 72°C for 5 min.

4. Prepare restriction endonuclease reaction by adding 10 µl of PCR product, 1 µl of BsaI (*see* **Subheading 2.1., item 14**), 2.5 µl of 10× NEBuffer 3 (*see* **Subheading 2.1., item 14**) and 11.5 µl of H_2O.

5. Incubate reaction at 50°C for 3 h.
6. Electrophorese 10 μl of digested PCR product on a 2% ethidium bromide-stained agarose gel by using 5 μl of HyperLadder IV (*see* **Subheading 2.1., item 4**) size standard and view using a Bio-Rad Gel Doc 1000 transilluminator and Quantity One 4.5.1 software.
7. Digested positive and negative controls are run on each gel to visualize mutant DNA (1015-bp) and wild-type DNA (318- and 697-bp) fragment patterns.

3.1.5. Characterization of Insertion Mutations

3.1.5.1. Generation of Amplicon to Be Cloned

1. Prepare a premix of MegaMix Blue (*see* **Subheading 2.1., item 3**) containing primers at 500 μM sufficient for 25-μl reactions on a 96-well plate. PCR primers are 5'-GAC CTG GGC CTC TGC AAG AAG CGC-3' (forward) and 5'-GGC ACT TCC CAG CAT GTA GCC G-3' (reverse). (Note that MegaMix Blue contains an enzyme that has 3-prime terminal adenosine triphosphate transferase activity that ensures that the amplicon has "A" overhangs to anneal to the "T" overhangs of the cloning vector).
2. Add 1 μl of genomic DNA at 20 ng/μl.
3. Thermal cycling is performed on an MJ Research Tetrad 1 PCR machine or similar using the following cycling parameters:

   a. 94°C for 5 min.
   b. 94°C for 30 s.
   c. 65°C for 30 s.
   d. 72°C for 1 min.
   e. Repeat steps b–d 34 times.
   f. 72°C for 10 min.

4. Electrophorese 5 μl of PCR product on a 2% ethidium bromide-stained agarose gel by using 5 μl of HyperLadder IV (*see* **Subheading 2.1., item 4**) size standard and view using a Bio-Rad Gel Doc 1000 transilluminator and Quantity One 4.5.1 software.
5. To preserve the "A" overhangs, before cloning, carry out as little manipulation as possible and use fresh product.

3.1.5.2. Ligation and Cloning

1. Perform TOPO TA cloning as described in the online user manual version O, 10 April 2006 (25-0276). Updates of this protocol are available at www.invitrogen. com. Use Invitrogen cat. no. K4575-01 (*see* **Subheading 2.1., item 15**) (TOP10, Chemically Competent *E. coli*, 20 reactions).

3.1.5.3. Analysis and Sequencing of Recombinant Clones

1. There is no need to miniprep possible positive clones. Pick white and light blue clones (color enhancement can be obtained by leaving the plate at 4°C overnight if this is preferred) and inoculate wells of a prewarmed 96-well tissue culture or storage plate containing 150–200 μl of LB broth (*see* **Subheading 2.1., item 16**) containing the appropriate antibiotic.
2. Incubate at 37°C for about 5–6 h or until the wells are opaque.
3. Transfer 50-μl aliquots to a 96-well PCR plate.
4. Seal the plate and place on a PCR machine for 10 min at 99°C to lyse the bacteria. Aliquots (1 μl) of these crude DNA preparations can then be used to produce amplicons by using the same methods used to produce the original amplicon by simple transfer to a fresh 96-well PCR plate containing the appropriate PCR mix.
5. Electrophorese 5 μl of PCR product on a 2% ethidium bromide-stained agarose gel by using 5 μl of HyperLadder IV (*see* **Subheading 2.1., item 4**) size standard. The gel is viewed using a Bio-Rad Gel Doc 1000 transilluminator and Quantity One 4.5.1 software.
6. Amplicons from positive wells can be purified and sequenced according to the automated sequencing protocol for PCR products (*see* **Subheadings 3.1.2.2.–3.1.2.6.**).
7. Amplicons will contain Taq polymerase artefacts preserved in the cloning process. Therefore, at least three clones should be sequenced to obtain a consensus sequence, preferably on both strands.

3.2. Immunoblotting

3.2.1. Biosafety

1. All procedures are performed within a class 1 microbiological safety cabinet situated within an ACDP level III containment laboratory with strict adherence to local rules of safe working practice. Informed consent for the analysis of samples must be in place before investigation.
2. No unsealed biological material (tissue or derivative sample thereof) is manipulated outside of the class 1 microbiological safety cabinet. Disposable gloves, safety gown, and safety glasses are worn at all times.
3. 1.5-ml screw-top microfuge tubes containing a rubber O-ring are used.
4. Guidelines for decontamination of human prions are available (*63*). All disposable plasticware (e.g., tubes, tips, and so on) and solutions containing biological material are decontaminated in 50% (v/v) sodium hypochlorite solution (containing >20,000 ppm available chlorine prepared in water) for at least 1 h before disposal of the liquid phase down designated laboratory sinks within the containment laboratory. Sharps (needles and scalpels) are disposed of immediately after use into a sharps bin and autoclaved at 136°C for 20 min before incineration.
5. Decontaminated plasticware is transferred to a sharps bin and autoclaved at 136°C for 20 min before incineration.

3.2.2. Preparation of Tissue Homogenate

1. Tissue specimens, stored frozen in sealed pots within the ACDP level III containment laboratory, are transferred into a class 1 microbiological safety cabinet and partially thawed and placed on a petri dish.
2. A suitable quantity of tissue is excised using a scalpel and sealed in a disposable plastic pot and weighed. The tissue is then prepared as a 10% (w/v) homogenate in Dulbecco's sterile PBS lacking Ca^{2+} and Mg^{2+} ions (*see* **Subheading 2.2., item 1**). The amount of PBS to add in microliters is equal to 9 times wet weight of tissue in milligrams. This calculation will produce a homogenate very close to a true 10% (w/v) ratio without the necessity of having to accurately measure the total volume of tissue in PBS before the homogenization process.
3. Homogenization of brain tissue is achieved by serial passage of tissue through syringe needles of decreasing diameter (needle gauges 19, 21, 23, and 25).
4. Homogenization of peripheral tissue is achieved through the use of glass Duall tissue grinders (*see* **Subheading 2.2., item 2**).
5. The homogenate is stored as aliquots in 1.5 ml screw-top microfuge tubes at $-80°C$.

3.2.3. Proteinase K Digestion and Electrophoresis

1. 10% brain homogenate is thawed, thoroughly vortexed, and then centrifuged at $100\,g$ (800 rpm) for 1 min in a microfuge (*see* **Note 1**).
2. 20-µl aliquots of the resultant supernatant are adjusted to a final concentration of 50 µg/ml proteinase K (*see* **Subheading 2.2., item 3**) by addition of 1.05 µl of a 1 mg/ml proteinase K stock solution (*see* **Note 2**).
3. Samples are incubated at 37°C for 1 h, followed by centrifugation at $16,100\,g$ (13,200 rpm) for 1 min in a microfuge.
4. The digestion is terminated by resuspension of the sample with an equal volume (21 µl) of 2× working SDS sample buffer (*see* **Subheading 2.2., item 4**) and *immediate* transfer to a 100°C heating block for 10 min.
5. Samples for analysis in the absence of proteinase K treatment are treated directly with an equal volume of 2× working SDS sample buffer (*see* **Subheading 2.2., item 4**) and heated similarly.
6. All samples are centrifuged at $16,100\,g$ (13,200 rpm) for 1 min in a microfuge, thoroughly vortexed, and then recentrifuged $16,100\,g$ for 1 min before electrophoresis of the supernatant.
7. 10 µl of the supernatant is loaded on an Invitrogen 16% Tris-glycine polyacrylamide mini gel (*see* **Subheading 2.2., item 5**) (*see* **Note 3**). The remainder of the sample can be stored at $-80°C$. Then, 10 µl of 1× working SDS sample buffer (prepared by mixing 2× working SDS sample buffer with an equal volume of water) should be added in any blank lane. Ten microliters of Seeblue prestained molecular mass markers (*see* **Subheading 2.2., item 6**) is used to calibrate the gel.
8. Gels are run at a constant voltage of 200 V for 80 min in SDS-PAGE running buffer (*see* **Subheading 2.2., item 7**) (*see* **Note 3**).

9. Gels are electroblotted (one gel per Invitrogen blot module) on to polyvinyli-dene difluoride membrane (*see* **Subheading 2.2., item 8**) in electroblotting buffer (*see* **Subheading 2.2., item 9**) at a constant voltage of 35 V for 2 h or 15 V overnight. Immobilon P membrane is soaked for 2 min in 100% methanol and then rinsed in electroblotting buffer immediately before use.

3.2.4. High-Sensitivity Chemiluminescence (ECL)

1. Blots are blocked with 5% (w/v) nonfat milk powder in PBST (*see* **Subheading 2.2., item 10**) for 1 h followed by brief rinsing with PBST.
2. Blots are incubated with anti-PrP monoclonal antibody 3F4 (*see* **Subheading 2.2., item 11**) at a final concentration of 0.2 μg/ml in PBST containing 0.1% (w/v) sodium azide for either 90 min or overnight.
3. Blots are washed for a minimum of 30 min and up to 60 min with at least six changes of PBST.
4. Blots are incubated for 1 h with a 1:10,000 dilution of goat anti-mouse IgG-phosphatase conjugate (*see* **Subheading 2.2., item 12**) in PBST.
5. Blots are washed for a minimum of 30 min and up to 60 min with at least six changes of PBST.
6. Blots are washed 2 × 5 min with 20 mM Tris, pH 9.8, containing 1 mM $MgCl_2$.
7. Blots are developed with chemiluminescent substrate CDP-Star (*see* **Subheading 2.2., item 13**) and visualized on Kodak Biomax MR film (*see* **Subheading 2.2., item 14**) (*see* **Note 4**).

3.2.5. Standard Enhanced Chemifluorescence (ECF)

1. Blots are blocked with 5% (w/v) nonfat milk powder in PBST (*see* **Subheading 2.2., item 10**) for 1 h followed by brief rinsing with PBST.
2. Blots are incubated with anti-PrP monoclonal antibody 3F4 (*see* **Subheading 2.2., item 11**) at a final concentration of 0.2 μg/ml in PBST containing 0.1% (w/v) sodium azide for either 90 min or overnight.
3. Blots are washed for a minimum of 30 min and up to 60 min with at least six changes of PBST.
4. Blots are incubated for 1 h with a 1:5,000 dilution of goat anti-mouse IgG-phosphatase conjugate (*see* **Subheading 2.2., item 12**) in PBST.
5. Blots are washed for a minimum of 30 min and up to 60 min with at least six changes of PBST.
6. Blots are washed 2 × 5 min with 20 mM Tris, pH 9.8, containing 1 mM $MgCl_2$.
7. Blots are developed with chemifluorescent substrate AttoPhos (*see* **Subheading 2.2., item 15**) and visualized on a Storm 840 PhosphorImager (GE Healthcare, Little Chalfont, Buckinghamshire, UK). PrP glycoforms are quantified with ImageQuaNT software (GE Healthcare) (*see* **Note 4**).

3.2.6. Sodium Phosphotungstic Acid Precipitation

Methods are adapted from the original procedure of Safar et al. *(64)* as described by Wadsworth et al. *(36)*.

1. 10% (w/v) homogenates from human brain or peripheral tissues prepared in Dulbecco's PBS lacking Ca^{2+} and Mg^{2+} ions (*see* **Subheading 2.2., item 1**) are centrifuged at 100g (800 rpm) for 1 min in a microfuge.
2. 500 µl of the resultant supernatant is mixed with an equal volume of 4% (w/v) sodium lauroylsarcosine (*see* **Subheading 2.2., item 16**) prepared in Dulbecco's PBS lacking Ca^{2+} and Mg^{2+} ions (*see* **Subheading 2.2., item 1**) and incubated for 10 min at 37°C with constant agitation.
3. Samples are adjusted to final concentrations of 50 U/ml Benzonase (*see* **Subheading 2.2., item 17**) (add 2 µl of 25 U/µl Benzon nuclease, purity 1) and 1 mM $MgCl_2$ (add 0.5 µl of 2 M $MgCl_2$ prepared in water) and incubated for 30 min at 37°C with constant agitation.
4. Samples are adjusted with 81.3 µl of a sodium phosphotungstic acid stock solution (*see* **Subheading 2.2., item 18**) to give a final concentration in the sample of 0.3% (w/v) sodium phosphotungstic acid. This stock solution is pre-warmed to 37°C before use, and both the sample and the stock solution should be at 37°C upon mixing to avoid formation of insoluble magnesium salts.
5. Samples are incubated at 37°C for 30 min with constant agitation before centrifugation at 16,100g (13,200 rpm) for 30 min in a microfuge. The microfuge rotor can be prewarmed to 37°C before use, because this helps to avoid salt precipitation during centrifugation.
6. After careful isolation of the supernatant, the sample is recentrifuged at 16,100g (13,200 rpm) for 2 min, and the residual supernatant is discarded. New tops are placed on the microfuge tubes.
7. Pellets are resuspended to a 20-µl final volume with Dulbecco's PBS lacking Ca^{2+} and Mg^{2+} ions containing 0.1% (w/v) sodium lauroylsarcosine and proteinase K digested and processed for immunoblotting as described in **Subheading 3.2.3, steps 2–9**) (*see* **Note 5**).

3.3. Immunohistochemistry

3.3.1. Procurement

3.3.1.1. Biosafety

For samples suspected to contain infectious prions, all procedures are performed within a class 1 microbiological safety cabinet situated within an ACDP level III containment laboratory with strict adherence to local rules of safe working practice. Informed consent for the analysis of samples must be in place before investigation. Samples are kept in a category III laboratory before decontamination with formic acid. Guidelines for decontamination of human prions are available

(63). Liquids that have been in contact with infected samples are decontaminated by mixing with an equal volume of 2 M sodium hydroxide for at least 1 h. For certain reagents, specialist disposal is preferred due to chemical incompatibilities *(63)*.

3.3.1.2. Whole Brain, Brain Hemispheres, or Whole Internal Organs

1. Large specimens of tissue (whole brain, brain hemispheres, whole internal organs) are suspended in 10% buffered formal-saline (*see* **Subheading 2.3.1., item 1**). The volume added should be approx 5 times the volume of tissue.
2. If there is excess blood within the sample, the 10% buffered formal-saline should be exchanged until it remains clear.
3. Tissue is left for up to 3 weeks to ensure adequate fixation and hardening.
4. After fixation, samples of tissue are excised with dimensions suitable for histology cassettes (*see* **Subheading 2.3.1., item 2**).

3.3.1.3. Small Specimens of Brain or Peripheral Tissues

1. Smaller pieces of brain (frontal cortex, temporal cortex, parietal cortex, occipital cortex, cerebellum) or samples of other peripheral tissues (tonsil, spleen lymph nodes), with dimensions no larger than approx 3 cm × 3 cm × 1 cm, are commonly provided for investigation.
2. Tissue samples are immersed in approx 5 volumes of 10% buffered formal-saline (*see* **Subheading 2.3.1., item 1**).
3. If there is excess blood within the sample, the 10% buffered formal-saline should be exchanged until it remains clear.
4. Fixation of the samples is achieved after 2 days.
5. After fixation, samples of tissue are excised with dimensions suitable for histology cassettes (*see* **Subheading 2.3.1., item 2**).

3.3.2. Prion Deactivation with Formic Acid

1. All brain tissue must be of a size suitable for processing. Generally, this is considered the size and thickness of the histology cassettes (*see* **Subheading 2.3.2., item 1**). Care must be taken not to overfill the cassettes, because this will result in poor processing and distortion.
2. After being encased in labeled cassettes, the samples are immersed in 98% formic acid (*see* **Subheading 2.3.2., item 2**) for 1 h.
3. Formic acid is decanted into a waste pot half filled with 2 M sodium hydroxide (*see* **Subheading 2.3.2., item 3**).
4. Specimens are treated with approx 5 volumes of 10% buffered formal-saline (*see* **Subheading 2.3.2., item 4**) for 1 h.

5. The 10% buffered formal-saline (*see* **Subheading 2.3.2., item 4**) is exchanged at least once to ensure any excess of formic acid has been removed before tissue processing.
6. Samples are removed from the ACDP level III containment laboratory.
7. Samples are placed on a tissue processor in an ACDP level II containment laboratory.

3.3.3. Tissue Processing

10% buffered formal saline (*see* **Subheading 2.3.3., item 1**) is an aqueous fixative; therefore, the samples are treated through a series of processing stages before wax embedding. Each stage needs to be of sufficient length to ensure impregnation. The stages are as follows:
1. Dehydration: The samples are taken through a series of industrial methylated spirits (IMS) (*see* **Subheading 2.3.3., item 2**) (70, 90, 100%) to remove water (**Table 1**).
2. Clearing: The alcohol is replaced by xylene (*see* **Subheading 2.3.3., item 3**), a fluid miscible with IMS and paraffin wax (*see* **Subheading 2.3.3., item 4**) (**Table 1**).
3. Impregnation: the xylene is replaced with molten paraffin wax (*see* **Subheading 2.3.3., item 4**) (**Table 1**).
4. Embedding: The samples are embedded in the desired orientation in molten paraffin wax (*see* **Subheading 2.3.3., item 4**). Once the wax has hardened, the samples are ready for sectioning.

3.3.4. Tissue Sectioning

1. The microtome (*see* **Subheading 2.3.4., item 1**) is set at 8 µm for tissue sectioning, although this measure can be varied.
2. The sample block, now in wax, is mounted on to the microtome chuck and serial sections of the sample are taken.

Table 1 Protocol for overnight processing of tissue samples for immunohistochemistry

Solution	Time (min)	Temperature
10% buffered formal-saline	30	Ambient
IMS 70%	75	Ambient
IMS 70%	75	Ambient
IMS 70%	75	Ambient
IMS 90%	60	Ambient
IMS 90%	60	Ambient
IMS 100%	75	Ambient
IMS 100%	75	Ambient
Xylene	75	Ambient
Xylene	75	Ambient
Molten paraffin wax	50	60°C
Molten paraffin wax	50	60°C
Molten paraffin wax	50	60°C
Molten paraffin wax	50	60°C

3. Sections are floated out on a water bath set at 40°C.
4. The sections are mounted on SuperFrost microscope slides (*see* **Subheading 2.3.4., item 2**) and left to air dry at 37°C for approx 2 h.
5. Slides are dried at 60°C for a minimum of 2 h, after which they are ready to be stained.
6. Tonsil sections require cutting just before staining. Immunoreactivity is markedly reduced if sections are exposed to air for long periods of time.

3.3.5. Tissue Staining

3.3.5.1. Staining with Hematoxylin and Eosin (H&E)

1. Rehydrate the sections by deparaffinising in three changes of xylene (*see* **Subheading 2.3.5., item 1**), followed by sequential washing for 1–2 min with graded alcohol (*see* **Subheading 2.3.5., item 2**) (100% × 2, 90%, and 70%) and final washing in running tap water.
2. Place the slides in filtered Harris hematoxylin solution (*see* **Subheading 2.3.5., item 3**) for 5 min.
3. Wash briefly in running tap water and differentiate in 1% acid alcohol (*see* **Subheading 2.3.5., item 4**) for 30 s.
4. Wash well in running tap water and allow the color to develop. Check microscopically. Nuclei look dark blue, whereas background shows a weak residual hematoxylin coloration.
5. Wash briefly in running water and stain with Eosin Y solution (*see* **Subheading 2.3.5., item 5**) for 2–3 min.
6. Wash sections sequentially for ~1–2 min with water, 70% ethanol, 90% ethanol, 100% ethanol, and xylene.
7. Mount sections in a xylene-based mounting medium, Pertex (*see* **Subheading 2.3.5., item 6**) (*see* **Note 6**).

3.3.5.2. Staining for Glial Fibrillary Acidic Protein (GFAP)

1. The sections to be stained are placed in plastic racks and deparaffinized, as described in **Subheading 3.3.5.1.**
2. The slides are placed on the Benchmark XT Staining Machine (*see* **Subheading 2.3.5., item 7**) (Ventana Medical Systems) with a 4-min pretreatment with Protease 1 (*see* **Subheading 2.3.5., item 8**).
3. The slides are incubated with a GFAP antibody (*see* **Subheading 2.3.5., item 9**) diluted 1:1000 in antibody diluent (*see* **Subheading 2.3.5., item 9**).
4. The slides are stained using the staining kit IViewDAB (*see* **Subheading 2.3.5., item 10**) and counterstained using hematoxylin (*see* **Subheading 2.3.5., item 11**) and a bluing reagent (*see* **Subheading 2.3.5., item 12**).
5. Once the run is finished, the slides are washed in hot soapy water (diluted washing up liquid) and dehydrated through alcohol and xylene.
6. Mount sections in a xylene-based mounting medium, Pertex (*see* **Subheading 2.3.5., item 6**) (*see* **Note 7**).

3.3.5.3. Staining for PrP

1. The sections to be stained are placed in plastic racks and deparaffinzed as described in **Subheading 3.3.5.1.**
2. The pretreatment for detection of abnormal PrP deposition is dependent upon the tissue and the length of fixation. For human brain samples that have been fixed for up to ~2 weeks, the microwave heat retrieval method is preferred. If the brain samples are fixed for longer periods, the pressure cooker method is used. If tonsil or other secondary lymphoid tissue is being examined, the autoclaving heat retrieval method is used.

3.3.5.3.1. Microwave Method

3. After deparaffinisation (*see* **Subheading 2.5.1.**), the slides are placed in 1 liter of Tris-EDTA-citrate buffer (*see* **Subheading 2.3.5., item 13**), and then they were placed in a microwave for 25 min at 800-W power.
4. The slides are washed in running cold tap water for 3 min.
5. The samples are covered with 98% formic acid (*see* **Subheading 2.3.5., item 14**), incubated for 5 min, and then washed in running cold tap water for 5 min to remove excess formic acid.

3.3.5.3.2. Pressure Cooker Method

6. After deparaffinization (*see* **Subheading 2.5.1.**), the slides are placed in a pressure cooker containing 1.5 liters of boiling Tris-EDTA-citrate buffer (*see* **Subheading 2.3.5., item 13**) for 5 min at high pressure and 5 min at low pressure.
7. Place the slides under running cold water for 5 min and treat with 98% formic acid (*see* **Subheading 2.3.5., item 14**) for a further 5 min. Wash the slides in running water for 5 min to remove excess formic acid.

3.3.5.3.3. Autoclaving Heat Retrieval

8. After deparaffinization (*see* **Subheading 3.3.5.1.**), the slides are placed in an autoclave resistant tub containing 1 liter of citrate buffer (*see* **Subheading 2.3.5., item 15**).
9. The tub is covered with aluminium foil and run in an autoclave at 121°C for 20 min. Allow autoclave to return to low pressure before removing the tub. Place slides under running cold tap water.
10. Treat with 98% formic acid (*see* **Subheading 2.3.5., item 14**) for a further 5 min. Wash the slides in running water for 5 min to remove excess formic acid.

3.3.5.3.4. Automated Staining

11. The monoclonal anti-PrP antibody ICSM35 (*see* **Subheading 2.3.5., item 16**) (1 mg/ml stock) is used at a 1:1,000 dilution in antibody diluent (*see* **Subheading 2.3.5., item 9**).

12. Automated staining is carried out on the Benchmark Staining Machine (*see* **Subheading 2.3.5., item 7**) from Ventana Medical Systems.

13. The slides are subjected to further pre-treatment with Protease 3 (*see* **Subheading 2.3.5., item 17**) for 4 min, and then 10 min with Superblock (*see* **Subheading 2.3.5., item 18**) , a blocking agent. The slides are stained using the staining kit IViewDAB (*see* **Subheading 2.3.5., item 10**) and counterstained using hematoxylin (*see* **Subheading 2.3.5., item 11**) and a bluing reagent (*see* **Subheading 2.3.5., item 12**).

14. Once the staining process is complete, the slides are washed in hot soapy water (diluted washing-up liquid) and dehydrated through alcohol and xylene. They were mounted in a xylene-based mountant as described in **Subheading 3.3.5.1.** (*see* **Note 8**).

3.3.5.3.5. Manual Staining

15. Sections are deparaffinized as far as 100% alcohol (*see* **Subheading 3.3.5.1.**).

16. Block endogenous peroxidase activity on the sections by treatment with 2.5% (v/v) hydrogen peroxide in methanol (*see* **Subheading 2.3.5., item 19**) for 30 min.

17. Wash sections in running tap water for 5 min and then in purified water for 5 min. Next, they were transferred to an appropriate container for autoclaving. Autoclave at 121°C for 20 min in Tris-buffered saline, pH 7.6 (*see* **Subheading 2.3.5., item 20**).

18. Cool slides in running tap water. Treat the slides in 98% formic acid (*see* **Subheading 2.3.5., item 14**) for 5 min, wash in running tap water for 5–10 min.

19. Treat sections with 4 M guanidine thiocyanate (*see* **Subheading 2.3.5., item 21**) at 4°C for 2 h and then wash in tap water and transfer to TBS (*see* **Subheading 2.3.5., item 20**).

20. Block nonspecific immunoglobulin staining with normal rabbit serum (*see* **Subheading 2.3.5., item 22**) diluted 1:10 in TBS for 30 min. Do not wash off.

21. Apply primary antibody ICSM35 (*see* **Subheading 2.3.5., item 16**) (1 mg/ml stock) at 1:1500 dilution in TBS containing 1:100 normal rabbit serum (*see* **Subheading 2.3.5., item 22**) , overnight at 4°C.

22. Wash in several changes of TBS.

23. Incubate in biotinylated rabbit anti-mouse immunoglobulins (*see* **Subheading 2.3.5., item 23**) 1:200 in TBS for 45 min.

24. Wash in several changes of TBS.

25. Incubate in ABC complex (*see* **Subheading 2.3.5., item 24**) for 45 min.

26. Wash in several changes of TBS.

27. Develop in 3,3 diaminobenzidine tetrachloride (*see* **Subheading 2.3.5., item 25**) (25 mg/100 ml of TBS) plus 30 μl of hydrogen peroxide (*see* **Subheading 2.3.5., item 19**) (added just before use) for 5–15 min. Check microscopically. Once chromagen has developed to satisfaction, wash slides in running tap water for 10 min.

28. Counterstain in Harris hematoxylin (*see* **Subheading 2.3.5., item 3**) for 3 min.
29. Differentiate in 1% acid alcohol (*see* **Subheading 2.3.5., item 4**) for 5 s.
30. Allow blue coloration to develop in tap water, 5 min.
31. Dehydrate, clear, and mount as described in Subheading 3.3.5.1. (*see* **Note 8**).

Acknowledgements We especially thank all patients and their families for generously consenting to use of human tissues in this research, and the UK neuropathologists who have kindly helped in providing these tissues. We are grateful to R. Young for preparation of the figures. This work was funded by the UK Medical Research Council and the European Commission, and it was performed under approval from the Institute of Neurology/National Hospital for Neurology and Neurosurgery Local Research Ethics Committee.

4. Notes

1. Whole brain homogenate can be analyzed by identical procedures; however problems of high sample viscosity due to nucleic acid aggregation are often encountered. For processing of 20 μl of whole brain homogenate, preincubation with 0.5 μl of Benzonase for 10 min at 20°C is recommended before further sample analysis by using appropriately adjusted volumes of subsequent reagents.

2. PrP^{Sc} is covalently indistinguishable from PrP^C, but it can by differentiated from PrP^C by its partial resistance to proteolysis and its marked insolubility in detergents *(1, 65)*. Under conditions in which PrP^C is completely degraded by the nonspecific protease, proteinase K, PrP^{Sc} in sporadic and acquired forms of human prion disease exists in an aggregated form with the C-terminal two thirds of the protein showing marked resistance to proteolysis, leading to the generation of amino terminally truncated fragments of di-, mono- and nonglycosylated PrP *(1, 65)* (**Fig. 4**).

3. The procedures described here have been optimized for use with Invitogen 16% acrylamide precast Tris-glycine gels. Variation in the resolution of the system may occur if other gel systems are used or if reagent compositions are varied from those listed here. Optimal resolution of PrP^{Sc} fragment size is achieved after electrophoresis for 80 min at 200 V. For improved separation of PrP glycoforms for densitometry analysis, electrophoresis is performed for 90 min at 200 V.

4. To date, we have identified four major types of human PrP^{Sc} associated with sporadic and acquired human prion diseases that can be differentiated by their fragment size on immunoblots after limited proteinase K digestion of brain homogenates *(8, 10, 12)* (**Fig. 4**). These types can be further classified by the ratio of the three PrP bands seen after protease digestion, corresponding to amino-terminally truncated cleavage products generated from di-, mono-, or nonglycosylated PrP^{Sc} (**Figs. 4, 5**). PrP^{Sc} types 1–3 are seen in classical (sporadic or iatrogenic) CJD brain, whereas type 4 PrP^{Sc} is uniquely seen in vCJD brain *(8, 10,*

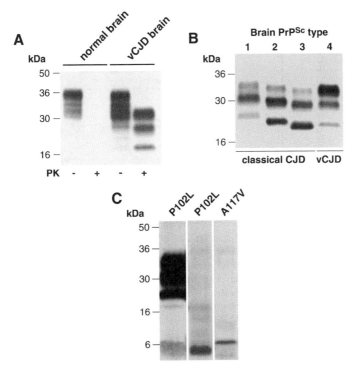

Fig. 4 Immunoblot analysis of human PrP. (**A**) Immunoblot analysis of normal human brain and vCJD brain homogenate before and after treatment with proteinase K (PK). PrP^C in both normal and vCJD brain is completely degraded by PK, whereas PrP^{Sc} present in vCJD brain shows resistance to proteolytic degradation leading to the generation of amino terminally truncated fragments of di-, mono-, and nonglycosylated PrP. (**B**) Immunoblot of PK digested brain homogenate with monoclonal antibody 3F4 showing PrP^{Sc} types 1–4 in human brain. Types 1–3 PrP^{Sc} are seen in the brain of classical forms of CJD (either sporadic or iatrogenic CJD), whereas type 4 PrP^{Sc} is uniquely seen in vCJD brain. Classification according to Hill et al. (*12*). (**C**) Immunoblots of PK digested brain homogenate from cases of inherited prion disease with *PRNP* mutations showing protease-resistant PrP fragments of ~6–8 kDa. The *PRNP* point mutation is designated above each immunoblot. Immunoblots were developed with anti-PrP monoclonal antibody 3F4

12). An earlier classification of PrP^{Sc} types seen in classical CJD described only two banding patterns (*9*) with PrP^{Sc} types 1 and 2 that we describe corresponding with the type 1 pattern of Gambetti and colleagues, and our type 3 fragment size corresponding to their type 2 pattern (*11, 66*). Although type 4 PrP^{Sc} is readily distinguished from the PrP^{Sc} types seen in classical CJD by a predominance of the diglycosylated PrP glycoform, type 4 PrP^{Sc} also has a distinct proteolytic fragment size (*12*) (**Fig. 4**), although this is not recognized by the alternative classification, which designates type 4 PrP^{Sc} as type 2b (*66*). Although proteinase K-resistant PrP fragments of ~21–30 kDa seen in inherited prion disease caused by *PRNP* P102L, D178N, and E200K mutations have molecular masses similar in size to those seen in classical CJD (*15, 67–69*), the glycoform ratio is distinct from PrP^{Sc} fragments

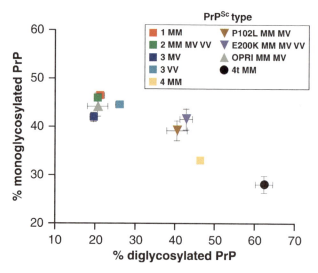

Fig. 5 PrP glycoform ratios in human prion disease. PK digestion of brain homogenate and analysis by enhanced chemifluorescence with anti-PrP monoclonal antibody 3F4 enables calculation of the proportions of di-, mono-, and nonglycosylated PrP. The plot shows the protease-resistant PrP glycoform ratio seen classical CJD (PrP^{Sc} types 1–3), vCJD (PrP^{Sc} type 4 in brain and type 4t PrP^{Sc} in tonsil) and in cases of inherited prion disease. The key shows PrP^{Sc} type or mutation and *PRNP* codon 129 genotype (methionine [M] and valine [V]). Classification according to Hill et al. *(12, 15)*. Data points represent the mean relative proportions of di- and mono-PrP as percentage ± S.E.M. In some cases the error bars were smaller than the symbols used

seen in both classical CJD *(15, 67–69)* and vCJD *(15)* (**Fig. 5**). Individuals with these mutations also propagate PrP^{Sc} with distinct fragment sizes *(15, 67, 68)*. The fragment sizes and glycoform ratios of PrP^{Sc} seen in 2-, 4-, and 6-OPRI cases are indistinguishable from those of PrP^{Sc} seen in classical CJD *(15)*. Importantly detection of PrP^{Sc} in the molecular mass range of ~21–30 kDa is by no means a consistent feature in inherited prion disease; and some cases, in particular those in which amyloid plaques are a prominent feature, show smaller protease resistant PrP fragments of ~7–15 kDa derived from the central portion of PrP *(15, 22, 67, 68, 70–72)* (**Fig. 4**).

5. Sodium phosphotungstic acid precipitation facilitates highly efficient recovery and detection of PrP^{Sc} from human tissue homogenate, even when present at levels 10^4–10^5-fold lower than found in brain *(14, 36)*. This procedure is now the preferred method for diagnostic analysis of tonsil in cases of suspected vCJD, and it should detect PrP^{Sc} in tonsil if levels reach 0.1% or above the maximum levels seen in necropsy vCJD tonsil *(34, 36)*. A distinctive PrP^{Sc} type, designated type 4t, is seen in both ante-mortem and post-mortem tonsil from patients with vCJD *((36, 56)*; *see* Figs. 5 and 7), including secondary vCJD infection resulting from blood transfusion *(39)*. Type 4t PrP^{Sc} in tonsil differs in the proportions of the PrP glycoforms from type 4 PrP^{Sc} seen in vCJD brain *((36, 56)*; *see* Figs. 5

sCJD   vCJD

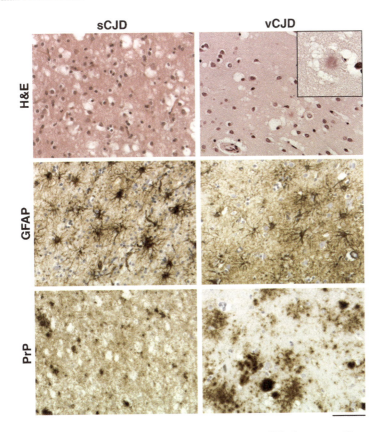

Fig. 6 Prion disease pathology. Brain sections from sCJD and vCJD show spongiform neurode-generation after H&E staining (*H&E*), proliferation of reactive astrocytes after immunohistochem-istry using anti-GFAP antibodies (*GFAP*), and abnormal PrP immunoreactivity after immunohistochemistry using anti-PrP monoclonal antibody ICSM35 (*PrP*). Abnormal PrP deposi-tion in sCJD most commonly presents as diffuse, synaptic staining, whereas vCJD is distinguished by the presence of florid PrP plaques consisting of a round amyloid core surrounded by a ring of spongiform vacuoles. Bar = 100 μm. Inset, high-power magnification of a florid PrP plaque

and 7), implying the superimposition of tissue and strain specific effects on PrP glycosylation *(32, 56)*.

6. On H&E-stained sections nuclei are stained deep blue, and the cytoplasm is stained pink. The cortex and subcortical white matter can be readily distin-guished. In the cortex of a patient with prion disease, there may be variable degrees of spongiosis, accompanied astroglial proliferation (**Fig. 6**). Neuronal loss also may be evident. Although synaptic PrP deposition is generally not rec-ognizable on H&E sections, amyloid PrP plaques as seen in GSS and vCJD may be a prominent feature (**Fig. 6**). In the cerebellum, spongiosis is generally less evident; however, PrP plaques may be observed particularly in GSS.

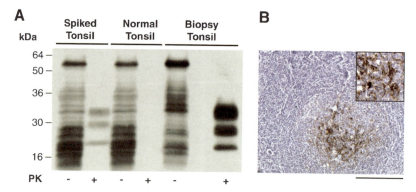

Fig. 7 Abnormal PrP in vCJD tonsil. (**A**) Diagnostic PrP^{Sc} analysis of tonsil biopsy tissue. Aliquots (0.5 ml) of 10% tonsil biopsy homogenate from a patient with suspected vCJD or 10% normal human tonsil homogenate, either lacking or containing a spike of 50 nl of 10% vCJD brain homogenate, were subjected to sodium phosphotungstic acid precipitation. Then, 20 µl aliquots of whole samples isolated before centrifugation were analyzed in the absence of PK digestion (−) and compared with PK digestion products (+) derived from the entire sodium phosphotungstic acid pellets. The immunoblot was analyzed with anti-PrP monoclonal antibody 3F4 and high-sensitivity enhanced chemiluminescence. (**B**) Immunohistochemical analysis of necropsy vCJD tonsil. Abnormal PrP immunoreactivity in vCJD tonsil is confined to lymphatic follicles with deposition mainly in dendritic cells. Anti-PrP monoclonal antibody ICSM 35. Bar = 160 µm. Inset, high-power magnification of PrP deposits

7. Reactive astrocytes are readily visualized by GFAP immunohistochemistry. They are characterized by prominent processes (**Fig. 6**). In the white matter, there may be a diffuse fibrillary gliosis.

8. Abnormal PrP deposition can present with a multitude of intensities, shapes, and distributions. The *synaptic* pattern is characterized by a fine, dispersed distribution, and it is the predominant pattern of abnormal PrP staining seen in sporadic CJD (**Fig. 6**). In contrast, PrP amyloid plaques are a predominant feature in GSS, kuru, and vCJD. The histopathologic features of vCJD are remarkably consistent and distinguish it from other human prion diseases with large numbers of PrP-positive amyloid plaques that differ in morphology from the plaques seen in kuru and GSS in that the surrounding tissue takes on a microvacuolated appearance, giving the plaques a florid appearance (**Fig. 6**). Abnormal PrP immunoreactivity in vCJD tonsil is confined to lymphatic follicles with deposition mainly in dendritic cells (**Fig. 7**).

References

1. Prusiner SB. (1998) Prions. Proc Natl Acad Sci U S A;95:13363–13383.
2. Collinge J. (2001) Prion diseases of humans and animals: their causes and molecular basis. Annu Rev Neurosci;24:519–550.

3. Collinge J. (2005) Molecular neurology of prion disease. J Neurol Neurosurg Psychiatry; 76:906–919.

4. Wadsworth JD, Hill AF, Beck JA, Collinge J. (2003) Molecular and clinical classification of human prion disease. Br Med Bull;66:241–254.

5. Wadsworth JD, Collinge J. (2007) Update on human prion disease. Biochim Biophys Acta; 1772:598–609.

6. Weissmann C. (2004) The state of the prion. Nat Rev Microbiol;2:861–871.

7. Telling GC, Parchi P, DeArmond SJ, et al. (1996) Evidence for the conformation of the pathologic isoform of the prion protein enciphering and propagating prion diversity. Science;274:2079–2082.

8. Collinge J, Sidle KCL, Meads J, Ironside J, Hill AF. (1996) Molecular analysis of prion strain variation and the aetiology of 'new variant' CJD. Nature;383:685–690.

9. Parchi P, Castellani R, Capellari S, et al. (1996) Molecular basis of phenotypic variability in sporadic Creutzfeldt-Jakob disease. Ann Neurol;39:767–778.

10. Wadsworth JDF, Hill AF, Joiner S, Jackson GS, Clarke AR, Collinge J. (1999) Strain-specific prion-protein conformation determined by metal ions. Nat Cell Biol;1:55–59.

11. Parchi P, Giese A, Capellari S, et al. (1999) Classification of sporadic Creutzfeldt-Jakob Disease based on molecular and phenotypic analysis of 300 subjects. Ann Neurol; 46:224–233.

12. Hill AF, Joiner S, Wadsworth JD, et al. (2003) Molecular classification of sporadic Creutzfeldt-Jakob disease. Brain;126:1333–1346.

13. Zanusso G, Farinazzo A, Prelli F, et al. (2004) Identification of distinct N-terminal truncated forms of prion protein in different Creutzfeldt-Jakob disease subtypes. J Biol Chem;79: 38936–38942.

14. Safar JG, Geschwind MD, Deering C, et al. (2005) Diagnosis of human prion disease. Proc Natl Acad Sci U S A;102:3501–3506.

15. Hill A, Joiner S, Beck J, et al. (2006) Distinct glycoform ratios of protease resistant prion protein associated with *PRNP* point mutations. Brain;129:676–685.

16. Lloyd SE, Onwuazor ON, Beck JA, et al. (2001) Identification of multiple quantitative trait loci linked to prion disease incubation period in mice. Proc Natl Acad Sci U S A;98:6279–6283.

17. Asante EA, Linehan JM, Desbruslais M, et al. (2002) BSE prions propagate as either variant CJD-like or sporadic CJD-like prion strains in transgenic mice expressing human prion protein. EMBO J;21:6358–6366.

18. Collinge J. (1997) Human prion diseases and bovine spongiform encephalopathy (BSE). Hum Mol Genetics;6:1699–1705.

19. Collinge J, Palmer.M.S. (1997) Prion Diseases. 1st ed. Oxford: Oxford University Press.

20. Brown P, Cathala F, Raubertas RF, Gajdusek DC, Castaigne P. (1987) The epidemiology of Creutzfeldt-Jakob disease: conclusion of a 15-year investigation in France and review of the world literature. Neurology;37:895–904.

21. Collins SJ, Sanchez-Juan P, Masters CL, et al. (2006) Determinants of diagnostic investigation sensitivities across the clinical spectrum of sporadic Creutzfeldt-Jakob disease. Brain;129:2278–2287.

22. Kovacs GG, Trabattoni G, Hainfellner JA, Ironside JW, Knight RS, Budka H. (2002) Mutations of the prion protein gene phenotypic spectrum. J Neurol; 249:1567–1582.

23. Mead S. (2006) Prion disease genetics. Eur J Hum Genet;14:273–281.

24. Brown P, Preece MA, Will RG. (1992) "Friendly fire" in medicine: hormones, homografts, and Creutzfeldt-Jakob disease. Lancet;340:24–27.

25. Brown P, Preece M, Brandel JP, et al. (2000) Iatrogenic Creutzfeldt-Jakob disease at the millennium. Neurology;55:1075–1081.

26. Alpers MP. Epidemiology and clinical aspects of kuru. (1987) In: Prusiner SB, McKinley MP, editors. Prions: Novel Infectious Pathogens Causing Scrapie and Creutzfeldt-Jakob Disease. San Diego: Academic Press, 451–465.

27. Mead S, Stumpf MP, Whitfield J, et al. (2003) Balancing selection at the prion protein gene consistent with prehistoric kurulike epidemics. Science;300:640–643.

28. Collinge J, Whitfield J, McKintosh E, et al. (2006) Kuru in the 21st century-an acquired human prion disease with very long incubation periods. Lancet;367:2068–2074.
29. Will RG, Ironside JW, Zeidler M, et al. (1996) A new variant of Creutzfeldt-Jakob disease in the UK. Lancet;347:921–925.
30. Hill AF, Desbruslais M, Joiner S, et al. (1997) The same prion strain causes vCJD and BSE. Nature;389:448–450.
31. Bruce ME, Will RG, Ironside JW, et al. (1997) Transmissions to mice indicate that 'new variant' CJD is caused by the BSE agent. Nature;389:498–501.
32. Collinge J. (1999) Variant Creutzfeldt-Jakob disease. Lancet;354:317–323.
33. Hilton DA, Ghani AC, Conyers L, et al. (2004) Prevalence of lymphoreticular prion protein accumulation in UK tissue samples. J Pathol;203:733–739.
34. Frosh A, Smith LC, Jackson CJ, et al. (2004) Analysis of 2000 consecutive UK tonsillectomy specimens for disease-related prion protein. Lancet;364:1260–1262.
35. Hilton DA. (2005) Pathogenesis and prevalence of variant Creutzfeldt-Jakob disease. J Pathol;208:134–141.
36. Wadsworth JDF, Joiner S, Hill AF, et al. (2001) Tissue distribution of protease resistant prion protein in variant CJD using a highly sensitive immuno-blotting assay. Lancet;358:171–180.
37. Llewelyn CA, Hewitt PE, Knight RS, et al. (2004) Possible transmission of variant Creutzfeldt-Jakob disease by blood transfusion. Lancet;363:417–421.
38. Peden AH, Head MW, Ritchie DL, Bell JE, Ironside JW. (2004) Preclinical vCJD after blood transfusion in a PRNP codon 129 heterozygous patient. Lancet; 364:527–529.
39. Wroe SJ, Pal S, Siddique D, et al. (2006) Clinical presentation and pre-mortem diagnosis of variant Creutzfeldt-Jakob disease associated with blood transfusion: a case report. Lancet;368:2061–2067.
40. Collinge J, Palmer MS, Dryden AJ. (1991) Genetic predisposition to iatrogenic Creutzfeldt-Jakob disease. Lancet;337:1441–1442.
41. Palmer MS, Dryden AJ, Hughes JT, Collinge J. (1991) Homozygous prion protein genotype predisposes to sporadic Creutzfeldt-Jakob disease. Nature; 352:340–342.
42. Windl O, Dempster M, Estibeiro JP, et al. (1996) Genetic basis of Creutzfeldt-Jakob disease in the United Kingdom: a systematic analysis of predisposing mutations and allelic variation in the PRNP gene. Hum Genet;98:259–264.
43. Lee HS, Brown P, Cervenáková L, et al. (2001) Increased susceptibility to Kuru of carriers of the PRNP 129 methionine/methionine genotype. J Infect Dis;183:192–196.
44. Wadsworth JD, Asante EA, Desbruslais M, et al. (2004) Human prion protein with valine 129 prevents expression of variant CJD phenotype. Science;306:1793–1796.
45. Collinge J, Harding AE, Owen F, et al. (1989) Diagnosis of Gerstmann-Sträussler syndrome in familial dementia with prion protein gene analysis. Lancet;2:15–17.
46. Collinge J, Owen F, Poulter M, et al. (1990) Prion dementia without characteristic pathology. Lancet;336:7–9.
47. Collinge J, Brown J, Hardy J, et al. (1992) Inherited prion disease with 144 base pair gene insertion: II: clinical and pathological features. Brain;115:687–710.
48. Mallucci GR, Campbell TA, Dickinson A, et al.(1999) Inherited prion disease with an alanine to valine mutation at codon 117 in the prion protein gene. Brain;122:1823–1837.
49. Mead S, Poulter M, Beck J, et al. (2006) Inherited prion disease with six octapeptide repeat insertional mutation–molecular analysis of phenotypic heterogeneity. Brain; 129:2297–2317.
50. World Health Organisation. (2003) WHO manual for surveillance of human transmissible spongiform encephalopathies. http://www.who.int/bloodproducts/TSE-manual2003.pdf
51. Collinge J, Poulter M, Davis MB, et al. (1991) Presymptomatic detection or exclusion of prion protein gene defects in families with inherited prion diseases. Am J Hum Genet;49:1351–1354.
52. Budka H, Aguzzi A, Brown P, et al. (1995) Neuropathological diagnostic criteria for Creutzfeldt-Jakob disease (CJD) and other human spongiform encephalopathies (prion diseases). Brain Pathol;5:459–466.
53. Budka H. Neuropathology of prion diseases. (2003) Br Med Bull;66:121–130.

54. Hainfellner JA, Brantner-Inthaler S, Cervenáková L, et al. (1995) The original Gerstmann-Sträussler-Scheinker family of Austria: divergent clinicopathological phenotypes but constant PrP genotype. Brain Pathol;5:201–211.
55. Ironside JW, Head MW. Neuropathology and molecular biology of variant Creutzfeldt-Jakob disease. (2004) Curr Top Microbiol Immunol;284:133–159.
56. Hill AF, Butterworth RJ, Joiner S, et al. (1999) Investigation of variant Creutzfeldt-Jakob disease and other human prion diseases with tonsil biopsy samples. Lancet;353:183–189.
57. Glatzel M, Abela E, Maissen M, Aguzzi A. (2003) Extraneural pathologic prion protein in sporadic Creutzfeldt-Jakob disease. N Engl J Med;349:1812–1820.
58. Hilton DA, Sutak J, Smith ME, et al. (2004) Specificity of lymphoreticular accumulation of prion protein for variant Creutzfeldt-Jakob disease. J Clin Pathol;57:300–302.
59. Head MW, Ritchie D, Smith N, et al. (2004) Peripheral tissue involvement in sporadic, iatrogenic, and variant Creutzfeldt-Jakob disease: an immunohistochemical, quantitative, and biochemical study. Am J Pathol;164:143–153.
60. Joiner S, Linehan JM, Brandner S, Wadsworth JD, Collinge J. (2005) High levels of disease related prion protein in the ileum in variant Creutzfeldt-Jakob disease. Gut;54:1506–1508.
61. Wadsworth JD, Joiner S, Fox K, et al. (2007) Prion infectivity in vCJD rectum. Gut;56:90–94.
62. Siddique D, Kennedy A, Thomas D, et al. (2005) Tonsil biopsy in the investigation of suspected variant Creutzfeldt-Jakob disease–a cohort study of 50 patients. J Neurol Sci;238:S1–S570.
63. Advisory Committee on Dangerous Pathogens and the Spongiform Encephalopathy Advisory Committee. (2003) Transmissible spongiform encephalopathy agents: safe working and the prevention of infection. UK Department of Health. http://www.advisorybodies.doh.gov.uk/acdp/tseguidance/Index.htm
64. Safar J, Wille H, Itri V, et al. (1998) Eight prion strains have PrP^{Sc} molecules with different conformations. Nat Med; 4:1157–1165.
65. Riesner D. (2003) Biochemistry and structure of PrP^C and PrP^{Sc}. Br Med Bull;66:21–33.
66. Parchi P, Capellari S, Chen SG, et al. (1997) Typing prion isoforms. Nature;386:232–233.
67. Piccardo P, Dlouhy SR, Lievens PMJ, et al. (1998) Phenotypic variability of Gerstmann-Sträussler-Scheinker disease is associated with prion protein heterogeneity. J Neuropathol Exp Neurol;57:979–988.
68. Parchi P, Chen SG, Brown P, et al. (1998) Different patterns of truncated prion protein fragments correlate with distinct phenotypes in P102L Gerstmann-Sträussler-Scheinker disease. Proc Natl Acad Sci U S A;95:8322–8327.
69. Furukawa H, Doh-ura K, Kikuchi H, Tateishi J, Iwaki T. (1998) A comparative study of abnormal prion protein isoforms between Gerstmann-Sträussler-Scheinker syndrome and Creutzfeldt-Jakob disease. J Neurol Sci;158:71–75.
70. Piccardo P, Liepnieks JJ, William A, et al. (2001) Prion proteins with different conformations accumulate in Geustmann-Sträussler-Scheinker disease caused by A117V and F198S mutations. Am J Pathol;158:2201–2207.
71. Tagliavini F, Lievens PMJ, Tranchant C, et al. (2001) A 7-kDa prion protein (PrP) fragment, an integral component of the PrP region required for infectivity, is the major amyloid protein in Gerstmann-Sträussler-Scheinker disease A117V. J Biol Chem;276:6009–6015.
72. Wadsworth JD, Joiner S, Linehan JM, et al. (2006) Phenotypic heterogeneity in inherited prion disease (P102L) is associated with differential propagation of protease-resistant wild-type and mutant prion protein. Brain;129:1557–1569.

Chapter 15
Analysis of Endogenous PrP^C Processing in Neuronal and Non-neuronal Cell Lines

Victoria Lewis and Steven J. Collins

Summary Numerous transmembrane and glycosylphosphatidylinositol (GPI)-anchored proteins, covering a vast range of structural and functional classes, are recognized to undergo proteolytic cleavage or shedding from the plasma membrane. Although this widespread phenomenon seems fundamental to normal cellular biology, proteolytic processing also seems to play a central role in the pathogenesis of some neurodegenerative disorders such as Alzheimer's disease. An analogous situation may exist in prion disorders. The GPI-anchored cellular prion protein (PrP^C) may be endoproteolytically cleaved at two different sites: one at the C-terminal end of the octameric repeat region and the other within a potentially neurotoxic and amyloidogenic region of the protein. The relevance of these alternative proteolytic events to normal cell function and pathogenesis is incompletely resolved. Study and characterization of the constitutive processing of PrP^C will provide insight into the biological relevance of alternative cleavages in terms of normal PrP^C function, and also into the potential role, if any, to disease causation.

Keywords Endogenous cleavage; endoproteolytic processing; prion protein (PrP); PrP^C; PrP^{res}.

1. Introduction

The cellular prion protein, PrP^C, is ubiquitously expressed, with highest levels found in neurons (*1–3*). Despite uncertainties regarding the precise biological function of PrP^C, expression is an absolute requirement for successful disease transmission (*4, 5*). After synthesis, PrP^C undergoes a number of posttranslational modifications, including cleavage of the amino- and carboxy-terminal signal peptides (the latter resulting in the attachment of a glycosylphosphatidylinositol [GPI] anchor), asparagine-linked glycosylation, and formation of a single disulfide bond (*6, 7*). PrP^C is among many membrane-bound proteins, encompassing those linked through a GPI-moiety and integral transmembrane proteins, that are known to undergo constitutive proteolytic processing or shedding (*8*). Because of the extensive structural

From: *Prion Protein Protocols.*
Methods in Molecular Biology, Vol. 459.
Edited by: A. F. Hill © Humana Press, Totowa, NJ

and functional classes of membrane-associated proteins to which this processing applies, cleavage is undoubtedly of fundamental biological significance, allowing functional diversity of a protein without the requirement for separate biosynthetic pathways *(8)*.

The mature PrP^C protein is predominantly endoproteolytically cleaved at either of two cleavage sites (**Fig. 1A**). One site is at residues 111/112 (human sequence), producing the C1 fragment *(9, 10)*, and the second is at around amino acid residue 90, producing the C2 fragment *(10, 11)*. Experimental evidence has revealed at least some of the proteases or mechanisms responsible for these cleavages, namely, *a d*isintegrin *a*nd *m*etalloprotease (ADAMs) 10 and 17 (also

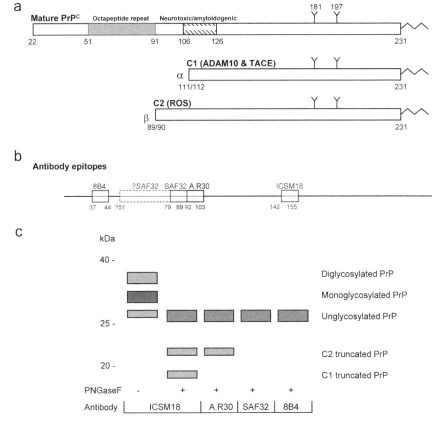

Fig. 1 (**A**) Schematic representation of cleavage sites of PrP^C resulting in the C1 and C2 PrP fragments. (**B**) Diagrammatic representation of the epitopes for antibodies used to detect the cleavage products of PrP. (**C**) Cartoon showing the expected results and mobilities of different PrP fragments when examined by Western blotting with a range of PrP specific antibodies

known as TACE) in C1 cleavage *(12)*, and reactive oxygen species (ROS) *(13, 14)* in C2 cleavage. ADAM10 has been identified as important for constitutive PrP processing, whereas TACE seems to have inducible processing activity *(12)*. There is also evidence suggesting the participation of calpains in C2-type cleavage, although this may relate more to cleavage of PrP^res at the C2 site *(11)*, and not PrP^C *(14)*. The physiological significance of endogenous PrP^C processing is not yet understood. It is possible that the resulting soluble N-terminal fragments may behave as signaling molecules, or that cleavage of PrP is generating an altered biological function of the residual membrane anchored, N-terminally truncated PrP species.

There are some interesting parallels between the prion diseases and Alzheimer's disease (AD), another neurodegenerative disease associated with protein accumulation and aggregation. AD pathogenesis is associated with generation of the toxic amyloid-β (Aβ) peptide, which is produced as the result of sequential β-secretase, then γ-secretase cleavage of the amyloid precursor protein (APP) *(15)*. Normal APP processing by α-secretases, of which ADAM10 and TACE have been implicated *(16)*, followed by γ-secretase cleavage is described as the non-amyloidogenic pathway, because it subserves APP cleavage at a site preventing generation of the toxic Aβ peptide and produces a nontoxic p3 fragment *(15)*. Analogous to APP processing, it is hypothesised that the C1 cleavage of PrP^C, which occurs within the potentially neurotoxic and amyloidogenic region of 106–126 *(17)*, may be protective by disrupting this segment *(18, 19)*. There is also some evidence to suggest C2 cleavage is protective against oxidative stress *(14)*. It is, therefore, possible that any aberrant or compromised processing of PrP^C within the cell could lead to a cascade of oxidative or other cellular damage that may play a role in the neurodegeneration observed in prion diseases, or perhaps influence susceptibility to prion infection by providing a better template for conversion of PrP^C to PrP^res.

2. Materials

2.1. Cell Culture

1. Cell lines are maintained in either Dulbecco's modified Eagle's medium or Opti-MEM (Invitrogen, Carlsbad, CA) containing 10% (v/v) fetal bovine serum (JRH Biosciences, Sigma-Aldrich, St. Louis, MO) and 1% (v/v) penicillin-streptomycin (Invitrogen).
2. Sterile cell scrapers were used to harvest cells for passage, and for collection of cell pellet for experimentation.
3. 1× phosphate-buffered saline (PBS): 140 mM NaCl, 2.7 mM KCl, 1.8 mM KH_2PO_4, 10 mM Na_2HPO_4. Adjust pH to 7.4 with HCl. Sterilize by autoclaving at 121°C for 20 minutes. Store at 4°C.

2.2. Cell Lysis and Protein Determination

1. Lysis buffer stock: 25 mM Tris-HCl, 150 mM NaCl, 5 mM EDTA, 1% Triton X-100 (v/v). Adjust pH to 7.5. Store at 4°C
2. Protease inhibitor cocktail: 1 Complete Mini™ tablet (Roche Applied Science, Penzberg, Germany) is dissolved in 1 ml of water to make a 10× stock. Subaliquot 100 µl volumes. Store at −20°C (*see* **Note 1**).
3. Protein determination: BCA Protein Assay kit (Pierce Chemical, Rockford, IL), reagents A and B, and albumin standards are provided. Store reagents A and B at room temperature. From the 2 mg/ml albumin standard, dilute with water to make 1, 0.5, 0.4, 0.2, and 0.1 mg/ml. Store albumin standards at 4°C.

2.3. Deglycosylation of PrP^C

1. Incubation buffer: 50 mM EDTA, 0.02% (w/v) sodium azide, made up in 1× PBS. Adjust pH to 8.0. Store at 4°C.
2. 10× denaturation buffer: 5% (w/v) sodium dodecyl sulfate (SDS) and 5% (v/v) β-mercaptoethanol (BME) made up in incubation buffer. Store at 4°C.
3. Ten percent solution of Nonidet P-40 (also called Igepal CA-630, Sigma-Aldrich) in water. Store at 4°C.
4. *N*-Glycosidase F (PNGase F) (Roche Applied Science). Store at −20°C.
5. 4× sample buffer stock: final concentration 170 mM Tris-HCl, pH 6.8, 8% SDS, 4% glycine, 0.02% bromophenol blue. Store at room temperature. To make 4× sample buffer that is ready to use and mix with samples, add 120 µl of BME (12%, v/v, final concentration) and 160 µl of glycerol with 720 µl of 4× stock. Store at 4°C.

2.4. SDS-Polyacrylamide Gel Electrophoresis (PAGE)

1. Resolving gel buffer: 2 M Tris-HCl, pH 8.8. Store at 4°C.
2. Stacking gel buffer: 0.5 M Tris-HCl, pH 6.8. Store at 4°C.
3. Other gel components: 30% acrylamide/Bis, 37.5:1 solution (Bio-Rad, Hercules, CA), *N*,*N*,*N*′,*N*′-tetra-methyl-ethylenediamine (TEMED) (Bio-Rad), ammonium persulfate (APS) (Sigma-Aldrich) made to 10% (w/v) solution with water. Store at 4°C.
4. 10× running buffer: 250 mM Tris, 2 M glycine, 1% SDS. To make 1× running buffer, dilute 100 ml of 10× stock with 900 ml of water. Both can be stored at room temperature.
5. Benchmark™ prestained protein ladder (Invitrogen). Store at −20°C.

2.5. *Western Blotting*

1. 10× transfer buffer (no methanol): 250 mM Tris, 2 M glycine. Store 10× stock at room temperature. To make up 1× transfer buffer, dilute 100 ml of 10× stock with 700 ml of water and add 200 ml (20%, v/v) methanol. Mix well and store at 4°C (*see* **Note 2**).
2. Immobilon-P polyvinylidene fluoride (PVDF 0.45-μm pore) membrane (Millipore Corporation, Billerica, MA).
3. 10× PBST: 1.4 M NaCl, 27 mM KCl, 18 mM KH$_2$PO$_4$, 100 mM Na$_2$HPO$_4$, 0.5% (v/v) Tween-20. To make 1× PBST, dilute 100 ml of 10× stock with 900 ml of water. Both can be stored at room temperature.
4. Block buffer: 5% (w/v) skim-milk powder in 1× PBST, prepared immediately before use (*see* **Note 3**).
5. Antibodies (**Fig. 1B**).

   a. ICSM18 (monoclonal antibody, 1.4 mg/ml, residues 142–155, generous gift from Professor John Collinge, MRC Prion Unit, Institute of Neurology, London, UK).
   b. SAF32 (monoclonal antibody, 200 μg/ml, residues 79–92, Cayman Chemical, Ann Arbor, MI).
   c. 8B4 (monoclonal, residues 37–44, Alicon, Schlieren, Switzerland).
   d. A.R30 (polyclonal antibody, residues 89–103, generous gift from Dr. Victoria Lawson, Department of Pathology and Mental Health Research Institute of Victoria, University of Melbourne, Melbourne, Australia).
   e. Sheep anti-mouse-horseradish peroxidase (HRP) (GE Healthcare, Little Chalfont, Buckinghamshire, UK).
   f. Sheep anti-rabbit-HRP (Dako Denmark A/S, Glostrup, Denmark).

6. ECL Plus Western Blotting Detection System (GE Healthcare). Store at 4°C.

3. Methods

To simplify visualization of C1 and C2, which like full-length PrP^C are glycosylated and GPI-anchored, the PrP proteins can be enzymatically stripped of their sugars. The C1 and C2 fragments can then be differentiated using epitope mapping by Western blotting with antibodies to specific regions of PrP (**Fig. 1B**). Depending on the immunoreactivity profile (**Fig. 1C**), assumptions can be made about the PrP^C region where cleavage has occurred.

3.1. *Cell Lysis and Protein Determination*

1. Cells are cultured in T75 flasks until confluent, at which point they are rinsed twice with approx 5 ml of ice-cold 1× PBS, and then scraped in approximately a further

10 ml of PBS and transferred into a 15-ml Falcon tube. Tubes are centrifuged at 720 rcf, for 3 min in a benchtop centrifuge, and the supernatant is removed. The cell pellet is then resuspended in 500 μl of PBS and transferred to an appropriately labeled 1.5-ml microfuge tube. Tubes are then centrifuged at 600 rcf for 3 min in a microcentrifuge. Again, the supernatant is removed (*see* **Note 4**).

2. Make up fresh cell lysis buffer by adding 900 μl of lysis buffer stock to 100 μl of 10× Complete inhibitor cocktail (final concentration of 1× protease inhibitors).

3. To lyse cells, add 100 μl of cell lysis buffer to cell pellet and triturate through end of pipette tip to break up pellet. Incubate crude lysate on ice, at 4°C for 30 min, triturate again, and then spin at 100 rcf, for 3 min at 4°C to pellet cellular debris. Remove the supernatant (crude lysate) to a new labeled microfuge tube and discard pellet.

4. Perform a BCA assay on the crude lysate to determine protein concentration. Briefly, pipette in triplicate, 10 μl each bovine serum albumin (BSA) standard and 10 μl of a 1:4 dilution of lysate into wells in 96-well plate. Then, mix BCA reagents A:B at a ratio of 50:1 and add 200 μl/well. Cover plate and incubate at 37°C for 20 min, and then read absorbance at 560 nm. Draw a standard curve based on the absorbance of the BSA standards, and then use the curve to calculate the concentration of the lysate in the well. Multiply by the dilution factor (i.e., x4) to get the concentration of total protein in the crude lysate.

3.2. Deglycosylation of PrP^C

1. Turn on heat blocks: set at 100°C and 37°C.
2. Pipette 10 μl of lysate (dilute crude lysate with appropriate volume of fresh lysis buffer to final total protein content of 20–80 μg, depending on expression levels of PrP^C) into microfuge tube.
3. Add 1 μl of 10× denaturation buffer and boil sample for 10 min.
4. Spin sample (max speed on microcentrifuge) briefly to bring down any condensation, and cool lysate to room temperature approximately 10 min.
5. To denatured sample, add 2.5 μl of 10% NP-40, 2 μl (2 units) PNGase of F, and 4.5 μl of 1× incubation buffer (*see* **Note 5**).
6. Incubate sample overnight (approx 14–16 h) at 37°C.
7. Spin sample (max speed in microcentrifuge) briefly to bring down any condensation.
8. Add 7 μl of 4× sample buffer to lysate (*see* **Note 6**).

3.3. SDS-PAGE

1. The following instructions assume the use of the Bio-Rad Mini-Protean® II Electrophoresis Cell, although adaptation to other mini-gel systems is possible.

2. Set up a gel case by placing together two clean glass plates (one large, one small, wiped with 70% ethanol and a Kimwipe, and allowed to air dry), held apart with 1.5-mm-thick spacers, onto the clamp assembly (larger plate at back). Tighten the clamp assembly screws to hold the glass plate together securely, and then attach the clamp assembly to the casting stand (*see* **Note 7**).

3. Prepare 15% resolving gel/s by mixing (per mini-gel) in an appropriate sized tube: 3.75 ml of acrylamide/bis, 2.25 ml of water, 1.41 ml of resolving gel buffer, and 75 µl of 10% SDS. Then, add 25 µl of 10% APS and 3.75 µl of TEMED immediately before pouring gel. Mix by inverting the tube approx three times (*see* **Note 8**). Pour gel solution between glass plates, allowing a gap of approx 2 cm from the top of the smaller glass plate for the stacking gel. Using a transfer pipette, overlay gel with water-saturated butan-1-ol and ensure gel setup is placed on a level surface by placing a "bulls-eye" spirit level on top of the casting stand. Allow resolving gel to set for 1 h at room temperature.

4. Pour off the butan-1-ol, and rinse the top of the gel thoroughly (at least five times) with water. Dry the inside of the glass plates by absorbing any water droplets onto a small piece of blotting paper placed between the plates (*see* **Note 9**).

5. Prepare stacking gel(s) by mixing (per mini-gel) in a 15-ml tube: 562.5 µl of acrylamide/bis, 1.95 ml of water, 705 µl of stacking gel buffer, and 56.25 µl of 10% SDS. Add 28.12 µl of 10% APS and 2.812 µl of TEMED last. Mix gently by inverting the tube, and then using a transfer pipette, pour the stacking gel on top of the resolving gel. Insert the comb into the liquid stacking gel and allow at least 5 mm between the bottom of the comb teeth and the top of the set resolving gel. Allow the stacking gel to set for 30 min at room temperature (*see* **Note 10**).

6. After stacking gel has set, remove clamp assembly/gel from casting stand, and then carefully remove the comb and rinse out wells twice with water, then twice with 1× running buffer.

7. Set up gel electrophoresis system, using two prepared gels, or one gel and a buffer dam, by pushing the clamp assembly/gel/dam into the locator slots of the electrode/core and place this into the buffer tank, such that two chambers are formed (inner, anode; outer, cathode), and fill both chambers with 1× running buffer.

8. Tubes containing samples for electrophoresis are placed into a heat block set at 100°C, and boiled for 10 min to denature the proteins. Pulse spin the samples in a microfuge to bring down any condensation.

9. Load 27 µl (i.e., all that is in the tube) of prepared samples carefully into wells, avoiding overflow from well. Allow one well for loading 5 µl of Benchmark (or other suitable) molecular weight standard, usually at one end of the gel to allow for orientation purposes (*see* **Note 11**).

10. Attach the lid of the electrophoresis tank and connect to power supply. Run gel at 40 mA/gel for approx 1 h, or until the dye front has reached the bottom of the gel.

3.4. Western Blotting

1. The following instructions assume the use of the Bio-Rad Mini Trans-Blot Electrophoretic Transfer Cell, but can be adapted to other Western blotting systems.
2. Just before the end of the SDS-PAGE, set up for Western blotting by presoaking a piece of PVDF membrane, just larger than the size of one gel, in 100% methanol. Cut four pieces of blotting/filter paper, just larger than the size of the PVDF membrane. In a tray, large enough for one open transfer cassette add enough 1× transfer buffer to presoak two sponges.
3. Disconnect the gel electrophoresis unit and disassemble the core/clamp assembly/ gel case carefully, to avoid tearing the gel. Cut off the stacking gel and discard.
4. Set up transfer sandwich in the tray, on top of the black/gray side of the open transfer cassette as follows, taking care not to allow any bubbles to form between each consecutive layer: sponge, two pieces of filter paper, gel, PVDF membrane, two pieces of filter paper, sponge. Close transfer cassette, and place cassette into electrode module such that the black side of the cassette is alongside the black wall of the electrode module. This ensures that the gel and membrane are correctly positioned so that the cathode (negative, black) to anode (positive, red) direction of electricity flow transfers proteins resolved in the gel electrophoretically onto the membrane.
5. Add the electrode module, magnetic stirrer bar and ice pack to the buffer tank, and then fill tank (approx 800 ml) with cold 1× transfer buffer. Place tank on magnetic stirrer plate, and turn on to activate the bar, so as to ensure continued circulation of the cold transfer buffer around the tank. Place the lid on the buffer tank, connect to power supply and transfer for 45 min at 380 mA.
6. Immediately before the end of the transfer, prepare 20 ml of blocking buffer, and pour into blotting tub/container.
7. After transfer is finished, disassemble transfer sandwich, and immediately place membrane into tub (containing blocking solution) such that the proteins will be on top of the membrane, not touching the bottom of the container (*see* **Note 12**). The Benchmark molecular weight marker should be clearly visible on the membrane. Gel and filter paper can be discarded.
8. Block PVDF membrane for 1 h at room temperature.
9. Immediately before the end of the block, prepare the appropriate dilution of primary antibody in freshly made blocking solution (**Table 1**).

Table 1 Optimized Antibody Dilutions

Primary antibody	Secondary antibody
ICSM18 1:35,000	Anti-mouse-HRP 1:10,000
A.R30 1:5,000	Anti-rabbit-HRP 1:10,000
SAF32 1:3,000	Anti-mouse-HRP 1:10,000
8B4 1:10,000	Anti-mouse-HRP 1:10,000

10. After 1 h has passed for blocking, remove block buffer and discard, and then immediately add the primary antibody solution, place lid on container, and ensure airtight seal (so as to avoid evaporation). Place the container on rocking platform at 4°C to incubate overnight (approx 16 h).

11. Wash off primary antibody with 1× PBST. Specifically, discard used primary antibody, and then rinse/wash as follows: three rinses, 10-min wash, 2 rinses, 3-min wash, 1 rinse, and two 3-min washes. A "rinse" is defined as adding approx 15 ml of 1× PBST to container, agitating briefly before discarding PBST, whereas a "wash" is adding at least 30 ml of 1× PBST and placing container on a rocking platform at room temperature for a specified period.

12. During the last 3-min wash, prepare the appropriate dilution of secondary antibody in freshly made block buffer (**Table 1**).

13. Incubate the PVDF membrane in secondary antibody for 1 h at room temperate on rocking platform.

14. Wash off secondary antibody with 1× PBST, following the method for washing off the primary antibody (*see* **step 11**).

15. During the last wash, prepare ECL-Plus chemiluminescence reagents (as per the manufacturer's instructions; approx 1.5 ml/mini-gel is required). In dark room (red safelight turned on) remove membrane from container and place on clean, dry surface (e.g., overhead transparency). Immediately add ECL-Plus reagents, and leave for 5 min.

16. Remove blot from ECL-Plus reagents and place between two pieces (cut to size of X-ray film cassette) of overhead transparency, and place this into film cassette. Expose membrane to film (try a 1-min exposure first, and based on the intensity of signal, expose again as appropriate) before developing and fixing (*see* **Note 13**).

Acknowledgements   We thank Professor John Collinge for the ICSM18, Professor Man Sun-Sy for the 8B4 monoclonal antibodies, and Dr. Victoria Lawson for the A.R30 polyclonal antibody. This work was supported by Australian NHMRC Program grants 208978 and 400202.

4. Notes

1. Complete Mini tablets are stable when stored at 4°C.
2. 1× transfer buffer can be reused approximately five times.
3. For one mini-gel, 20 ml (i.e., 1 g of skim milk powder in 20 ml of PBST) is sufficient.
4. Cell pellets can be used immediately for experimentation, or they can be stored at −80°C for subsequent use.
5. If needed, one can prepare an "untreated" control by omitting the PNGase F and adding an extra 2 μl of 1× incubation buffer. Untreated samples are otherwise treated identically to those containing PNGase F.

6. **Subheading 3.2., steps 7** and **8** are best performed to coincide with the setting of the stacking gel (*see* **Subheading 3.3., steps 5–8**), but if needed, samples can be boiled for 10 min, spun, and stored at −20°C until required for Western blotting. In such cases, samples should be thawed and reboiled before loading into gel.
7. To avoid later difficulties, ensure the glass gel case does not leak by adding a few milliliters of water and watch for leaks. Pour water out of the top of the glass gel case, and dry the inside of the glass plate with filter paper.
8. It is important at this stage to avoid the formation of bubbles in the tube, so do not shake vigorously.
9. Do not let this blotting paper touch the surface of the already set resolving gel.
10. When positioning the comb, if any bubbles are present at the bottom of the comb teeth, remove the comb, rinse, and dry. Then, replace in the liquid stacking gel. This is so as to avoid bubbles being at the bottom of the well in which the samples will be loaded.
11. If there are any empty lanes, then prepare some 1× sample buffer (mix 25 μl of 4× sample buffer and 75 μl of water) and add the same volume (27 μl) of this to the well.
12. It is important to not let the PVDF membrane dry at this stage.
13. Instead of, or in addition to, using film and developer, one can expose membrane in digital imaging machine/system to record and retain an image.

References

1. Bendheim PE, Brown HR, Rudelli RD, et al. (1992) Nearly ubiquitous tissue distribution of the scrapie agent precursor protein. Neurology;42(1):149–156.
2. Horiuchi M, Yamazaki N, Ikeda T, Ishiguro N, Shinagawa M. (1995) A cellular form of prion protein (PrPC) exists in many non-neuronal tissues of sheep. J Gen Virol;76(10):2583–2587.
3. Ford MJ, Burton LJ, Morris RJ, Hall SM. (2002) Selective expression of prion protein in peripheral tissues of the adult mouse. Neuroscience;113(1):177–192.
4. Sakaguchi S, Katamine S, Shigematsu K, et al. (1995) Accumulation of proteinase K-resistant prion protein (PrP) is restricted by the expression level of normal PrP in mice inoculated with a mouse-adapted strain of the Creutzfeldt-Jakob disease agent. J Virol;69(12):7586–7592.
5. Bueler H, Aguzzi A, Sailer A, et al. (1993) Mice devoid of PrP are resistant to scrapie. Cell;73(7):1339–1347.
6. Collins SJ, Lawson VA, Masters CL. (2004) Transmissible spongiform encephalopathies. Lancet;363(9402):51–61.
7. Lawson VA, Collins SJ, Masters CL, Hill AF. (2005) Prion protein glycosylation. J Neurochem;93(4):793–801.
8. Ehlers MR, Riordan JF. (1991) Membrane proteins with soluble counterparts: role of proteolysis in the release of transmembrane proteins. Biochemistry;30(42):10065–10074.
9. Chen SG, Teplow DB, Parchi P, Teller JK, Gambetti P, Autilio-Gambetti L. (1995) Truncated forms of the human prion protein in normal brain and in prion diseases. J Biol Chem;270(32):19173–19180.
10. Jimenez-Huete A, Lievens PM, Vidal R, et al. (1998) Endogenous proteolytic cleavage of normal and disease-associated isoforms of the human prion protein in neural and non-neural tissues. Am J Pathol;153(5):1561–1572.

11. Yadavalli R, Guttmann RP, Seward T, Centers AP, Williamson RA, Telling GC. (2004) Calpain-dependent endoproteolytic cleavage of PrP^Sc modulates scrapie prion propagation. J Biol Chem;279(21):21948–21956.

12. Vincent B, Paitel E, Saftig P, et al. (2001) The disintegrins ADAM10 and TACE contribute to the constitutive and phorbol ester-regulated normal cleavage of the cellular prion protein. J Biol Chem;276(41):37743–37746.

13. McMahon HE, Mange A, Nishida N, Creminon C, Casanova D, Lehmann S. (2001) Cleavage of the amino terminus of the prion protein by reactive oxygen species. J Biol Chem;276(3):2286–2291.

14. Watt NT, Taylor DR, Gillott A, Thomas DA, Perera WS, Hooper NM. (2005) Reactive oxygen species-mediated beta-cleavage of the prion protein in the cellular response to oxidative stress. J Biol Chem;280(43):35914–35921.

15. Wilquet V, De Strooper B. (2004) Amyloid-beta precursor protein processing in neurodegeneration. Curr Opin Neurobiol;14(5):582–588.

16. Buxbaum JD, Liu KN, Luo Y, et al. (1998) Evidence that tumor necrosis factor alpha converting enzyme is involved in regulated alpha-secretase cleavage of the Alzheimer amyloid protein precursor. J Biol Chem;273(43):27765–27767.

17. Forloni G, Angeretti N, Chiesa R, et al. (1993) Neurotoxicity of a prion protein fragment. Nature;362(6420):543–546.

18. Checler F, Vincent B. (2002) Alzheimer's and prion diseases: distinct pathologies, common proteolytic denominators. Trends Neurosci;25(12):616–620.

19. Vincent B. (2004) ADAM proteases: protective role in Alzheimer's and prion diseases? Curr Alzheimer Res;1(3):165–174.

Chapter 16
Molecular Typing of PrP^{res} in Human Sporadic CJD Brain Tissue

Victoria Lewis, Genevieve M. Klug, Andrew F. Hill, and Steven J. Collins

Summary Within the spectrum of sporadic human transmissible spongiform encephalopathies (TSEs), there is considerable diversity of disease phenotypes. At least part of this variation is thought to be on the basis of different "strains" of prions (the infectious agent). Tissue deposition of PrP^{res} (the abnormal disease-associated conformation of the prion protein) is considered a hallmark of TSE pathology, and it can be visualized by Western blotting typically as three bands depicting the diglycosylated, monoglycosylated, and unglycosylated species. It is the mobility of the unglycosylated PrP^{res}, and the relative abundance of the two glycosylated bands, along with the prion protein gene (*PRNP*) codon 129 genotype, that seem to correlate with distinct clinico-pathological profiles of sporadic Creutzfeldt–Jakob disease.

Keywords Glycotyping; prion protein; transmissible spongiform encephalopathy (TSE); Western blotting.

1. Introduction

The etiology of human transmissible spongiform encephalopathies (TSEs) includes those with a genetic basis (genetic Creutzfeldt–Jakob disease [CJD], fatal familial insomnia, and Gerstmann–Sträussler–Scheinker syndrome), those acquired through accidental horizontal transmission (iatrogenic CJD, Kuru and variant CJD), and for the majority of human CJD cases (85–90%), those with no known attributable cause (sporadic CJD; sCJD) *(1)*. Within the heterogeneous sCJD group are a number of phenotype subtypes or "strains." These correlate with the patient's molecular PrP^{res} type, in combination with the PRNP genotype at codon 129.

PrP^{res} is defined as the protease-*res*istant, disease-associated conformer of the normal prion protein PrP^C. The altered structure and higher β-sheet content of PrP^{res} confer poorer solubility and a relative protease resistance of the C-terminal residues of the protein, which allows for recognition in biochemical assays. The distinct patterns of PrP^{res}, as visualized on Western blots, defines the PrP^{res} molecular types. Specifically, the relative electrophoretic mobility of the unglycosylated PrP^{res} band

From: *Prion Protein Protocols.*
Methods in Molecular Biology, Vol. 459.
Edited by: A. F. Hill © Humana Press, Totowa, NJ

after proteinase K (PK) digestion can be used to differentiate between different PrP^{res} types with relative abundance of the di- and monoglycosylate isoforms, adding further complexity to clarification schemes. Different prion strains are in part envisaged due to PrP^{res} molecules folding into slightly different conformations, allowing predominant digestion of the N-terminal amino acids up to alternative residues, resulting in a protein with a distinct size *(2)*.

There is good evidence that the different PrP^{res} types visualized by Western blot represent different prion strains *(2–4)*, and there has been congruence with strain reporting in various geographical regions, suggesting not environmental but endogenous determinants of TSE strain type *(3)*.

2. Materials

2.1. Preparation of Brain Homogenate and PK Digestion

1. Sterile scalpels and screw-capped tubes.
2. Phosphate buffered saline (PBS): Prepare 10× stock from 100 mM Na$_2$HPO$_4$, 31 mM KH$_2$PO$_4$, 1.23 M NaCl.

   a. Adjust to pH 7.4 with HCl if necessary.
   b. Autoclave before storage at room temperature.
   c. Prepare working stock by dilution in sterile, distilled water (1 part 10× PBS:9 parts distilled water).

3. Hybaid Ribolyser™ (Bio-Rad, Hercules, CA) or dounce tissue grinder, size 21 (Kimble/Kontes Glass, Vineland, NJ).
4. Zirconium beads (1.0 mm; BioSpec Products, Bartlesville, OK).
5. PK (Invitrogen, Carlsbad, CA): 10 mg/ml in distilled water.

   a. Store at −20°C.
   b. Prepare working stocks of 1 mg/mL in distilled water.

6. Pefabloc (Roche Molecular Biochemicals, Mannheim, Germany): 100 mM working stock in distilled water.

   a. Store at −20°C.

7. 2× sample buffer: 125 mM Tris-HCl, pH 6.8, 4% (w/v) sodium dodecyl sulphate (SDS), 6% (v/v) 2-mercaptoethanol, 20% (v/v) glycerol, 0.005% (w/v) bromophenol blue.

2.2. SDS-Polyacrylamide Gel Electrophoresis (PAGE)

1. Novex 16% Pre-cast Tris-glycine gels, 1.0 mm, 10-well (Invitrogen).
2. Pre-cast gel/Sure-Lock system (Invitrogen).

3. 10× running buffer: 250 mM Tris, 1.92 M glycine, 1% (w/v) SDS.

   a. Store at room temperature.
   b. Prepare 1× working stock by dilution in distilled water (1 part 10× running buffer:9 parts distilled water).

4. Benchmark prestained protein ladder (Invitrogen).

2.3. Western Blotting

1. 10× transfer buffer: 250 mM Tris, 1.92 M glycine.

   a. Store at room temperature.
   b. Prepare working stock by dilution of 100 ml of 10× transfer buffer, 20% (v/v) methanol in distilled water.
   c. Store at 4°C until ready for use.

2. Immobilon-P transfer membrane (polyvinylidene diflouride; PVDF) (Millipore Corporation, Billerica, MA).
3. Phosphate-buffered saline with Tween-20 (PBS-T): Prepare working stock by diluting 100 ml of 10× PBS with 900 ml of distilled water, and addition of 0.05% (v/v) Tween-20.

   a. Store at room temperature.

4. Block buffer: 5% (w/v) nonfat dry milk powder in PBS-T.
5. Primary antibody: 3F4 monoclonal mouse antibody (kind gift from Dr. V. Lawson, Department of Pathology, The University of Melbourne, Melbourne, Australia).
6. Secondary antibody: Sheep anti-mouse horseradish peroxidase-conjugated whole antibody (GE Healthcare, Little Chalfont, Buckinghamshire, UK).
7. Enhanced chemiluminescence (ECL) reagents (GE Healthcare).
8. Biomax light film (Sigma-Aldrich, St. Louis, MO).

3. Methods

All steps up to and including loading of samples onto the SDS-PAGE gel are carried out in a biohazard safety cabinet class II for operator protection and minimization of sample cross-contamination.

3.1. Brain Sampling and Homogenization

1. Frozen (stored at −80°C) brain tissue is partially thawed overnight at −20°C. Using a long-handled, sterile scalpel remove a small sample of cerebral cortex and place in an appropriately labeled screw-cap tube, and weigh.

2. Add 500 mg of zirconium beads to a second sterile screw-capped tube. Add pre-weighed tissue to prepared tube and the appropriate volume of sterile PBS to give a 10% (w/v) brain homogenate preparation.
3. Sealed samples are then transferred to a Hybaid Ribolyser and homogenized with a 45-s pulse at maximum speed (6.5). A repeat homogenization may be required to achieve a uniform homogenate. If this is the case, a 10-min incubation at 4°C is necessary between homogenization steps to avoid the sample overheating. Store homogenate at −80°C until ready for analysis.

3.2. PK Digestion

1. Add 1 µl of 1 mg/ml PK to sterile screw-capped tube and then add 9 µl of 10% (w/v) brain homogenate (test samples and known type 1–type 4 controls), giving a final concentration of 100 µg/ml.
2. Incubate samples for 1 h at 37°C with no agitation.
3. Add 100 mM Pefabloc stock to a final concentration of 10 mM to inhibit enzyme activity.
4. Immediately, add the appropriate volume of 2× sample buffer, and boil the sample for 10 min. Spin briefly to condense sample 100 rcf, 10 s in Eppendorf centrifuge (Eppendorf, Hamburg, Germany).

3.3. SDS-PAGE

The following instructions are to be used in conjunction with the Sure-Lock system (Invitrogen).

1. Open 16% Tris-glycine gel packet, decant buffer solution, remove comb and adhesive strip, and rinse lanes gently with tap water. Assemble Safe-Lock gel tank, and precast gel, by using buffer dam as appropriate.
2. Prepare a 1× running buffer solution by mixing 100 ml of 10× running buffer with 900 ml of distilled water. Fill both the inner and outer chambers of the Sure-Lock tank with 1× running buffer, up to the level of approx 5–10 mm above the wells.
3. Load appropriate volume of each test sample and controls to wells (*see* **Notes 1** and **2**). This will usually vary between 5 and 20 µl of PK-digested brain tissue, depending on the relative amounts of PrP^{res} in the samples. The aim is to visually balance the amounts of PrP^{res} for optimal resolution and comparison of the unglycosylated bands (*see* **Note 3**).
4. Connect the gel tank to a power supply and run the gel at 200 V for approx 90 min to allow for adequate resolution (*see* **Note 4**).

3.4. Western Blotting

1. Immediately before the end of the SDS-PAGE, cut a gel-sized piece of PVDF transfer membrane filter. Pre-wet the membrane in 100% methanol for 1–2 s, rinse in distilled water, and then equilibrate in precooled (4°C) 1× transfer buffer 10–15 min.
2. Upon completion of PAGE, prepare a transfer "sandwich" in a tray containing precooled (4°C) 1× transfer buffer (*see* **Note 5**). Add transfer cassette to tray, and immerse cassette sponges in transfer buffer. Add one sponge and two pieces precut filter paper to the cassette.
3. Using the spatula tool provided with the Sure-Lock system, open the gel cassette, and carefully excise the stacking gel/wells (approximately top 1.5 cm of gel), and the lower 0.5 mm of the gel.
4. Carefully load the gel slab to the sandwich and overlay the prewet PVDF. Use the roller tool provided with the Sure-Lock system, to gently remove air bubbles between the gel slab and membrane. Add two pieces of precut filter paper and the second pre-wet sponge. Close cassette and add to transfer tank.
5. Add ice-block and magnetic stirrer to ensure buffer circulation. Fill tank with precooled (4°C) 1× transfer buffer.
6. Connect to power supply and transfer proteins from gel to membrane at 0.38 A for 60 min.
7. After transfer, turn off power supply and disassemble the transfer cassette. Carefully remove the membrane from the cassette and any excess membrane can be trimmed and discarded.

3.5. Immunoprobing

1. Add PVDF membrane to an incubation tray with 20 ml of block buffer and incubate at room temperature for 1 h, with agitation on rocking or orbital shaking platform.
2. After incubation, the block buffer is discarded. Add the primary antibody (3F4) in block buffer at a 1:5,000 dilution and incubate overnight at 4°C with agitation.
3. Discard primary antibody and wash membrane with PBS-T; three brief rinses, then 10-min wash, two brief rinses, 3-min wash, one brief rinse, 3-min wash, 3-min wash). Typically, the volume per wash is 15–30 ml.
4. During the final wash, dilute the secondary antibody in block buffer to a 1:15,000 dilution. After discarding the last wash, the secondary antibody preparation is added to the membrane and incubated at room temperature for 1 h with agitation.
5. After the secondary antibody solution is discarded, the membrane is washed with PBS-T as in **step 3**.

6. During the wash step, the ECL reagents are equilibrated to room temperature. During the final wash, ECL reagents A and B are combined according to manufacturer's guidelines.

7. Using tweezers, the membrane is removed, blotted onto Kim-Wipes to remove excess wash buffer, and then applied to a clean, dry acetate sheet. To avoid drying of the membrane, the ECL reagents are added directly to the membrane and agitated as necessary to ensure even coverage of the entire membrane. Incubate 5 min at room temperature.

8. The membrane is transferred to a second appropriately sized acetate sheet and then carefully overlaid with a sheet, avoiding air bubbles forming between the membrane and sheeting. Excess ECL fluid can be blotted with Kim-Wipes.

9. The acetate membrane-PVDF sandwich is transferred to an X-ray film cassette and placed in a dark room under photographic safe lighting.

10. Biomax light film is then trimmed to an appropriate size and then exposed to the acetate membrane-PVDF sandwich for a sufficient period of time to enable a clear comparison of the glycotypes. Routine exposure times range from 30 s to 10 min. Often a shorter exposure period provides the best clarity of images for distinguishing and classifying the four glycotypes (*see* **Note 6**).

11. Develop and fix film as per manufacturer's instructions.

Acknowledgements The ANCJDR is funded by the Commonwealth Department of Health and Ageing. We thank the families and clinicians for ongoing support in the surveillance of CJD in Australia. Original T1–T4 standards were supplied to the ANCJDR by the National Institute for Biological Standards and Control (NIBSC) as part of the World Health Organization Collaborative Nomenclature Study (2001–2002). Since then, Australian cases of T1–3 glycotypes have been determined, and they are used as control samples for future glycotyping gels. We thank Dr. Victoria Lawson for the kind gift of the monoclonal mouse antibody 3F4.

4. Notes

1. To ensure optimal transfer and resolution of glycotypes, avoid using the first and last lanes of the 16% polyacrylamide gels. It is preferable to use these lanes for molecular weight markers or 1× sample buffer.

2. Often samples will need to be retested to ensure accurate typing of PrP^{res}. When this is necessary, it is useful to adjust the SDS-PAGE loading pattern so that the sample is positioned between the most relevant control glycotypes.

3. The concentrations of PrP^{res} vary between brain samples. It is preferable to have approximately equivalent protein loadings for the controls and test samples. To achieve this, PK-digested samples can be diluted using 1× sample buffer and then reanalyzed by loading equivalent sample volumes for PAGE.

4. To improve the resolution of samples that prove difficult to glycotype, the running time for SDS-PAGE can be extended to 100 min.

5. 1× Transfer buffer can be reused four to five times without a loss of transfer efficiency, and it should be stored at 4°C.

6. It is useful to prepare at least one short- and one long-exposure blot. Short exposures provide visualization of greater detail of the glycotype patterns, which is especially important where a brain sample may comprise of more than one PrP^res glycotype. Longer exposures may be necessary for the visualization of unexpectedly lower concentrations of PrP^res.

References

1. Collins S, Boyd A, Lee JS, et al. Creutzfeldt-Jakob disease in Australia 1970–1999. Neurology 2002;59(9):1365–71.
2. Hill AF, Joiner S, Wadsworth JD, et al. Molecular classification of sporadic Creutzfeldt-Jakob disease. Brain 2003;126(6):1333–46.
3. Lewis V, Hill AF, Klug GM, Boyd A, Masters CL, Collins SJ. Australian sporadic CJD analysis supports endogenous determinants of molecular-clinical profiles. Neurology 2005;65(1):113–8.
4. Parchi P, Giese A, Capellari S, et al. Classification of sporadic Creutzfeldt-Jakob disease based on molecular and phenotypic analysis of 300 subjects. Ann Neurol 1999;46:224–33.

Chapter 17
Transgenic Mouse Models of Prion Diseases

Glenn C. Telling

Summary Prions represent a new biological paradigm of protein-mediated information transfer. In mammals, prions are the cause of fatal, transmissible neurodegenerative diseases, often referred to as transmissible spongiform encephalopathies. Many unresolved issues remain, including the exact molecular nature of the prion, the detailed mechanism of prion propagation, and the mechanism by which prion diseases can be both genetic and infectious. In addition, we know little about the mechanism by which neurons degenerate during prion diseases. Tied to this, the physiological function of the normal form of the prion protein remains unclear, and it is uncertain whether loss of this function contributes to prion pathogenesis. The factors governing the transmission of prions between species remain unclear, in particular the means by which prion strains and PrP primary structure interact to affect interspecies prion transmission. Despite all these unknowns, dramatic advances in our understanding of prions have occurred because of their transmissibility to experimental animals and the development of transgenic mouse models has done much to further our understanding about various aspects of prion biology. In this chapter, I review recent advances in our understanding of prion biology that derive from this powerful and informative approach.

Keywords Bovine spongiform encephalopathy (BSE); chronic wasting disease (CWD); Creutzfeldt–Jakob disease (CJD); prions; scrapie; transgenic mice; transmissible spongiform encephalopathies (TSEs).

1. Introduction

The prion disorders of animals and humans include scrapie in sheep, bovine spongiform encephalopathy (BSE), chronic wasting disease (CWD) of deer and elk, and human Creutzfeldt–Jakob disease (CJD). They share a number of common features, the most consistent being the neuropathologic changes that accompany disease in the central nervous system (CNS). These features typically consist of neuronal vacuolation and degeneration, a reactive proliferation of astroglia, and, in some

From: *Prion Protein Protocols.*
Methods in Molecular Biology, Vol. 459.
Edited by: A. F. Hill © Humana Press, Totowa, NJ

conditions such as kuru, Gerstmann–Straussler–Scheinker (GSS) syndrome, and variant CJD (vCJD), the deposition of amyloid plaques.

Central to all the prion diseases is the abnormal metabolism of the prion protein (PrP). The normal form of PrP, referred to as PrP^C, is a sialoglycoprotein of molecular weight 33–35 kDa that is attached to the surface of neurons and other cell types by means of a glycophoshphatidylinositol (GPI) anchor. PrP^C is sensitive to protease treatment and soluble in detergents. The disease-associated isoform, referred to as PrP^{Sc}, is found only in infected brains, and it is partially resistant to protease treatment, insoluble in detergents, and tends to aggregate. In most examples of prion disease, protease cleavage of the N-terminal 66, or so, amino acids of PrP^{Sc} gives rise to a protease-resistant core, originally referred to as PrP$^{27-30}$. Considerable evidence now supports the once unorthodox hypothesis that prions lack nucleic acid and are composed largely, if not entirely, of PrP^{Sc} and that during disease PrP^{Sc} imposes its conformation on PrP^C, resulting in the exponential accumulation of PrP^{Sc}.

2. Prion Transmission Barriers

An important aspect of prion diseases is their transmissibility. Inoculation of diseased brain material into individuals of the same species will typically reproduce the disease. In contrast, the passage of prions from one species to another is generally inefficient, and it is referred to as the species barrier *(1)*. The barrier may be absolute in which case no transmission is recorded, or partial, in which case primary transmission is either characterized by long incubation times and low attack rates followed by greatly reduced incubation times and high attack rates on secondary passage, or by high attack rates on both primary and secondary passage with reduced incubation times on secondary passage.

3. PrP Primary Structure, Prion Strains, and the Species Barrier

Although the elements controlling interspecies prion transmission are not completely understood, seminal studies in transgenic Tg mice *(2, 3)* and cell-free systems *(4)* suggested that the sequences of PrP^{Sc} in the inoculum and PrP^C in the host should be isologous for optimal progression of disease. The influence of intraspecies PrP polymorphisms on prion disease susceptibility in mice *(5)*, sheep *(6)*, and humans *(7–9)* supported the concept that the primary structure of PrP was the foremost molecular determinant of prion species barriers. Curiously, the rules governing PrP primary structure control over prion transmission seem not to apply in the case of the bank voles, which are susceptible to prions from a number of mammals *(10)*.

An equally important component affecting prion transmission is the strain of prion. Mammalian prions strains are classically defined in terms of their incubation times in susceptible animals and by the profile of lesions they produce in the CNS. Because numerous studies indicate that strain diversity is encoded in the tertiary structure of PrP^{Sc} *(11–15)*, assessment of PrP^{Sc} conformation, the neuroanatomic distribution of PrP^{Sc}, and the extent of PrP^{Sc} glycosylation are parameters that also have been used to characterize prion strains *(14, 16–20)*.

The unexpectedly wide host range of BSE prions, exemplified by transmission to humans as variant CJD (vCJD), and the restricted transmission of vCJD prions in human PrP Tg mice *(13, 21)* compared with bovine PrP Tg mice *(22)* exemplify the capacity of particular prion strains to overcome the influence of PrP primary structure. Although transmission of vCJD prions in Tg mice is relatively inefficient, generation of the vCJD phenotype requires expression of human PrP with methionine at residue 129 *(23)*. Interestingly, transmission studies of BSE in Tg mice consistently reveal the existence of more than one molecular type of PrP^{Sc} *(13, 21, 23)*. Thus, in addition to the usual type 4, or 19-kDa PrP^{Sc} form, a second type 2 or 21-kDa form of PrP^{Sc} is detected under these conditions, suggesting that more than one BSE-derived prion strain might infect humans.

4. Various Effects of Endogenous Mouse PrP Expression in Tg Mouse Models

The availability of *Prnp$^{0/0}$* knockout mice and the characterization of increasing numbers of transgenic mice expressing different PrP alleles have revealed interesting protective functions of wild-type PrP on various PrP-related pathologies. In some cases, disease can be partially or fully suppressed by coexpression of wild-type PrP. This effect was first observed during the characterization of Tg mice expressing human PrP, referred to as Tg(HuPrP) mice, which only became susceptible to CJD prions when endogenous wild-type mouse PrP was eliminated by crossing the transgene array to *Prnp$^{0/0}$* knockout mice *(24, 25)*. For this reason, subsequent transgenic mouse models expressing various foreign PrP coding sequences have generally been either produced in or ultimately crossed with *Prnp$^{0/0}$* knockout mice. Coincidentally, these observations also formed the partial basis for the model of prion replication involving an auxiliary factor, referred to as protein X, which was proposed to bind to PrP^{C} and facilitate conversion to PrP^{Sc} *(13, 25, 26)*.

Interactions between wild-type and mutant prion proteins also have been shown to modulate neurodegeneration in Tg mouse models expressing mutated PrP, the seminal observations being made in Tg mouse models of GSS *(27)*. In contrast, neurodegenertion induced by expression of a nine-octapeptide insertion associated with familial CJD, designated Tg(PG14), is unaffected by wild-type mouse PrP expression *(28)*. Subsequently, coexpression of wild-type mouse PrP was shown to rescue the neurodegenerative phenotype of mice expressing amino-terminal deletions *(29–31)*. Similar to the protein X model, these studies also are consistent with a model in which mutated and wild-type PrP compete for a hypothetical binding partner that, in this case, serves to transduce neurotoxic signals. Interestingly, although Tg mice expressing PrP tagged at its amino terminus with green fluorescent protein (GFP) supported compromised prion replication, prion propagation was facilitated by coexpression of wild-type PrP, suggesting that wild-type PrP rescued an altered amino-terminal function in the tagged PrP *(32)*. In contrast, the effect of tagging PrP at the C terminus with GFP was that Tg mice expressing this construct were incapable of sustaining prion infection and the PrP-GFP chimera acted as a dominant-negative inhibitor of wild-type PrP conversion to PrP^{Sc} *(33)*.

5. Transgenic Mouse Models of Mammalian Prion Diseases

The discovery that PrP primary structure was an important determinant of interspecies prion transmission *(2, 34)* paved the way for the development of a variety of facile Tg mouse models in which to study the biology of various mammalian prion diseases. In addition to the aforementioned Tg models of human prion disease, Tg mice expressing bovine PrP, referred to as Tg(BoPrP) mice, were developed that were susceptible to BSE prions *(35–37)*, whereas Tg mice expressing the ovine PrP coding sequence with alanine, arginine, and glutamine, or valine, arginine, and glutamine at codons 136, 154, and 171 were developed to study sheep scrapie prions *(38–40)*. Tg mouse models with susceptibility to transmissible mink encephalopathy prions also have been generated *(41)*.

Transgenic mice overexpressing bovine PrP (Tgbov XV) have been used to characterize an atypical neuropathological and molecular form of BSE, referred to as bovine "amyloidotic" spongiform encephalopathy (BASE). BASE differs from classical BSE with respect to PrP^{Sc} type and the presence of PrP amyloid plaques. Although BSE and BASE transmitted readily to Tgbov XV mice, they produced different clinical, neuropathologic, and molecular disease phenotypes *(42)*. Interestingly, the same study indicated that BASE prions were able to convert into BSE prions upon serial transmission in inbred mice. The relationship of this finding to the apparently protean nature of BSE prions in the aforementioned transmission studies *(13, 21, 23)* remains to be determined.

Ovine Tg mice have been shown to be a useful tool for discriminating scrapie strains *(43)* and in particular for differentiating BSE in sheep from natural scrapie isolates *(38, 44, 45)*. Tg mouse models are at the forefront of characterizing newly emerging "atypical" scrapie strains *(46)* related to so-called Nor98 cases first identified in Norwegian sheep.

In recent years, attention has focused on another emerging prion disease, namely, CWD of deer, elk, and moose. Of all the prion diseases, CWD is perhaps the most enigmatic, and it is unique in being the only known prion disease of wild animals. It is unclear whether CWD originated from the transmission of, for example, sheep scrapie to deer and elk, or whether it represents a naturally occurring TSE of cervids. Although the natural route of CWD transmission is unknown, lateral transmission *(47)* via ingestion of forage or water contaminated by secretions, excretions, or other sources, for example, CWD-infected carcasses *(48)*, is the most plausible natural route. Long thought to be limited in the wild to a relatively small endemic area in northeastern Colorado, southeastern Wyoming, and southwestern Nebraska, the disease has since been found in states east of the Mississippi in free-ranging deer in Wisconsin *(49)*, Illinois, New York, and West Virginia, and in the Canadian provinces of Saskatchewan and Alberta. In addition to its distribution in an increasingly wide geographic area of North America, outbreaks of CWD have occurred in South Korea as a result of importation of subclinically infected animals *(50, 51)*.

To address many of the unknown features of CWD pathogenesis, and to provide information about prion species barriers, strains, and the mechanisms of CWD prion propagation, Tg mice expressing cervid PrP, referred to as Tg(CerPrP), were developed. The prototype Tg(CerPrP) mice developed signs of neurologic dysfunction ~230 days after intracerebral inoculation with four CWD isolates *(52)*. The brains of sick Tg mice recapitulated the cardinal neuropathological features of CWD. As part of a larger study of CWD pathogenesis, Tg(CerPrP) mice were used as a sensitive means to show that skeletal muscles of CWD-infected deer harbor infectious prions, demonstrating that humans consuming or handling meat from CWD-infected animals are at risk to prion exposure *(53)*. Similar analyses of skeletal muscle BSE-affected cattle in a larger study of BSE pathogenesis by using Tgbov XV mice did not reveal high levels of BSE infectivity *(54)*. Since the seminal reports of accelerated CWD transmission from deer and elk to Tg(CerPrP) mice *(52)*, several other groups have reported similar results by using comparable Tg mouse models *(55–59)*. CWD also has been transmitted, albeit with less efficiency, to Tg mice expressing mouse *(60)* or Syrian hamster PrP *(61)*.

6. Assessing the Potential for Interspecies Prion Transmission by Using Transgenic Mouse Models

Investigating the susceptibility of Tg mice to prions from other species provides an approach for addressing the possible origins of prion diseases and a sensitive means of addressing the potential for interspecies prion transmission. The absence of a transmission barrier for Suffolk sheep scrapie prions in Tg mice expressing bovine PrP suggested that cattle may be highly susceptible to sheep scrapie *(15)*. The uncertain origin of CWD and its apparently contagious means of transmission, uncertain strain prevalence, and environmental persistence raise the possibility of uncontrolled prion dissemination and bring into question the risk to other species of developing a novel CWD-related disease, for example, via shared grazing of CWD-contaminated rangeland. These are issues that are currently being addressed using the aforementioned Tg mouse models. Moreover, contamination of the human food chain with BSE prions and the resulting deaths of young adults and teenagers have raised concern over the uncertain zoonotic potential of CWD.

To address the zoonotic potential of CWD prions, brain homogenates from diseased elk were inoculated into Tg mice expressing human PrP. In this limited study, CWD prions from elk did not induce disease *(58)*; however, it is worth noting that failure of BSE prion transmission to Tg(HuPrP) mice was cited as early evidence for a substantial transmission barrier for BSE prions in humans *(62)*. As mentioned, subsequent studies indicated that polymorphism at 129 had a significant effect on the transmissibility of BSE prions in Tg(HuPrP) mice *(21, 23)*. In the CWD transmissions to humanized mice, only mice expressing HuPrP with methionine at 129 were tested for their susceptibility to CWD elk prions. The effects of the valine 129 polymorphism and the susceptibility of Tg(HuPrP) mice to CWD prions from deer are, therefore, currently unknown. Along these lines, the coding sequence of the elk PrP gene *(PRNP)* is also polymorphic at codon 132 encoding either methionine or leucine *(63, 64)*. Because residue 132 in elk is equivalent to the human codon 129 polymorphism, it seems likely that the CerPrP 132 polymorphism would similarly influence CWD prion pathogenesis and possibly affect interspecies prion transmission.

7. Transgenic Modeling of Inherited Human Prion Diseases

Approximately 10–20% of human prion diseases exhibit an autosomal dominant mode of inheritance resulting from missense or insertion mutations in the coding sequence of the human *PRNP*. Five of these mutations are genetically linked to loci controlling familial CJD, GSS syndrome, and fatal familial insomnia (FFI), which are inherited human prion diseases that can be transmitted to experimental animals. GSS syndrome, which is characterized clinically by ataxia and dementia and neuropathologically by the deposition of PrP amyloid, most commonly results from mutation at codon 102 of *PRNP* resulting in the substitution of leucine (L) for proline (P) *(65)*. GSS linked to this mutation is transmissible to nonhuman primates *(66)*, wild-type mice *(67)*, Tg mice expressing a chimeric mouse–human PrP gene expressing the GSS mutation *(25)*, and *Prnp* gene-targeted mice referred to as 101LL *(68)*.

Initial studies in Tg(GSS) mice that attempted to understand how an inherited disease also could be infectious suggested that prions in the brains of spontaneously sick Tg(GSS) mice could be transmitted to Tg mice expressing lower levels of mutant protein, referred to as Tg196 mice *(27, 69)*. Disease also was induced in Tg196 mice by a mutant synthetic peptide composed of MoPrP residues 89–103 refolded into a β-sheet conformation *(70)*, and this disease was subsequently propagated to additional Tg196 mice *(71)*. Although these studies lent support to the prion hypothesis, because they suggested that pathogenic PrP gene mutations resulted in the spontaneous formation of PrP^{Sc} and *de novo* production of prions *(72)*, this explanation was controversial for several reasons. Although protease-sensitive forms of PrP^{Sc} (sPrP^{Sc}) have been identified using biochemical and immunologic methods *(18, 73)*, the lack of protease-resistant PrP^{Sc} (rPrP^{Sc}) in the brains of spontaneously sick or recipient mice eliminated a property that, to some, was synonymous with prion infectivity. Moreover, *Prnp* gene-targeted 101LL mice expressing MoPrP-P101L failed to spontaneously develop neurodegenerative disease *(68)*. Finally, disease transmission from spontaneously sick mice to wild-type mice did not occur, and spontaneous disease was eventually registered in aged Tg196 mice *(70, 71)* complicating the interpretation of the original transmission experiments.

To address the apparent dissociation of prion infectivity and PrP^{Sc} in this well-established Tg model, subsequent studies attempted to use means other than differential resistance to proteinase K treatment to detect disease-associated forms of PrP in spontaneously sick Tg mice expressing MoPrP-P101L. Using the prototype PrP^{Sc}-specific monoclonal antibody reagent referred to as 15B3 *(74)*, Nazor and coworkers showed that disease in Tg mice overexpressing MoPrP-P101L results from the spontaneous conversion of mutant PrPC to protease-sensitive MoPrPSc-P101L, defined by its reactivity with 15B3, that accumulates as aggregates in the brains of sick Tg mice *(75)*. To understand the influence of mutant PrP expression levels on the transmissibility of spontaneously generated pathogenic MoPrP-P101L, they produced mice in which transgene copy numbers and levels of MoPrP-P101L expression were carefully defined. Although inoculation of disease-associated MoPrP-P101L accelerated disease in Tg mice expressing MoPrP-P101L from multiple transgenes, disease transmission neither occurred to wild-type nor Tg mice expressing MoPrP-P101L from two transgene copies that did not develop disease spontaneously in their natural life span.

Because disease transmission from spontaneously sick Tg(GSS) mice depended on recipient mice expressing MoPrP-P101L at levels greater than that produced by two transgene copies, and, because such levels of overexpression ultimately resulted in spontaneous disease in older uninoculated recipients, these results suggest that the phenomenon of disease transmission from spontaneously sick Tg(GSS) mice might be more appropriately viewed as disease acceleration whereby inoculation of disease-associated MoPrP-P101L promotes the aggregation of precursors of pathologic MoPrP-P101L that result from transgene overexpression. Such a scheme is consistent with a nucleated polymerization mechanism of prion replication originally postulated from cell-free conversion systems *(4)* and subsequently demonstrated to be the basis of prion propagation in lower eukaryotes *(76)*. According to this model, PrP^C is in equilibrium with PrP^{Sc}, or its precursor, and the equilibrium normally favors PrP^C. Also, PrP^{Sc} is stable only in its aggregated form which can "seed" polymerization of additional PrP^C, thus converting it into additional PrP^{Sc}.

The role of PrP overexpression in the production of synthetic mammalian prions (SMPs) *(77)* originating from *Escherichia coli*-derived recombinant MoPrP remains to be determined. Although the transmission properties and protease resistance of MoPrP(89-230) SMPs are clearly different from disease-associated MoPrP-P101L, it may be significant that the Tg mice in which these SMPs were initially derived expressed MoPrP(89-231) at levels 16 times higher than normal.

Tg mice expressing a mouse PrP version of a nine-octapeptide insertion associated with familial CJD, designated Tg(PG14), exhibited a slowly progressive neurological disorder characterized by apoptotic loss of cerebellar granule cells, gliosis, but no spongiosis *(78)*. Whether the brains of sick Tg(PG14) mice, like sick Tg(GSS) mice, contain 15B3-immunoprecipitable PrP has not yet been reported; however, in both models, mutated PrP adopts different pathologic conformations either spontaneously or after inoculation with authentic prions *(75, 79)*. Like Tg(GSS) mice, brain homogenates from spontaneously sick Tg(PG14) mice failed to transmit disease to Tg mice that express low levels of mutated PrP that do not become sick spontaneously. Whether differences in the state of aggregation of PG14^{spon} compared with MoPrP-P101L will affect its ability to accelerate disease progression in overexpressor Tg(PG14) mice remains to be determined. Transgenic mice expressing a mutation at codon 117 associated with a telencephalic form of GSS *(80)* also spontaneously develop neurodegenerative disease, and they accumulate an aberrant, neurotoxic form of PrP termed ^{Ctm}PrP, which seems to be distinct from conventional protease-resistant PrP^{Sc} *(81)*.

8. PrP Knockout Mice: Physiological and Pathophysiological Functions of PrP

Knockout mouse models were crucial in elucidating the precursor–product relationship between PrP^C and PrP^{Sc}. Although the prion resistant phenotype of $Prnp^{0/0}$ knockout mice and their inability to replicate infectivity (82–84) was in accordance with the prion hypothesis, $Prnp^{0/0}$ knockout mice have not been as informative in defining the physiological and possible pathological functions of PrP^C. Because the gene encoding PrP is highly conserved among species and it is ubiquitously expressed, it was anticipated that $Prnp^{0/0}$ knockout mice would display overt phenotypic deficits that would provide information about essential physiologic functions of PrP^C. In contrast to expectations, initial studies of $Prnp^{0/0}$ knockout mice produced in Zurich and Edinburgh indicated that they developed and behaved normally *(85, 86)*. Although subtle phenotypic defects in the Zurich and Edinburgh $Prnp^{0/0}$ knockout mouse strains suggested roles for PrP^C in maintaining normal circadian rhythms *(87)*, superoxide dismutase activity and protection from oxidative stress *(88)* and copper metabolism *(89)*, the lack of overt phenotypic deficits raised the possibility that adaptive developmental changes might compensate for loss of PrP^C function in $Prnp^{0/0}$ mice. Arguing against this possibility, Tg mice in which expression of PrP^C was suppressed in adult mice by using a tetracycline-responsive transactivator *(90)*, or mice in which neuronal PrP was postnatally ablated *(91)*, remained free of any abnormal phenotype.

Although $Prnp^{0/0}$ knockout mice produced in Nagasaki developed progressive ataxia and Purkinje cell degeneration at ~70 weeks of age *(92)*, these defects were found to be the result of inappropriate and neurotoxic expression of the PrP-like protein Doppel (Dpl) in the CNS rather than loss of PrP function *(93, 94)*. Dpl has 24% sequence homology, and it shares many structural similarities with PrP^C: both proteins are GPI-anchored to the external face of the plasma membrane, they have two *N*-gylcosylation sites and a globular C terminus with three α-helices, and two short strand β-strands and a disulfide bond *(95)*. Both are copper-binding proteins, despite the lack of the octapeptide repeats in Dpl *(96)*. Dpl is primarily expressed by spermatid cells, and it is not normally expressed in the CNS *(95)*.

Interestingly, the structure of Dpl resembles the aforementioned mutant N-terminally truncated PrP between codons 32 and 121 or 32 and 134, which, when expressed in Tg mice, also cause ataxia and degeneration of the granule layer of the cerebellum, and, like Dpl, is overcome by the coexpression of wild-type PrP *(29)*. Targeting N-terminally truncated PrP to Purkinje cells also leads to PrP-reversible Purkinje cell loss and ataxia, further substantiating the notion that Dpl and truncated PrP cause Purkinje cell degeneration by the same mechanism *(97)*. Recent studies by two groups have demonstrated that Tg mice expressing PrP deleted between residues 105 and 125 spontaneously develop a severe, lethal neurodegenerative phenotype in the absence of endogenous PrP *(31)*, whereas Tg mice expressing PrP deleted between residues 94 and 134 exhibit a rapidly progressive, lethal phenotype with extensive central and peripheral myelin degeneration that also is rescued by wild-type PrP or PrP lacking all octarepeats *(30)*.

Other lines of evidence suggest that PrP^C may provide a neuroprotective signaling function *(98, 99)*, raising the possibility that impairment of this function is a feature of prion pathogenesis. Grafts of PrP^C-expressing neuroectodermal tissue into *Prnp$^{0/0}$* knockout mouse brains and Tg mice in which neuronal PrP^C expression was eliminated but that harbored substantial PrP^{Sc} deposits in non-neuronal cells convincingly demonstrated that PrP^{Sc} is not toxic to cells that do not express PrP^C, even after long-term exposure *(100, 101)*. In accordance with a neuroprotective role for PrP, Dpl-induced neurodegeneration is rescued by coexpression of wild-type PrP *(94, 102)*. To address the mechanism by which PrP protects neurons from Dpl-induced toxicity, mutant PrP transgenes were introduced into Nagasaki *Prnp$^{0/0}$* knockout mice. Only mutant PrP lacking amino acid residues 23–88 failed to rescue Dpl-induced ataxia and Purkinje cell death *(103)*.

Recent transgenic mouse studies, which confirm longstanding observations that the amyloidogenic properties of PrP are unrelated to prion infectivity, also may be consistent with a neuroprotective signaling function for PrP^C. Transgenic mice expressing PrP lacking the signal sequence for GPI modification that anchors PrP to the external surface of the cell membrane replicate prions and accumulate PrP^{Sc}, but they do not develop clinical prion disease *(104)*. When inoculated with prions, the mice accumulated mutant PrP in the form of abundant amyloid plaques, but they failed to develop neurologic symptoms up to 600 days later. Although the dissociation of clinical disease and accumulation of disease-associated PrP is consistent with a mechanism in which neurons die in response to neurotoxic signaling events mediated by GPI-anchored PrP^C, the data also supported an alternative hypothesis in which both benign amyloid-forming, protease-resistant PrP, and neurotoxic PrP^{Sc} are produced in the brains of prion-infected transgenic mice *(105)*. In support of this interpretation, prion titers in the brain tissues of asymptomatic transgenic mice were 10-fold lower those found in sick wild-type mice, and prion titers decreased as amyloid plaques accumulated, suggesting that the lack of clinical disease in transgenic mice resulted from relatively inefficient conversion of mutant protein to PrP^{Sc}.

In spite of mounting evidence consistent with a neuroprotective role for PrP^C, it has been difficult to reconcile the absence of pathology and apparently normal behavior of *Prnp$^{0/0}$* knockout mice, with a mechanism of prion pathogenesis involving progressive loss of PrP^C-mediated neuroprotection. Initial behavioral tests of the Zurich *Prnp$^{0/0}$* mice, including swimming navigation, Y-maze discrimination, and the two-way avoidance shuttle box test, indicated that elimination of PrP expression did not compromise learning ability. Subsequent studies of learning and memory and associated hippocampal function by using *Prnp$^{0/0}$* knockout mice provided variable results *(106–109)*. Electrophysiological studies suggesting that GABA$_A$ receptor-mediated fast inhibition and long-term potentiation were impaired in hippocampal slices from Zurich *Prnp$^{0/0}$* mice *(110, 111)* were not independently confirmed *(112, 113)*.

Although progressive cognitive dysfunction is a feature of prion diseases in humans, physical problems such as ataxia, and changes in gait and posture also are typical characteristics. Recent reports indicate that *Prnp$^{0/0}$* knockout exhibit an age-dependent motor behavior deficit, suggesting a function for PrP^C in maintaining sensorimotor coordination *(114)*. This study also reported that the brains of *Prnp$^{0/0}$* knockout mice exhibited region-specific spongiform degeneration and reactive astrocytic gliosis that are neuropathologic hallmarks of prion diseases. A neuroprotective role for PrP^C raises the possibility that loss of normal PrP^C function in the brains of patients expressing mutant *PRNP* alleles is a component of the pathology of inherited prion disorders *(115)*. Cerebellar ataxia is an early symptom of GSS, preceding the onset of progressive dementia, and in some cases the clinical phase of disease may be as long as 10 years. The occurrence of sensorimotor deficits in Tg(GSS) mice similar to those observed in *Prnp$^{0/0}$* mice, but considerably earlier than the onset of end-stage neurodegeneration *(114)*, is consistent with a mechanism involving progressive loss of mutant PrP^C function. In patients with inherited prion diseases such as GSS, loss of PrP function is presumably exacerbated at later times by the dominant-negative effects of mutant PrP^{Sc}. Indeed, previous transgenic studies demonstrated that MoPrP^{Sc}-P101L accelerates prion disease in Tg(GSS) mice and that its accumulation is a late event in the course of disease *(75)*.

References

1. Pattison, I. H. (1965) Slow, Latent and Temperate Virus Infections, NINDB Monograph 2 (Gajdusek, D. C., Gibbs, C. J., Jr., and Alpers, M. P., Eds.), pp. 249–57, U.S. Government Printing, Washington, DC.
2. Prusiner, S. B., Scott, M., Foster, D., Pan, K.-M., Groth, D., Mirenda, C., Torchia, M., Yang, S.-L., Serban, D., Carlson, G. A., Hoppe, P. C., Westaway, D., and DeArmond, S. J. (1990) *Cell* **63**, 673–86.
3. Scott, M., Groth, D., Foster, D., Torchia, M., Yang, S.-L., DeArmond, S. J., and Prusiner, S. B. (1993) *Cell* **73**, 979–88.

4. Kocisko, D. A., Come, J. H., Priola, S. A., Chesebro, B., Raymond, G. J., Lansbury, P. T., Jr., and Caughey, B. (1994) *Nature* **370,** 471–74.
5. Westaway, D., Goodman, P. A., Mirenda, C. A., McKinley, M. P., Carlson, G. A., and Prusiner, S. B. (1987) *Cell* **51,** 651–62.
6. Hunter, N., Goldmann, W., Marshall, E., and O'Neill, G. (2000) *Arch Virol Suppl*, 181–8.
7. Baker, H. F., Poulter, M., Crow, T. J., Frith, C. D., Lofthouse, R., and Ridley, R. M. (1991) *Lancet* **337,** 1286.
8. Collinge, J., Palmer, M. S., and Dryden, A. J. (1991) *Lancet* **337,** 1441–42.
9. Palmer, M. S., Dryden, A. J., Hughes, J. T., and Collinge, J. (1991) *Nature* **352,** 340–42.
10. Nonno, R., Di Bari, M. A., Cardone, F., Vaccari, G., Fazzi, P., Dell'Omo, G., Cartoni, C., Ingrosso, L., Boyle, A., Galeno, R., Sbriccoli, M., Lipp, H. P., Bruce, M., Pocchiari, M., and Agrimi, U. (2006) *PLoS Pathog* **2,** e12.
11. Bessen, R. A., and Marsh, R. F. (1994) *J Virol* **68,** 7859–68.
12. Telling, G. C., Parchi, P., DeArmond, S. J., Cortelli, P., Montagna, P., Gabizon, R., Mastrianni, J., Lugaresi, E., Gambetti, P., and Prusiner, S. B. (1996) *Science* **274,** 2079–82.
13. Korth, C., Kaneko, K., Groth, D., Heye, N., Telling, G., Mastrianni, J., Parchi, P., Gambetti, P., Will, R., Ironside, J., Heinrich, C., Tremblay, P., DeArmond, S. J., and Prusiner, S. B. (2003) *Proc Natl Acad Sci U S A* **100,** 4784–9.
14. Peretz, D., Williamson, R. A., Legname, G., Matsunaga, Y., Vergara, J., Burton, D. R., DeArmond, S. J., Prusiner, S. B., and Scott, M. R. (2002) *Neuron* **34,** 921–32.
15. Scott, M. R., Peretz, D., Nguyen, H. O., Dearmond, S. J., and Prusiner, S. B. (2005) *J Virol* **79,** 5259–71.
16. Taraboulos, A., Jendroska, K., Serban, D., Yang, S.-L., DeArmond, S. J., and Prusiner, S. B. (1992) *Proc Natl Acad Sci U S A* **89,** 7620–24.
17. Hecker, R., Taraboulos, A., Scott, M., Pan, K.-M., Torchia, M., Jendroska, K., DeArmond, S. J., and Prusiner, S. B. (1992) *Genes Dev* **6,** 1213–28.
18. Safar, J., Wille, H., Itri, V., Groth, D., Serban, H., Torchia, M., Cohen, F. E., and Prusiner, S. B. (1998) *Nat Med* **4,** 1157–65.
19. Collinge, J., Sidle, K. C. L., Meads, J., Ironside, J., and Hill, A. F. (1996) *Nature* **383,** 685–90.
20. Hill, A. F., Desbruslais, M., Joiner, S., Sidle, K. C. L., Gowland, I., Collinge, J., Doey, L. J., and Lantos, P. (1997) *Nature* **389,** 448–50.
21. Asante, E. A., Linehan, J. M., Desbruslais, M., Joiner, S., Gowland, I., Wood, A. L., Welch, J., Hill, A. F., Lloyd, S. E., Wadsworth, J. D., and Collinge, J. (2002) *EMBO J* **21,** 6358–66.
22. Scott, M. R., Will, R., Ironside, J., Nguyen, H.-O. B., Tremblay, P., DeArmond, S. J., and Prusiner, S. B. (1999) *Proc Natl Acad Sci U S A* **26**.
23. Wadsworth, J. D., Asante, E. A., Desbruslais, M., Linehan, J. M., Joiner, S., Gowland, I., Welch, J., Stone, L., Lloyd, S. E., Hill, A. F., Brandner, S., and Collinge, J. (2004) *Science* **306,** 1793–6.
24. Telling, G. C., Scott, M., Hsiao, K. K., Foster, D., Yang, S. L., Torchia, M., Sidle, K. C., Collinge, J., DeArmond, S. J., and Prusiner, S. B. (1994) *Proc Natl Acad Sci U S A* **91,** 9936–40.
25. Telling, G. C., Scott, M., Mastrianni, J., Gabizon, R., Torchia, M., Cohen, F. E., DeArmond, S. J., and Prusiner, S. B. (1995) *Cell* **83,** 79–90.
26. Perrier, V., Kaneko, K., Safar, J., Vergara, J., Tremblay, P., DeArmond, S. J., Cohen, F. E., Prusiner, S. B., and Wallace, A. C. (2002) *Proc Natl Acad Sci U S A* **99,** 13079–84.
27. Telling, G. C., Haga, T., Torchia, M., Tremblay, P., DeArmond, S. J., and Prusiner, S. B. (1996) *Genes Dev* **10,** 1736–50.
28. Chiesa, R., Drisaldi, B., Quaglio, E., Migheli, A., Piccardo, P., Ghetti, B., and Harris, D. A. (2000) *Proc Natl Acad Sci U S A* **97,** 5574–9.
29. Shmerling, D., Hegyi, I., Fischer, M., Blattler, T., Brandner, S., Gotz, J., Rulicke, T., Flechsig, E., Cozzio, A., von Mering, C., Hangartner, C., Aguzzi, A., and Weissmann, C. (1998) *Cell* **93,** 203–14.

30. Baumann, F., Tolnay, M., Brabeck, C., Pahnke, J., Kloz, U., Niemann, H. H., Heikenwalder, M., Rulicke, T., Burkle, A., and Aguzzi, A. (2007) *EMBO J* **26**, 538–47.
31. Li, A., Christensen, H. M., Stewart, L. R., Roth, K. A., Chiesa, R., and Harris, D. A. (2007) *EMBO J* **26**, 548–58.
32. Bian, J., Nazor, K. E., Angers, R., Jernigan, M., Seward, T., Centers, A., Green, M., and Telling, G. C. (2006) *Biochem Biophys Res Commun* **340**, 894–900.
33. Barmada, S. J., and Harris, D. A. (2005) *J Neurosci* **25**, 5824–32.
34. Scott, M., Foster, D., Mirenda, C., Serban, D., Coufal, F., Wälchli, M., Torchia, M., Groth, D., Carlson, G., DeArmond, S. J., Westaway, D., and Prusiner, S. B. (1989) *Cell* **59**, 847–57.
35. Scott, M. R., Safar, J., Telling, G., Nguyen, O., Groth, D., Torchia, M., Koehler, R., Tremblay, P., Walther, D., Cohen, F. E., DeArmond, S. J., and Prusiner, S. B. (1997) *Proc Natl Acad Sci U S A* **94**, 14279–84.
36. Buschmann, A., Pfaff, E., Reifenberg, K., Muller, H. M., and Groschup, M. H. (2000) *Arch Virol Suppl*, 75–86.
37. Castilla, J., Gutierrez Adan, A., Brun, A., Pintado, B., Ramirez, M. A., Parra, B., Doyle, D., Rogers, M., Salguero, F. J., Sanchez, C., Sanchez-Vizcaino, J. M., and Torres, J. M. (2003) *Arch Virol* **148**, 677–91.
38. Crozet, C., Bencsik, A., Flamant, F., Lezmi, S., Samarut, J., and Baron, T. (2001) *EMBO Rep* **2**, 952–6.
39. Crozet, C., Flamant, F., Bencsik, A., Aubert, D., Samarut, J., and Baron, T. (2001) *J Virol* **75**, 5328–34.
40. Vilotte, J. L., Soulier, S., Essalmani, R., Stinnakre, M. G., Vaiman, D., Lepourry, L., Da Silva, J. C., Besnard, N., Dawson, M., Buschmann, A., Groschup, M., Petit, S., Madelaine, M. F., Rakatobe, S., Le Dur, A., Vilette, D., and Laude, H. (2001) *J Virol* **75**, 5977–84.
41. Windl, O., Buchholz, M., Neubauer, A., Schulz-Schaeffer, W., Groschup, M., Walter, S., Arendt, S., Neumann, M., Voss, A. K., and Kretzschmar, H. A. (2005) *J Virol* **79**, 14971–5.
42. Capobianco, R., Casalone, C., Suardi, S., Mangieri, M., Miccolo, C., Limido, L., Catania, M., Rossi, G., Fede, G. D., Giaccone, G., Bruzzone, M. G., Minati, L., Corona, C., Acutis, P., Gelmetti, D., Lombardi, G., Groschup, M. H., Buschmann, A., Zanusso, G., Monaco, S., Caramelli, M., and Tagliavini, F. (2007) *PLoS Pathog* **3**, e31.
43. Bencsik, A., Philippe, S., Debeer, S., Crozet, C., Calavas, D., and Baron, T. (2007) *Histochem Cell Biol* **127**, 531–9.
44. Cordier, C., Bencsik, A., Philippe, S., Betemps, D., Ronzon, F., Calavas, D., Crozet, C., and Baron, T. (2006) *J Gen Virol* **87**, 3763–71.
45. Baron, T., Crozet, C., Biacabe, A. G., Philippe, S., Verchere, J., Bencsik, A., Madec, J. Y., Calavas, D., and Samarut, J. (2004) *J Virol* **78**, 6243–51.
46. Le Dur, A., Beringue, V., Andreoletti, O., Reine, F., Lai, T. L., Baron, T., Bratberg, B., Vilotte, J. L., Sarradin, P., Benestad, S. L., and Laude, H. (2005) *Proc Natl Acad Sci U S A* **102**, 16031–6.
47. Williams, E. S., and Miller, M. W. (2002) *Rev Sci Tech* **21**, 305–16.
48. Miller, M. W., Williams, E. S., Hobbs, N.T., and Wolfe, L.L. (2004) Emerg Infect Dis **10**, 1003–6.
49. Joly, D. O., Ribic, C. A., Langenberg, J. A., Beheler, K., Batha, C. A., Dhuey, B. J., Rolley, R. E., Bartelt, G., Van Deelen, T. R., and Samuel, M. D. (2003) *Emerg Infect Dis* **9**, 599–601.
50. Sohn, H. J., Kim, J. H., Choi, K. S., Nah, J. J., Joo, Y. S., Jean, Y. H., Ahn, S. W., Kim, O. K., Kim, D. Y., and Balachandran, A. (2002) *J Vet Med Sci* **64**, 855–8.
51. Kim, T. Y., Shon, H. J., Joo, Y. S., Mun, U. K., Kang, K. S., and Lee, Y. S. (2005) *J Vet Med Sci* **67**, 753–9.
52. Browning, S. R., Mason, G. L., Seward, T., Green, M., Eliason, G. A., Mathiason, C., Miller, M. W., Williams, E. S., Hoover, E., and Telling, G. C. (2004) *J Virol* **78**, 13345–50.
53. Angers, R. C., Browning, S. R., Seward, T. S., Sigurdson, C. J., Miller, M. W., Hoover, E. A., and Telling, G. C. (2006) *Science* **311**, 1117.
54. Buschmann, A., and Groschup, M. H. (2005) *J Infect Dis* **192**, 934–42.

55. Meade-White, K., Race, B., Trifilo, M., Bossers, A., Favara, C., Lacasse, R., Miller, M., Williams, E., Oldstone, M., Race, R., and Chesebro, B. (2007) *J Virol* **81,** 4533–9.
56. LaFauci, G., Carp, R. I., Meeker, H. C., Ye, X., Kim, J. I., Natelli, M., Cedeno, M., Petersen, R. B., Kascsak, R., and Rubenstein, R. (2006) *J Gen Virol* **87,** 3773–80.
57. Tamguney, G., Giles, K., Bouzamondo-Bernstein, E., Bosque, P. J., Miller, M. W., Safar, J., DeArmond, S. J., and Prusiner, S. B. (2006) *J Virol* **80,** 9104–14.
58. Kong, Q., Huang, S., Zou, W., Vanegas, D., Wang, M., Wu, D., Yuan, J., Zheng, M., Bai, H., Deng, H., Chen, K., Jenny, A. L., O'Rourke, K., Belay, E. D., Schonberger, L. B., Petersen, R. B., Sy, M. S., Chen, S. G., and Gambetti, P. (2005) *J Neurosci* **25,** 7944–9.
59. Trifilo, M. J., Ying, G., Teng, C., and Oldstone, M. B. (2007) *Virology*.
60. Sigurdson, C. J., Manco, G., Schwarz, P., Liberski, P., Hoover, E. A., Hornemann, S., Polymenidou, M., Miller, M. W., Glatzel, M., and Aguzzi, A. (2006) *J Virol* **80,** 12303–11.
61. Raymond, G. J., Raymond, L. D., Meade-White, K. D., Hughson, A. G., Favara, C., Gardner, D., Williams, E. S., Miller, M. W., Race, R. E., and Caughey, B. (2007) *J Virol* **81,** 4305–14.
62. Collinge, J., Palmer, M. S., Sidle, K. C., Hill, A. F., Gowland, I., Meads, J., Asante, E., Bradley, R., Doey, L. J., and Lantos, P. L. (1995) *Nature* **378,** 779–83.
63. Schatzl, H. M., Wopfner, F., Gilch, S., von Brunn, A., and Jager, G. (1997) *Lancet* **349,** 1603–4.
64. O'Rourke, K. I., Baszler, T. V., Miller, J. M., Spraker, T. R., Sadler-Riggleman, I., and Knowles, D. P. (1998) *J Clin Microbiol* **36,** 1750–5.
65. Hsiao, K., Baker, H. F., Crow, T. J., Poulter, M., Owen, F., Terwilliger, J. D., Westaway, D., Ott, J., and Prusiner, S. B. (1989) *Nature* **338,** 342–45.
66. Brown, P., Gibbs, C. J., Jr., Rodgers-Johnson, P., Asher, D. M., Sulima, M. P., Bacote, A., Goldfarb, L. G., and Gajdusek, D. C. (1994) *Ann Neurol* **35,** 513–29.
67. Tateishi, J., and Kitamoto, T. (1995) *Brain Pathol* **5,** 53–59.
68. Manson, J. C., Jamieson, E., Baybutt, H., Tuzi, N. L., Barron, R., McConnell, I., Somerville, R., Ironside, J., Will, R., Sy, M. S., Melton, D. W., Hope, J., and Bostock, C. (1999) *EMBO J* **18,** 6855–64.
69. Hsiao, K. K., Groth, D., Scott, M., Yang, S.-L., Serban, H., Rapp, D., Foster, D., Torchia, M., DeArmond, S. J., and Prusiner, S. B. (1994) *Proc Natl Acad Sci U S A* **91,** 9126–30.
70. Kaneko K, B. H., Wille H, Zhang H, Groth D, Torchia M, Tremblay P, Safar J, Prusiner SB, DeArmond SJ, Baldwin MA, Cohen FE (2000) *J Mol Biol* **295,** 997–1007.
71. Tremblay, P., Ball, H. L., Kaneko, K., Groth, D., Hegde, R. S., Cohen, F. E., DeArmond, S. J., Prusiner, S. B., and Safar, J. G. (2004) *J Virol* **78,** 2088–99.
72. Cohen, F. E., Pan, K.-M., Huang, Z., Baldwin, M., Fletterick, R. J., and Prusiner, S. B. (1994) *Science* **264,** 530–31.
73. Tzaban, S., Friedlander, G., Schonberger, O., Horonchik, L., Yedidia, Y., Shaked, G., Gabizon, R., and Taraboulos, A. (2002) *Biochemistry* **41,** 12868–75.
74. Korth, C., Stierli, B., Streit, P., Moser, M., Schaller, O., Fischer, R., Schulz-Schaeffer, W., Kretzschmar, H., Raeber, A., Braun, U., Ehrensperger, F., Hornemann, S., Glockshuber, R., Riek, R., Billeter, M., Wuthrick, K., and Oesch, B. (1997) *Nature* **389,** 74–77.
75. Nazor, K. E., Kuhn, F., Seward, T., Green, M., Zwald, D., Purro, M., Schmid, J., Biffiger, K., Power, A. M., Oesch, B., Raeber, A. J., and Telling, G. C. (2005) *EMBO J* **24,** 2472–80.
76. Uptain, S. M., and Lindquist, S. (2002) *Annu Rev Microbiol* **56,** 703–41.
77. Legname, G., Baskakov, I. V., Nguyen, H. O., Riesner, D., Cohen, F. E., DeArmond, S. J., and Prusiner, S. B. (2004) *Science* **305,** 673–6.
78. Chiesa, R., Piccardo, P., Ghetti, B., and Harris, D. A. (1998) *Neuron* **21,** 1339–51.
79. Chiesa, R., Piccardo, P., Quaglio, E., Drisaldi, B., Si-Hoe, S. L., Takao, M., Ghetti, B., and Harris, D. A. (2003) *J Virol* **77,** 7611–22.
80. Hsiao, K. K., Cass, C., Schellenberg, G. D., Bird, T., Devine-Gage, E., Wisniewski, H., and Prusiner, S. B. (1991) *Neurology* **41,** 681–84.
81. Hegde, R. S., Mastrianni, J. A., Scott, M. R., DeFea, K. A., Tremblay, P., Torchia, M., DeArmond, S. J., Prusiner, S. B., and Lingappa, V. R. (1998) *Science* **279,** 827–34.

82. Büeler, H., Aguzzi, A., Sailer, A., Greiner, R.-A., Autenried, P., Aguet, M., and Weissmann, C. (1993) *Cell* **73,** 1339–47.
83. Büeler, H., Raeber, A., Sailer, A., Fischer, M., Aguzzi, A., and Weissmann, C. (1994) *Mol Med* **1,** 19–30.
84. Manson, J. C., Clarke, A. R., McBride, P. A., McConnell, I., and Hope, J. (1994) *Neurodegeneration* **3,** 331–40.
85. Büeler, H., Fischer, M., Lang, Y., Bluethmann, H., Lipp, H.-P., DeArmond, S. J., Prusiner, S. B., Aguet, M., and Weissmann, C. (1992) *Nature* **356,** 577–82.
86. Manson, J. C., Clarke, A. R., Hooper, M. L., Aitchison, L., McConnell, I., and Hope, J. (1994) *Mol Neurobiol* **8,** 121–27.
87. Tobler, I., Gaus, S. E., Deboer, T., Achermann, P., Fischer, M., Rülicke, T., Moser, M., Oesch, B., McBride, P. A., and Manson, J. C. (1996) *Nature* **380,** 639–42.
88. Brown, D. R., and Besinger, A. (1998) *Biochem J* **334,** 423–29.
89. Brown, D. R., Qin, K., Herms, J. W., Madlung, A., Manson, J., Strome, R., Fraser, P. E., Kruck, T., von Bohlen, A., Schulz-Schaeffer, W., Giese, A., Westaway, D., and Kretzschmar, H. (1997) *Nature* **390,** 684–87.
90. Tremblay, P., Meiner, Z., Galou, M., Heinrich, C., Petromilli, C., Lisse, T., Cayetano, J., Torchia, M., Mobley, W., Bujard, H., DeArmond, S. J., and Prusiner, S. B. (1998) *Proc Natl Acad Sci U S A* **95,** 12580–85.
91. Mallucci, G. R., Ratte, S., Asante, E. A., Linehan, J., Gowland, I., Jefferys, J. G., and Collinge, J. (2002) *EMBO J* **21,** 202–10.
92. Sakaguchi, S., Katamine, S., Nishida, N., Moriuchi, R., Shigematsu, K., Sugimoto, T., Nakatani, A., Kataoka, Y., Houtani, T., Shirabe, S., Okada, H., Hasegawa, S., Miyamoto, T., and Noda, T. (1996) *Nature* **380,** 528–31.
93. Moore, R. C., Lee, I. Y., Silverman, G. L., Harrison, P. M., Strome, R., Heinrich, C., Karunaratne, A., Pasternak, S. H., Chishti, M. A., Liang, Y., Mastrangelo, P., Wang, K., Smit, A. F. A., Katamine, S., Carlson, G. A., Cohen, F. E., Prusiner, S. B., Melton, D. W., Tremblay, P., Hood, L. E., and Westaway, D. (1999) *J Mol Biol* **292,** 797–817.
94. Rossi, D., Cozzio, A., Flechsig, E., Klein, M. A., Rulicke, T., Aguzzi, A., and Weissmann, C. (2001) *EMBO J* **20,** 694–702.
95. Silverman, G. L., Qin, K., Moore, R. C., Yang, Y., Mastrangelo, P., Tremblay, P., Prusiner, S. B., Cohen, F. E., and Westaway, D. (2000) *J Biol Chem* **275,** 26834–41.
96. Qin, K., Coomaraswamy, J., Mastrangelo, P., Yang, Y., Lugowski, S., Petromilli, C., Prusiner, S. B., Fraser, P. E., Goldberg, J. M., Chakrabartty, A., and Westaway, D. (2003) *J Biol Chem* **278,** 8888–96.
97. Flechsig, E., Hegyi, I., Leimeroth, R., Zuniga, A., Rossi, D., Cozzio, A., Schwarz, P., Rulicke, T., Gotz, J., Aguzzi, A., and Weissmann, C. (2003) *EMBO J* **22,** 3095–101.
98. Mouillet-Richard, S., Ermonval, M., Chebassier, C., Laplanche, J. L., Lehmann, S., Launay, J. M., and Kellermann, O. (2000) *Science* **289,** 1925–8.
99. Chiarini, L. B., Freitas, A. R., Zanata, S. M., Brentani, R. R., Martins, V. R., and Linden, R. (2002) *EMBO J* **21,** 3317–26.
100. Brandner, S., Isenmann, S., Raeber, A., Fischer, M., Sailer, A., Kobayashi, Y., Marino, S., Weissmann, C., and Aguzzi, A. (1996) *Nature* **379,** 339–43.
101. Mallucci, G., Dickinson, A., Linehan, J., Klohn, P. C., Brandner, S., and Collinge, J. (2003) *Science* **302,** 871–4.
102. Moore, R. C., Mastrangelo, P., Bouzamondo, E., Heinrich, C., Legname, G., Prusiner, S. B., Hood, L., Westaway, D., DeArmond, S. J., and Tremblay, P. (2001) *Proc Natl Acad Sci U S A* **98,** 15288–93.
103. Atarashi, R., Nishida, N., Shigematsu, K., Goto, S., Kondo, T., Sakaguchi, S., and Katamine, S. (2003) *J Biol Chem* **278,** 28944–9.
104. Chesebro, B., Trifilo, M., Race, R., Meade-White, K., Teng, C., LaCasse, R., Raymond, L., Favara, C., Baron, G., Priola, S., Caughey, B., Masliah, E., and Oldstone, M. (2005) *Science* **308,** 1435–9.
105. Telling, G. (2005) *N Engl J Med* **353,** 1177–9.

106. Lipp, H. P., Stagliar-Bozicevic, M., Fischer, M., and Wolfer, D. P. (1998) *Behav Brain Res* **95,** 47–54.
107. Roesler, R., Walz, R., Quevedo, J., de-Paris, F., Zanata, S. M., Graner, E., Izquierdo, I., Martins, V. R., and Brentani, R. R. (1999) *Brain Res Mol Brain Res* **71,** 349–53.
108. Coitinho, A. S., Roesler, R., Martins, V. R., Brentani, R. R., and Izquierdo, I. (2003) *Neuroreport* **14,** 1375–9.
109. Criado, J. R., Sanchez-Alavez, M., Conti, B., Giacchino, J. L., Wills, D. N., Henriksen, S. J., Race, R., Manson, J. C., Chesebro, B., and Oldstone, M. B. (2005) *Neurobiol Dis* **19,** 255–65.
110. Collinge, J., Whittington, M. A., Sidle, K. C., Smith, C. J., Palmer, M. S., Clarke, A. R., and Jefferys, J. G. R. (1994) *Nature* **370,** 295–97.
111. Colling, S. B., Collinge, J., and Jefferys, J. G. R. (1996) *Neurosci Lett* **209,** 49–52.
112. Herms, J. W., Kretzschmar, H. A., Titz, S., and Keller, B. U. (1995) *Eur J Neurosci* **7,** 2508–12.
113. Lledo, P.-M., Tremblay, P., DeArmond, S. J., Prusiner, S. B., and Nicoll, R. A. (1996) *Proc Natl Acad Sci U S A* **93,** 2403–07.
114. Nazor, K. E., Seward, T., and Telling, G. C. (2007) *Biochim Biophys Acta* **1772,** 645–53.
115. Samaia, H. B., and Brentani, R. R. (1998) *Mol Psychiatry* **3,** 196–7.

Chapter 18
Quantitative Bioassay of Surface-bound Prion Infectivity

Victoria A. Lawson

Summary The unconventional nature of the infectious agent of prion diseases poses a challenge to conventional infection control methodologies. The extra neural tissue distribution of variant and sporadic Creutzfeldt–Jakob disease has increased concern regarding the risk of prion disease transmission via general surgical procedures and highlighted the need for decontamination procedures that can be incorporated into routine processing. This chapter describe a quantitative method for assessing the prionocidal activity of chemical and physical decontamination methods against surface-bound prion infectivity.

Keywords Bioassay; decontamination; prion; quantitative; steel wire.

1. Introduction

Iatrogenic prion disease transmission *(1)* and laboratory studies *(2)* have highlighted the remarkable resistance of prion infectivity to chemical and physical decontamination. Conventional disinfection methods used in a medical setting are ineffective in the removal of prion infectivity *(2)*. The World Health Organization, therefore, recommends incineration of disposable instruments, materials, and wastes that have been exposed to high-infectivity tissues. Less effective methods, such as high heat sterilization or chemical deactivation (1 N sodium hydroxide or 20,000 ppm chlorine), may be applied to instruments and other materials subject to re-use. However, these chemicals are generally incompatible with the materials used in the construction of medical instruments *(3, 4)*.

Treatment of prion-infected tissue preparations with sodium hydroxide, chlorine, or high heat autoclaving have been shown to reduce the infectivity of several prion strains by more than 3 logs *(5)*. However, tissue preparations (homogenates or macerates) do not represent contamination in a medical setting where infected tissue is associated with the surface of medical instruments. Weissmann and colleagues demonstrated the ease and avidity with which prion infectivity becomes associated with stainless steel surfaces *(6, 7)*. Ensuing studies have assessed the efficacy of traditional

From: *Prion Protein Protocols.*
Methods in Molecular Biology, Vol. 459.
Edited by: A. F. Hill © Humana Press, Totowa, NJ

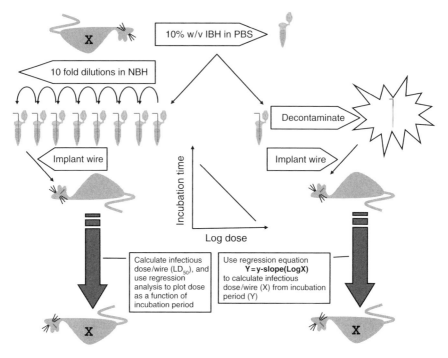

Fig. 1 Overview of quantitative bioassay to determine the efficacy of decontamination protocols by using prion contaminated surgical steel wires

and novel decontamination protocols against surface-bound prions *(7–11)*. Fichet et al. *(12)* reported a significant improvement in the method by demonstrating that the infectivity of prions associated with stainless steel wires can be titrated and that the efficacy of prion decontamination protocols could be determined by incubation time interval assay. The use of surface-bound models for the assessment of prion decontamination methods better models contamination in a surgical setting, circumvents problems associated with toxicity of the chemicals under investigation and recognizes the need for both cleaning and decontamination within the sterilization process *(16)*.

This chapter describes a method for the preparation, titration, and decontamination of surface-bound prions (**Fig. 1**).

2. Materials

2.1. Preparation of Brain Tissue Homogenates

1. Biosafety cabinet class II (BCII).
2. Prion-infected brain tissue (*see* **Note 1**).
3. Uninfected brain tissue (*see* **Note 2**).

4. Appropriately sized sterile container.
5. Dulbecco's phosphate-buffered saline (D-PBS; Invitrogen, Carlsbad, CA).
6. Blunt-ended cannulas (18-, 20-, 22-gauge; Monoject) and 10-ml syringes.
7. 26-gauge needles and 10-ml syringes.
8. Screw-capped tubes.

2.2. Preparation of Surface-bound prions

1. BCII.
3. Surgical steel wire (B. Braun stainless steel monofilament wire USP 4/0).
4. 2% (v/v) Triton X-100 prepared in distilled water.
5. Sterile petri dishes.
6. Sterile distilled water.
7. Phosphate-buffered saline (PBS; Invitrogen).
8. Forceps (*see* **Note 3**).
9. Scissors (*see* **Note 3**).
10. 50-ml tubes.
11. 5-ml screw-capped tubes (*see* **Note 4**).
12. 1.5-ml screw-capped tubes (*see* **Note 4**).
13. Parafilm.
14. Room temperature shaking platform.
15. 19-gauge needle or orange sticks cut to fit across petri dish.
16. 37°C dry oven.
17. Ultrasonication bath (Transonic 310, Elma).

2.3. Decontamination of Surface-bound Prions

1. BCII.
2. Forceps (*see* **Note 3**).
3. 1.5-ml screw-capped tubes.
4. Chemical decontamination reagent.
5. Thermomixer (Eppendorf, Hamburg, Germany).
6. Distilled water.
7. Sterile petri dishes.
8. 19-gauge needles or orange sticks cut to fit across petri dish.
9. Autoclave bags or similar (for testing physical decontamination protocol).
10. Autoclave (for testing physical decontamination protocol).

2.4. Bioassay of Surface-bound prions

1. BCII.
2. Prion-contaminated surgical steel wires prepared as described in **Subheading 3.2.**

3. Anaesthetic (as per institutional guidelines).
4. 26-gauge needles-1 for each indicator mouse.
5. Forceps (*see* **Note 3**).
6. Scissors (*see* **Note 3**).
7. Indicator mice (Prnp transgenic Tga20 mice) *(13)*.
8. Animal holding facilities.

3. Methods

All methods that describe the handling of infectious material should be performed in a BCII.

3.1. Preparation of 10% (w/v) Brain Homogenates

Homogenates 10% (w/v) of infected (IBH) and uninfected brain (UBH) tissue are prepared in the same manner. We prepare large stock of IBH and freeze 300-µl aliquots. Homogenates are prepared on ice.

1. Weigh container to be used to prepare tissue homogenate.
2. Remove infected or uninfected brain tissue from −80°C freezer and transfer to container.
3. Weigh tissue.
4. Add 9 volumes of D-PBS.
5. Using a graded series of blunt ended cannulas (19-20-22-gauge), homogenize tissue. Ensure complete homogenization of tissue at each gauge of needle before progressing to the next gauge.
6. Complete homogenization by passing homogenate through 26-gauge needle. Care should be taken to prevent injury by needle stick.
7. Prepare aliquots of tissue homogenate, snap-freeze in liquid nitrogen, and store at −80°C.

3.2. Preparation of Surface-bound Prions

3.2.1. Preparation of Surgical Steel Wires

1. Using a clean pair of forceps and scissors cut 30-mm lengths of surgical steel wire.
2. Bend surgical steel wire into an L-shape by bending one end to 90° approx 5 mm from end (*see* **Note 3**).
3. Submerge surgical steel wires in 2% (v/v) solution of TX-100 prepared in distilled water in a 50-ml tube.

4. Place 50-ml tubes in ultrasonication bath.
5. Clean wires by ultrasonication for 15 min.
6. Pour off Triton X-100 (TX-100) solution and rinse surgical steel wires 2 × 10 min at room temperature in 50-ml volume of sterile distilled water.
7. Remove surgical steel wires from tubes using clean forceps and place in a sterile petri dish.
8. Dry surgical steel wires (with lid off) at 37°C in a dry oven (1 h).
9. Close petri dish lid and seal with parafilm (*see* **Note 5**).

3.2.2. Contamination of Surgical Steel Wire

1. Thaw IBH.
2. Add appropriately diluted IBH to screw-capped tube to a depth of 10 mm (*see* **Note 4**). Handling surgical steel wires at the bent end insert up to 20 wires into each screw-capped tube. Ensure that the tip of each wire is submerged to a depth between 5 and 10 mm.
3. Incubate for 1 h at room temperature.
4. Carefully remove each surgical steel wire from tube handling the bent end of the wire with forceps.
5. Place surgical steel wires in sterile petri dish with clean end resting on a support (*see* **Note 6**).
6. Leave surgical steel wires in BCII hood to dry with lid of petri dish open (1 h), and then close lid of petri dish and leave overnight at room temperature.
7. Transfer surgical steel wires using forceps into a 50-ml tube containing PBS.
8. Wash surgical steel wires in PBS with agitation for 5 min at room temperature.
9. Carefully remove each surgical steel wire from the 50-ml tube handling the bent end of the wire with forceps.
10. Place wires in sterile petri dish with clean end resting on a support (*see* **Note 6**).
11. Leave wires in BCII hood to dry with lid of petri dish open (1 h).
12. Proceed to decontamination of surface-bound prions (*see* **Subheading 3.3.**) or bioassay of surface-bound prions (*see* **Subheading 3.4.**).
13. Prepared wires can be stored for bioassay in 5-ml screw-capped tubes, sealed with parafilm, at −80°C until implantation (*see* **Note 7**).

3.2.3. Dose Titration of Surface-bound Prion Infectivity

1. Prepare a 10-fold dilution series of IBH in UBH (*see* **Note 8**).
2. Expose clean surgical steel wires to dilutions of IBH as described for contamination of stainless steel wires in **Subheading 3.2.2.** (*see* **Note 9**).

3.3. Decontamination of Surface-bound Prions

For each decontamination procedure, prepare surgical steel wires contaminated with an undiluted (10%, w/v) IBH as described for contamination of surgical steel wires (*see* **Subheading 3.2.2.**). Leave untreated prion contaminated wires (positive control) in a sealed petri dish at room temperature while decontamination procedures are performed.

3.3.1. Chemical Treatment of Surface-bound Prion Infectivity

1. Prepare chemical decontamination reagents and aliquot 1-ml volumes into 1.5-ml screw-capped tubes.
2. Using forceps, transfer five prion-contaminated surgical steel wires into 1.5-ml screw-capped tube containing the appropriate chemical decontamination reagent (*see* **Note 10**).
3. Treat prion-contaminated surgical steel wires for the desired time and the desired temperature in thermomixer (300 rpm).
4. At the conclusion of the treatment, remove chemical decontamination reagent from surgical steel wires and rinse in 1-ml volume of distilled water.
5. Carefully remove each surgical steel wire from the 1.5-ml tube handling the bent end of the wire with forceps.
6. Place surgical steel wires in sterile petri dish with clean end resting on a support (*see* **Note 6**).
7. Leave surgical steel wires in BCII hood to dry with lid of petri dish open (30 min).
8. Proceed to physical treatment of surface-bound prion infectivity (*see* **Subheading 3.3.2.**) or transfer wires to 5-ml screw-capped tubes, seal with parafilm, and store at −80°C until implantation (*see* **Note 7**).

3.3.2. Physical Treatment of Surface-bound Prion Infectivity

1. Place surgical steel wires in an appropriate vessel for autoclaving (*see* **Note 11**).
2. Autoclave surgical steel wires.
3. After treatment, allow surgical steel wires to cool and transfer to 5-ml screw-capped tubes, seal with parafilm, and store at −80°C until implantation (*see* **Note 7**).

3.4. Bioassay of Surface-bound Prions

3.4.1. Implantation of Prion-contaminated Surgical Steel Wires

Seek institutional animal ethics approval before commencing this method.

1. For each surgical steel wire, prepare a 26-gauge needle with a pusher made from surgical steel wire (at least 2 cm longer than the needle and hub) inserted into the hub end of the needle.
2. Holding surgical steel wire for implantation by the contaminated end with a pair of forceps, cut a 5-mm length from the contaminated end (*see* **Note 3**).
3. Using forceps insert the surgical steel wire into the sharp end of the 26-gauge needle.
4. To implant surgical steel wire into the parietal region (*see* **Note 12**) of anesthetized indicator mouse (*see* **Note 13**) insert needle until contact is made with the base of the skull. Using very light pressure, use the pusher wire to hold the contaminated wire in place while gently withdrawing the needle.
5. Check for wire insertion by confirming that wire is no longer in needle (*see* **Note 14**).
6. Confirm that wire is completely embedded and not protruding from skull.
7. Monitor mice until completely recovered from anaesthetic.
8. Maintain mice and sacrifice when showing signs of clinical disease and before they become moribund as per institutional guidelines.
9. Confirm implantation of wire at time of sacrifice.
10. Confirm prion disease by Western blot analysis for protease-resistant prion protein in brain tissue.
11. At the conclusion of incubation period, confirm absence of nonclinical infection in asymptomatic mice by high sensitivity Western blot of brain tissue *(14, 15)*.

3.4.2. Determination of Dose

1. Estimate the lethal dose of surface-bound infectivity present on surgical steel wires (dose/wire) contaminated with an undiluted IBH (10%, w/v, homogenate) by using the method of Reed and Muench *(17)*.
2. Use regression analysis to plot the relationship between incubation period and infectivity titer (incubation time interval assay) and calculate the infectivity associated with treated wires using the following equation:

$$Y = y - slope(\log X)$$

where Y is incubation period, y is *y*-axis intercept, and X is dose (LD_{50}).

4. Notes

1. Prion-infected brain tissue was derived Balb/c mice inoculated with M1000 prions at the terminal stage of disease. Brain tissue was removed from euthanized animals, snap-frozen in liquid nitrogen, and stored at $-80°C$.
2. Uninfected bran tissue was derived from normal Balb/c mice. Brain tissue was removed from euthanized animals, snap-frozen in liquid nitrogen, and stored at $-80°C$.

3. Unless stated, surgical steel wires should be handled by the shorter arm of the "L" by using forceps. Clean forceps should be used to prepare wires. Once wires have been contaminated, a separate pair of forceps and scissors should be used for each dilution of IBH or treatment group to prevent cross-contamination.

4. Wires can be contaminated in 5-ml or 1.5-ml screw-capped tubes. A depth of 5–10 mm can be achieved by adding 1 ml of homogenate to a 5-ml screw-capped tube or 200 µl of homogenate to a 1.5-ml screw-capped tube. For storage of wires at −80°C before implantation, 5-ml screw-capped tubes will be needed.

5. Clean, dry surgical steel wires can be stored at room temperature in sterile petri dish until needed.

6. Use a 19-gauge needle or orange stick to support clean end of surgical steel wire to allow air to circulate around contaminated end.

7. All homogenates and surgical steel wires should undergo the same number of freeze-thaw cycles to ensure accuracy of incubation time interval assay.

8. To calculate the titer of infectivity (LD_{50}), the inoculum must be diluted beyond a dose at which 50% of inoculated animals succumb to disease. In our experience and the experience of others *(12)*, this occurs when surgical steel wires are exposed to a 1×10^{-6} dilution of IBH. To ensure accurate titration, homogenates should be diluted to 1×10^{-9}, because titer is dependent on the prion strain and indicator mouse strain used.

9. Include surgical steel wires exposed to UBH as negative control.

10. The number of surgical steel wires prepared will depend on the number of animals used for bioassay. The number of test animals used should be as few as required for statistical significance. Prepare at least one set (five surgical steel wires) more than required to allow for loss.

11. Wires should be placed in a container that approximates conditions of interest. Treatment of wires within sealed plastic containers will not be as effective as treatment in open containers or autoclave bags, which allow steam to circulate *(9, 12)*.

12. In 6- to 8-week-old Tga20 mice, implant wire approx 2 mm anterior to the back rim of the eye socket and 2 mm lateral to the midline, angling needle back towards the midline.

13. Mice should be sedated, in accordance with institutional guidelines, by using an anaesthetic combination such as xylazine/ketamine to induce a deep sedation of 20- to 60-min duration.

14. Care should be taken to account for all wires used to prevent "needle stick" injury from wires.

References

1. Brown, P., Preece, M., Brandel, J. P., Sato, T., McShane, L., Zerr, I., Fletcher, A., Will, R. G., Pocchiari, M., Cashman, N. R., d'Aignaux, J. H., Cervenakova, L., Fradkin, J., Schonberger, L. B., and Collins, S. J. (2000) Iatrogenic Creutzfeldt-Jakob disease at the millennium *Neurology* **55**, 1075–81.

2. Taylor, D. M. (2000) Inactivation of transmissible degenerative encephalopathy agents: a review. *Vet J* **159,** 10–7.
3. Brown, S. A., Merritt, K., Woods, T. O., and Busick, D. N. (2005) Effects on instruments of the World Health Organization–recommended protocols for decontamination after possible exposure to transmissible spongiform encephalopathy-contaminated tissue. *J Biomed Mater Res B Appl Biomater* **72,** 186–90.
4. McDonnell, G., and Burke, P. (2003) The challenge of prion decontamination. *Clin Infect Dis* **36,** 1152–4.
5. Rutala, W. A., and Weber, D. J. (2001) Creutzfeldt-Jakob disease: recommendations for disinfection and sterilization. *Clin Infect Dis* **32,** 1348–56.
6. Flechsig, E., Hegyi, I., Enari, M., Schwarz, P., Collinge, J., and Weissmann, C. (2001) Transmission of scrapie by steel-surface-bound prions. *Mol Med* **7,** 679–84.
7. Zobeley, E., Flechsig, E., Cozzio, A., Enari, M., and Weissmann, C. (1999) Infectivity of scrapie prions bound to a stainless steel surface. *Mol Med* **5,** 240–3.
8. Baxter, H. C., Campbell, G. A., Whittaker, A. G., Jones, A. C., Aitken, A., Simpson, A. H., Casey, M., Bountiff, L., Gibbard, L., and Baxter, R. L. (2005) Elimination of transmissible spongiform encephalopathy infectivity and decontamination of surgical instruments by using radio-frequency gas-plasma treatment. *J Gen Virol* **86,** 2393–9.
9. Jackson, G. S., McKintosh, E., Flechsig, E., Prodromidou, K., Hirsch, P., Linehan, J., Brandner, S., Clarke, A. R., Weissmann, C., and Collinge, J. (2005) An enzyme-detergent method for effective prion decontamination of surgical steel. *J Gen Virol* **86,** 869–78.
10. Lemmer, K., Mielke, M., Pauli, G., and Beekes, M. (2004) Decontamination of surgical instruments from prion proteins: in vitro studies on the detachment, destabilization and degradation of PrPSc bound to steel surfaces. *J Gen Virol* **85,** 3805–16.
11. Yan, Z. X., Stitz, L., Heeg, P., Pfaff, E., and Roth, K. (2004) Infectivity of prion protein bound to stainless steel wires: a model for testing decontamination procedures for transmissible spongiform encephalopathies. *Infect Control Hosp Epidemiol* **25,** 280–3.
12. Fichet, G., Comoy, E., Duval, C., Antloga, K., Dehen, C., Charbonnier, A., McDonnell, G., Brown, P., Lasmezas, C. I., and Deslys, J. P. (2004) Novel methods for disinfection of prion-contaminated medical devices. *Lancet* **364,** 521–6.
13. Fischer, M., Rulicke, T., Raeber, A., Sailer, A., Moser, M., Oesch, B., Brandner, S., Aguzzi, A., and Weissmann, C. (1996) Prion protein (PrP) with amino-proximal deletions restoring susceptibility of PrP knockout mice to scrapie. *EMBO J* **15,** 1255–64.
14. Collins, S. J., Lewis, V., Brazier, M. W., Hill, A. F., Lawson, V. A., Klug, G. M., and Masters, C. L. (2005) Extended period of asymptomatic prion disease after low dose inoculation: assessment of detection methods and implications for infection control. *Neurobiol Dis* **20,** 336–46.
15. Wadsworth, J. D., Joiner, S., Hill, A. F., Campbell, T. A., Desbruslais, M., Luthert, P. J., and Collinge, J. (2001) Tissue distribution of protease resistant prion protein in variant Creutzfeldt-Jakob disease using a highly sensitive immunoblotting assay. *Lancet* **358,** 171–80.
16. Lawson, V. A., Stewart, J. D., and Masters, C. L. (2007) Enzymatic detergent treatment protocol that reduces protease-resistant prion protein load and infectivity from surgical-steel monofilaments contaminated with a human-derived prion strain *J Gen Virol* **88,** 2905–14.
17. Reed, L. J., and Muench, H. (1938) A simple method of estimating fifty percent endpoints. *Am J Hyg* **27,** 493–497.

Index

Printed in the United States of America